Principles of Process Planning

Principles of Process Planning

A logical approach

Gideon Halevi

Adjunct Professor, Technion, Haifa, Israel

and

Roland D. Weill

Professor Emeritus of Mechanical Engineering,
Israel Institute of Technology, Haifa, Israel

SPRINGER-SCIENCE+BUSINESS MEDIA, B.V

First edition 1995

Originally published by Chapman & Hall in 1995
Softcover reprint of the hardcover 1st edition 1995

Typeset in 11/12pt Palatino by Cotswold Typesetting Ltd, Gloucester, England

ISBN 978-94-010-4544-5 ISBN 978-94-011-1250-5 (eBook)
DOI 10.1007/978-94-011-1250-5

A catalogue record for ths book is available from the British Library

∞ Printed on permanent acid-free text paper, manufactured in accordance with ANSI/NISO Z39.48-1992 and ANSI/NISO Z39.48-1984 (Permanence of Paper).

Contents

Preface

Process planning determines how a product is to be manufactured and is therefore a key element in the manufacturing process. It plays a major part in determining the cost of components and affects all factory activities, company competitiveness, production planning, production efficiency and product quality. It is a crucial link between design and manufacturing.

There are several levels of process planning activities. Early in product engineering and development, process planning is responsible for determining the general method of production. The selected general method of production affects the design constraints.

In the last stages of design, the designer has to consider ease of manufacturing in order for it to be economic. The part design data is transferred from engineering to manufacturing and process planners develop the detailed work package for manufacturing a part. Dimensions and tolerances are determined for each stage of processing of the workpiece. Process planning determines the sequence of operations and utilization of machine tools. Cutting tools, fixtures, gauges and other accessory tooling are also specified. Feeds, speeds and other parameters of the metal cutting and forming processes are determined.

In spite of the importance of process planning in the manufacturing cycle, there is no formal methodology which can be used, or can help to train personnel for this job. Process planning activities are predominantly labor intensive, depending on experience and the skill and intuition of production labor. Dependence on such methodologies often precludes a thorough analysis and optimization of the process plan and nearly always results in higher than necessary production costs, delays, errors and nonstandardization of processes.

Industry has also observed in recent years a decrease in the number of process planners. They are retiring from the field at a faster rate than new ones are being trained. This situation is endangering manufacturing's ability to perform efficiently. The above situation, coupled with the objective to increase quality and reduce lead time and cost or alternatively improve productivity, has led to the widespread interest in **CAPP – computer-aided process planning**.

CAPP systems may take the variant approach, the expert system approach, the artificial inteligence approach or the generative approach.

Unfortunately, because of the complexity of machined parts, process planning has resulted in CAPP remaining at the conceptual stage. The benefits are anticipated, but the path to follow has been unclear. All CAPP systems depend on expert process planners to set the **rules** or a **master** process plan, and on a skilled operator to evaluate the recommended process.

As a result, there is no substitute for a skilled process planner or other expert. The question is, who is an expert? By definition an expert is one 'having or manifesting the knowledge, skill and experience needed for success in a particular field or endeavor', or 'one who has acquired special skill in or knowledge and mastery of something'.

However, experience is obtained by practical work, where processes are defined, follow-up is made and corrective measures are taken during production. Experience is gained from problematic processes and rejected parts and corrections are made to obtain a successful result. Very little experience, or even the wrong kind of experience, can be gained from 'no problem' parts.

All research in the field of process planning has indicated that all experts have their own expertise and one expert's experience might be different from that of another. It is rare, therefore, for two planners to produce the same process. Each process will produce the part as specified, although different processes will result in different machine times and costs. In addition, expert rules are often contradictory and provision has to be made for conflict resolution which is difficult to implement.

This book is intended to enhance process planner expertise, to try to construct a 'master' expert, i.e. one who produces an effective process, accepted by any expert.

It is also intended to provide investigators in CAPP and production management fields with an understanding of process planning complexity, and supply them with objective technical data. In other words, to become independent of any particular expert opinion.

The book will also be useful for students in manufacturing who will gain a comprehensive and fundamental knowledge of the concept of 'process planning'.

The book is based upon the experience gained by working with over 250 process planners and studying their way of thinking. The conclusions are that experience is invaluable if it is interpreted correctly. However, usually it is not.

Nevertheless, it was found that process planners usually make their decisions based upon a global understanding without breaking it down into the individual parameters. They know the problem, they find a working solution, but they cannot pinpoint the controlling parameter that caused the problem. Therefore, they apply the same solution to many similar problems, even if the controlling parameters are not the same.

There are many parameters that impose constraints and affect the selection of a process: machine, tool, fixture, part geometry, tolerances, technology, economics.

The method used in this book is based on the belief that personal experience can be enriched by the exchange of practical experience between practitioners and by understanding the nature of the process. Using this method in workshops resulted in increased process planner productivity and plant productivity was improved by at least 30%.

The book guides the reader through process planning problems, while:

- comparing individual process planner's recommendations;
- analyzing the proposed processes;
- defining the controlling parameters;
- comparing practice to theoretical know-how;
- reaching conclusions, and defining the 'best' solution to the problem at hand.

Exercises are given at the end of the book.

The text is organized in such a way as to provide know-how in process planning. The first two chapters give an overview of the role and importance of process planning in the manufacturing cycle. Chapters 3 and 4 set the background of communication between product designer and process planner, i.e. interpretation of an engineering drawing. This is followed in Chapter 5 by a short review of the basic forming processes. A method of selecting these processes is proposed, taking into account engineering drawing specifications and economic considerations.

As forming by material removal is the most comprehensive process, the remaining chapters are devoted to gaining an in depth understanding of this and of metal cutting process planning and operation planning.

While every precaution has been taken in the preparation of this book, no liability is assumed by the writer and publisher with respect to the use of information contained within it. Process planners must exercise their judgement on the proposed processes and the decisions made.

Symbols and units

General

C = cost (\$)
T, t = time (s, min, hour)

Technical quantities (Excerpt from ISO 31, Quantities and Units in the various Fields of Science and Technology)

l = length (mm) (in mechanical engineering)
b = width (mm) (in mechanical engineering)
d = thickness (mm) (in mechanical engineering)
r = radius (mm) (in mechanical engineering)
$d, D, \varnothing$ = diameter (mm) (in mechanical engineering)
A = area $(mm)^2$
u, v, w = velocity (m/s), (m/min)
n = rotational speed (tour/min)
F = force (N)
T, M = moment (Nm)
μ = coefficient of friction
E = energy (J)
P = power (W)
t, δ = temperature (°C, K)
Q = heat quantity (J)

Basic quantities in cutting and grinding (Excerpt from ISO 3002)

r_ε = corner radius (mm)
v_c = cutting speed (m/min)
v_f = feed speed (m/min)
κ_r = tool cutting edge angle (radian)
λ_s = tool cutting edge inclination (radian)
κ_r' = minor cutting edge (radian)
γ_o = tool orthogonal rake (radian)
α_o = tool orthogonal clearance (radian)
f = feed (mm/tour)
f_z = feed per tooth (mm/tooth)
z = number of teeth
a = engagement of a cutting edge (depth of cut) (mm)

Q = material removal rate (mm^3/min)
V = material removal (mm^3)
R_a = arithmetic mean deviation of the profile (μm)
IT = interval of tolerance (mm)

CHAPTER 1

Introduction

The objective of the manufacturing process is to transform an idea into a saleable product.

The common methodology used to accomplish this objective is to divide the manufacturing process into several activities, arranged serially. Usually, each activity handles a different stage of the process, each stage representing a unique discipline and training.

Figure 1.1 describes the global structure of an industrial enterprise and points to the role of process planning as a part of production planning. This latter represents the general control of manufacturing activities in an enterprise. Whereas the general management of a company is concerned with long-range planning activities such as personnel policy, marketing, accounting, etc., it delegates the functions of product design and manufacture to special departments which are in charge of functions such as production planning, fabrication technologies, quality control, etc. The role of process planning itself is now explained.

1.1 THE PLACE OF PROCESS PLANNING IN THE MANUFACTURING CYCLE

Process planning is an important link in the manufacturing cycle. It defines in detail the process that transforms raw material into the desired form. The form is defined by the product designer, and is expressed in engineering drawings and GDT – geometric dimensioning and tolerances.

More precisely, process planning can be defined by a sequence of activities illustrated in Fig. 1.2, although not necessarily in the order shown. They comprise mainly:

- an interpretation of the specifications contained in the definition drawing of a part, including mainly dimensions and tolerances, geometric tolerances, surface roughness, material type, blank size, number of parts in a batch, etc.;
- a selection of processes and tools which are candidates for processing a part and its features by respecting the constraints imposed in the definition drawing;
- a determination of production tolerances and setting dimensions which ensure execution of the design tolerances, while choosing production

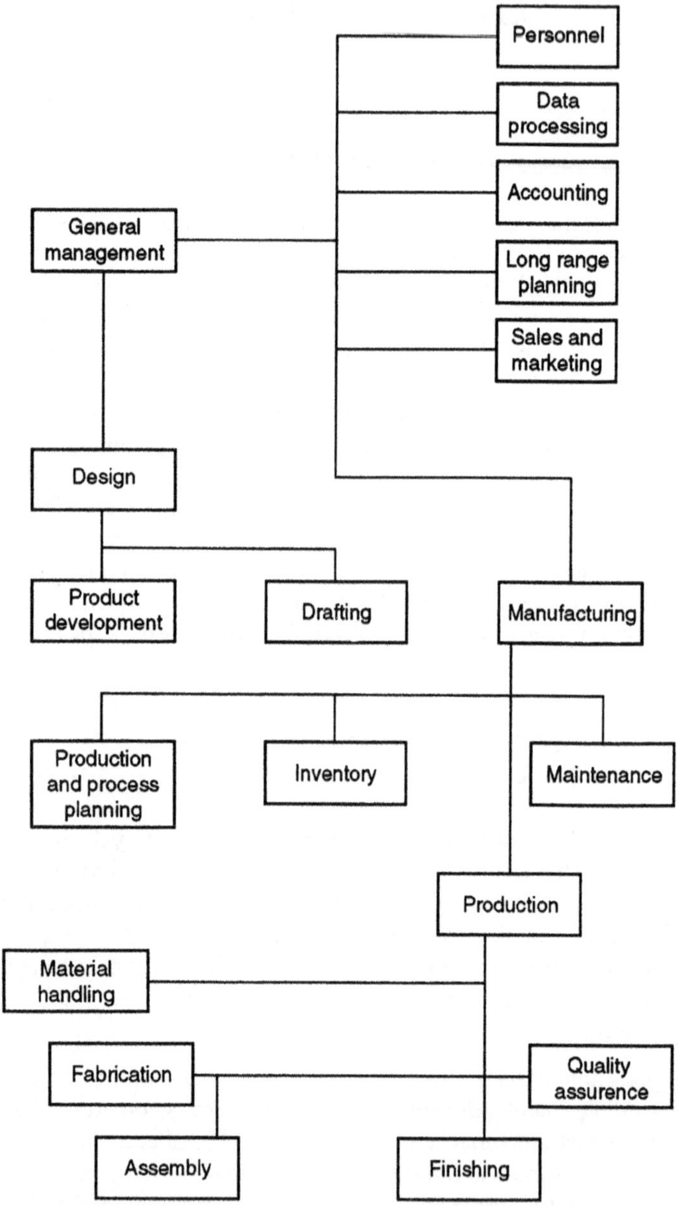

Fig. 1.1 Functionality of the global structure of an industrial enterprise.

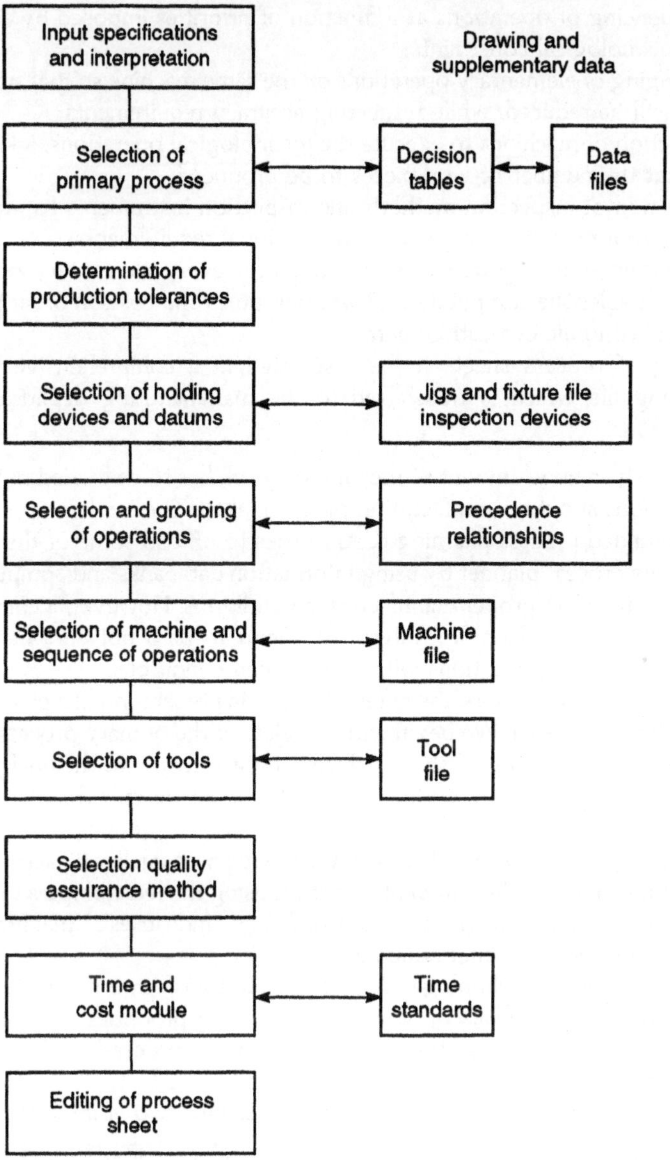

Fig. 1.2 Process planning activities.

dimensions for reasons of commodity and capability of manufacturing machinery;

- a selection of starting surfaces and datum surfaces to ensure precise execution of processing operations simultaneously with a selection of holding fixtures and checking of stability of a part by appropriate clamping;

- a sequencing of operations as a function of priorities imposed by accuracy and technological constraints;
- a grouping of elementary operations on the same machine so that operation time will be reduced, while respecting accuracy requirements;
- a selection of machines to execute the technological operations, taking into account the number of workpieces to be produced;
- a selection of inspection methods and inspection instruments to guarantee final conformity of component with functional requirements;
- a determination of processing conditions for every elementary operation which enables the computation of working times and costs in order to carry out an economic evaluation; and
- editing of process sheets to be assembled in a comprehensive process planning file which is transferred to the manufacturing department for execution.

Obviously, a huge amount of preparation work has to be carried out before final decisions about a manufacturing plan are taken. The modern approach of computer-aided process planning (CAPP) tries to offload some of the manual work of the process planner by using information databanks and computerized algorithms to select proper manufacturing conditions. However, a completely automated process planning system still seems to be a not-very-near ideal, except for special applications with well-defined conditions.

In the following chapters, the different stages in preparation of a process plan are developed in detail. We begin with a review of the primary processes used to produce a rough material shape, which is transformed into the final product by the methods which are outlined in this book. As described later, some of these primary processes can be used to deliver a final product, but normally, other processes have to be added to ensure the required quality (in accuracy and finish) of the product. The different industries using the technologies to deliver final products are known as 'transformation industries', 'manufacturing industries' or 'metalworking industries'.

A huge variety of technologies is employed to transform a raw material into a final product. A brief description of most forming processes is given, and a methodology of selecting the economic forming process is described in Chapter 5.

It is possible to divide the forming processes into two large groups:

1. the group of forming solids by material removal, such as turning, milling, boring, drilling, electrodischarge machining, etc.; and
2. all other processes groups, such as casting, forging, extrusion, punching, deep drawing, shearing, welding, assembling, etc.

In this book, only the first group will be analyzed in detail, for the following reasons:

- Metal removal processes represent the majority of transformation processes – about 80% in machines and production volume.

- Metal removal processes have an inherent flexibility which enables their use for a large spectrum of applications and batch volumes. This situation can only be reinforced in the future with the constant trend towards more variety in products and smaller quantities of products.
- On preparation of a metal removal process plan, one has to consider a much higher number of solutions than for a process plan for other forming processes. In this latter case, the constraints of material, shape, equipment capabilities and batch volume restrict the anticipated solutions to a fewer number of alternatives.

Therefore, most of the following chapters concentrate on technologies suitable for metal removal operations. To gain an in-depth knowledge of other forming technologies, which can be more productive in specific cases, it is recommended the reader consult the many good books available on the market.

1.2 PROCESS PLANNING AND THE ECONOMIC MANAGEMENT OF A COMPANY

The management of an enterprise is overwhelmingly based on economic considerations.

Managing of a company calls for many economic decisions such as the economics of manufacturing a certain product, capital investment and cash flow needs, type and number of machines needed, number of employees, due date of delivery, layout, etc. A decision implementation has to be based on intuition, on partially estimated data or accurate data. The better the data, the better the decision should be. In every case, process planning has to provide the background for economic evaluation.

For example, when a new product is introduced in the company, the finance department wants to know its manufacturing cost. To answer this question within reasonable accuracy, the bill of materials for the product has to be broken down, giving a list of all required parts and their quantity for a single product. For each part on the list, a process plan will be devised – listing the sequence of operations, the machines, the tools and fixtures used and machining time for each operation. The finance personnel will translate this data into costs.

Another example, in relation to data from process planning, is where management would like to know what capital investment has to be made in manufacturing facilities. To answer this question, a procedure similar to that used in the previous case has to be made. Then the data must be multiplied by the quantity of products to be manufactured per period. Where the same facility is used for several operations, the total time required for each facility is summed. When the total time per period is known, the number of required facilities of each type can be computed. Knowing the cost of each working station, management can transform this data into total investment.

Likewise, if management needs a breakdown of the workforce by profession, a similar evaluation must be made, but instead of summing facilities, there would be a summing of employees required for each facility.

Almost any industrial inquiry concerning the manufacturing process (floor space, due dates, lead time, work-in-process, etc.) addresses process planning as a data source. Process planning is the basis for the optimization of the whole production scenario and its alternatives, and not only for simple operations.

Finally, it is important to emphasize that process planning is required at any manufacturing plant, regardless of plant size, part complexity and batch size. The opinion that is often heard – that process planning is not suited for small batch sizes – is misleading. The problem with small batch sizes is not a process planning or manufacturing one; it is an economics problem. The difficulty here is to find a reasonable compromise between preparation (thinking) time and manufacturing time.

The following derivation supports this opinion using some realistic figures. Let the total cost of producing a part be as follows:

$$C = Q \times C_d \times T_d + C_t \times T_p + C_s \times T_s \tag{1.1}$$

where: C = total cost ($)
Q = batch quantity
C_d = hourly rate of direct work ($ per hour)
T_d = direct machining time (hours)
C_t = hourly rate of indirect labor ($ per hour)
T_p = indirect labor time (hours)
C_s = hourly rate for setup time ($ per hour)
T_s = time for setup (hours)

The indirect time T_p can be divided into two elements:

T_{pf} – a fixed time to handle an order and to arrive at an initial process plan; and
T_{pv} – the time to generate alternative process plans and evaluate them in order to arrive at an optimum process plan.

T_{pv} is a variable defining the thinking time; the longer it is, the better the process plan generated. However, total cost will increase. The value of this variable indicates the difference between the economics of handling a small batch size order and a large batch size. An evaluation of T_{pv} can be made as follows: If it is assumed that T_d (direct machining time) depends on T_{pv}, the thinking time, as shown in Fig. 1.3, can be expressed by the equation:

$$T_d = K_1 + \frac{K_2}{T_p} = K_1 + \frac{K_2}{T_{pf} + T_{pv}} \tag{1.2}$$

where K_1 and K_2 are constants, their values can be assigned by the following boundary conditions.

At $T_{pv} = \infty$

$$T_d = K_1 = T_{dmin}$$

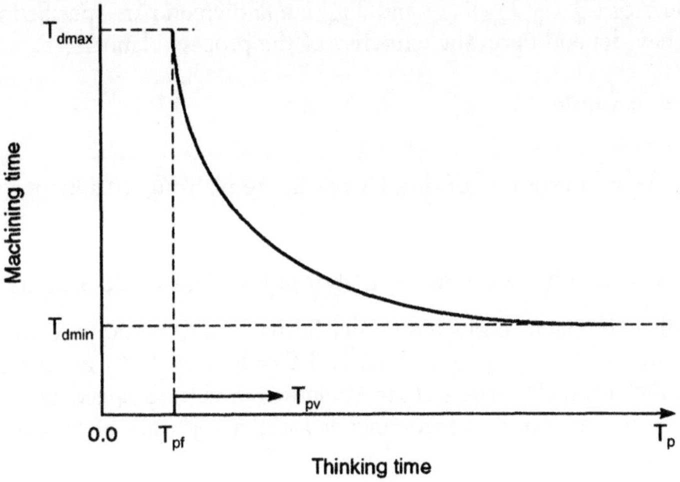

Fig. 1.3 Direct machining time as a function of thinking time.

At $T_{pv}=0$

$$T_d=K_1+\frac{K_2}{T_{pf}}=T_{dmax}$$

Hence $$K_2=(T_{dmax}-t_{dmin})\times T_{pf}$$

Thus

$$T_d=T_{dmin}+\frac{(T_{dmax}-T_{dmin})\times T_{pf}}{T_p} \tag{1.3}$$

Substituting T_d from (1.3) into (1.1) gives:

$$C=Q\times C_d\times\left\{T_{dmin}+\frac{(T_{dmax}-T_{dmin})\times T_{pf}}{T_p}\right\}+C_t\times T_p+C_s\times T_s \tag{1.4}$$

Differentiating (1.4) with respect to T_p and setting it equal to zero, will result in the minimum total cost of manufacturing a part with the quantity of products as a parameter and the thinking time as the variable.

$$\frac{dC}{dT_p}=-Q\times C_d\times\left\{\frac{(T_{dmax}-T_{dmin})\times T_{pf}}{T_p^2}\right\}+C_t=0$$

Hence

$$T_p^2=Q\frac{C_d}{C_t}T_{pf}(T_{dmax}-T_{dmin}) \tag{1.5}$$

The values of C_d, C_t, T_{pf}, T_{dmax} and T_{dmin} are known and are specific for every plant. They depend upon the expertise of the process planners.

Practical example
Taking

C_d = hourly rate of direct work in the plant @ 40 \$/hour

and

C_t = hourly rate of indirect labor in the plant @ 20 \$/hour

assume that the plant received an order to machine $Q = 100$ parts on a lathe, with a part $L = 20$ mm (7.87 in) long and $D = 30$ mm (1.181 in) in diameter.

As a first estimate, where $T_{pf} = 3$ minutes, a cutting speed of $v_c = 30$ m (196.85 ft) per minute was recommended and a feed rate of $f = 0.3$ mm per revolution.

Thus the machining time (in minutes) will be:

$$T_{max} = \frac{\pi DL}{v_c f 1000} = \frac{\pi \times 30 \times 200}{60 \times 0.3 \times 1000} = 1.047 \qquad (1.6)$$

However, by selecting the right tool shape and grade, the cutting speed can be as high as $v_c = 90$ m (295.25 ft)/min. By appropriate chucking and support, the feed rate can go up as high as $f = 0.5$ mm/rev.

Thus

$$T_{min} = \frac{\pi DL}{v_c f 1000} = \frac{\pi \times 30 \times 200}{90 \times 0.5 \times 1000} = 0.42 \qquad (1.7)$$

Using these values, the economic thinking time T_{pv} (in minutes) is computed.

$$T_p^2 = Q \frac{C_d}{C_t} T_{pf}(T_{dmax} - T_{dmin})$$

$$T_p^2 = 100 \frac{40}{20} 3(1.047 - 0.42) = 376.2$$

$$T_p = 19.4$$

$$T_{pv} = 19.4 - 3 = 16.4 \qquad (1.8)$$

This means that it is economic to allow only 16.4 minutes to compute the optimum process plan.

The machining time (in minutes) will be:

$$T_d = 0.42 + (1.047 - 0.42) \times 3/19.4 = 0.517 \qquad (1.9)$$

Similarly,

for a quantity of 50 parts, the economic T_d is 0.557 min
for a quantity of 10 parts, the economic T_d is 0.727 min

whereas the maximum time was $T_d = 1.047$ min.

This example merely demonstrates that the thinking time is an important parameter in determining a machining process. The process planner usually works under heavy preassure and lack of time, so that the above computation, which should be made for each operation, is generally not practical and should not be done for small quantities.

Today, two methods are used to try to overcome this dilemma and to shorten the process generation time. One is to use a computer, i.e. CAPP (computer-aided process planning). However, although research and development efforts in the field of computer-aided process planning over the past two decades have resulted in numerous experimental CAPP systems, they have had no significant effect on manufacturing planning practice.

Another method is to improve the process planner's intuition, knowledge and expertise. Thereby, the difference between T_{dmax} and T_{dmin} will be small, and T_{dmax} will be reached in less time T_p. From a study on process planners, it appears that they rely on experience and intuition; as different process planners have different experience, it is no wonder that for the same part, different process planners will devise different processes. Each process will produce the part as specified in the drawing, although different processes will have different machining times and costs.

Experience comes from practical work on the shop floor during production, where a process is defined and corrections are made. Experience is also gained from the rejected part and problematic processes. Very little experience, or even the wrong experience, can be gained from 'no problem' parts. But the process planner often neglects other considerations such as time and cost. The experienced process planner usually makes decisions based on comprehensive data without breaking it down to individual parameters; there is no time to analyze the problem, and the result is an empirical solution without justification.

It is important that the process planner should understand the process and the effect of each individual parameter of the process plan. Such understanding, and a methodic thinking flow will improve the performance of the process planner. One of the purposes of this book is to promote such systematic thinking among process planners.

To promote methodical thinking based on scientific knowledge, this book intends to analyze logically the process plan genesis and to present reliable information. It also explains the basic principles of process planning in order to give manufacturing engineering students a solid professional background.

1.3 PROCESS PLANNING AND PRODUCTION PLANNING

Process planning is the first step in the organization of a manufacturing plan. However, it does not take into account time and use of industrial facilities. Certainly without ensuring the feasibility of a process plan, it is pointless preparing production management and production control plans. Conversely, it would not be economical to design process plans which need equipment not available in the company, which would incur unnecessary expense or which would entail the use of machinery which is vital for other manufacturing operations. If one intends to produce economically, it is therefore extremely important to achieve an optimal balance between occupation of machinery and queuing times of in-process material.

Also, the wasting of time by long handling transfer periods between machines is undesirable. The purpose of production control is to supervise the flow of parts on the shop floor in order to minimize such losses of time and, at the same time, to respect the delivery dates of products. Scheduling in production management is therefore a very basic function in manufacturing and it has to be well matched with process planning. The trend in process planning developments is more and more to integrate the two functions of process planning and production planning in order to achieve better productivity.

A process planner must therefore be aware of the facilities in the shop and should know the load on the machines. But as there is no way of ascertaining which machine will be overloaded and which underloaded other than by estimates and assumptions, the process planner has to think about alternatives. Balancing machine loading is something the process planner has no responsibility for, or even has to consider. In fact, it is impossible to know the load if one does not know the process. It is a loop problem, and normally, the process planner will simply select a 'better' machine. Thus about 30% of the machines will be overloaded and the rest underloaded. This situation, together with added disruptions on shop floor, makes production management a difficult task.

One main difficulty lies with the fact that routing is fixed. Routing prescribes the flow of work in the plant and lists the sequence of workstations required to produce an item. It derives its information from process planning and presents it to production planning.

However, this methodology has one main drawback. Production planning uses technical information from process planning as an input. But this data does not reveal or reflect the basic intentions of the process planner, and therefore, does not result in an optimal production planning solution. This situation has been reconsidered recently with a view to gaining time for the production cycle and reducing manufacturing costs. This new approach, called 'concurrent engineering' (CE) is explained below.

1.4 PROCESS PLANNING AND CONCURRENT ENGINEERING

As mentioned before, technical data is incomplete at the production management phase; although it represents the final decision of the process planner, it does not reveal the process planner's intentions and imposed constraints (quantity, tooling, machinery, etc.).

Usually a process planner will evaluate several alternative solutions, such as going to the shop floor to consult with the foreman with regard to the machine load. Normally, the process planner applies innate knowledge and experience, but for the same machining requirements there could be different process alternatives. This means that process planning is more or less an iterative rather than a straight process. The iterations are however 'lost' when the process plan arrives at the production management stage.

Process planning is also a series of decisions, decisions that must uniquely specify the process, even if they are not mandatory to it. Once the process planner makes a decision, it becomes a constraint on all decisions which follow it. For example, a selected machine imposes constraints on the power available for the cutting operation, the torque at the spindle, the maximum depth of cut, the maximum cutting speed and the available speeds and feeds, the machining dimensions, the number of tools that can be used, the accuracy, the handling times, etc.

A single machining operation can be adjusted to comply with these constraints, but machining cost and time will be applied to the selected machine. Similarly, a selected tool imposes constraints on the maximum cutting speed, depth of cut, feed rate and tool life.

It is accepted that these constraints are artificial ones; they exist only because of the sequence of decisions made. Another sequence might result in a different set of constraints. Similarly, a decision made at the process planning stage will be a constraint in the production management stage. In summary, if the technical data could be available at the production management stage, savings in manufacturing a product mix could be realized.

New approaches have therefore been proposed to consider alternative routings in preparing production planning. One approach is based on a matrix format of alternative routings as a function of available machinery.

The matrix establishes a network of possible routings while deferring to a later stage the decision of which path to take. The choice of path may be changed after each operation. Hence the scheduling problem is to employ routing alternatives that are available to meet specifications in production, with no bottlenecks or disruptions, and achieving minimal operation costs.

A similar approach, called 'nonlinear process planning', has been proposed as an alternative. By this methodology, process planning is given a tree structure defining different routings, so that scheduling is only decided after a certain stage of advancement in the process plan is reached.

Concurrent engineering (CE) is based on the synergy between the design stages of a product and the manufacturing planning stages. CE involves the parallel processing of tasks and provides methods enabling different people to solve problems simultaneously according to their specific point of view. The term engineering in this instance does not just involve technical tasks such as design and manufacture, but includes such things as cost accounting, procurement, marketing and distribution and social behavior. People of different disciplines must work together in a cooperative manner and understand each other.

The common methodology used in manufacturing a product is to divide the industrial process into several activities, arranged serially. Usually, each activity handles a different stage of the process, each stage representing a unique discipline and training. The input to each activity is the output of the previous activity. Each activity optimizes its decisions according to its own criteria and the decisions are the output of this activity. Thus, the manufacturing process is a serial chain of activities (Fig. 1.4), as follows.

- Product design
- Process planning

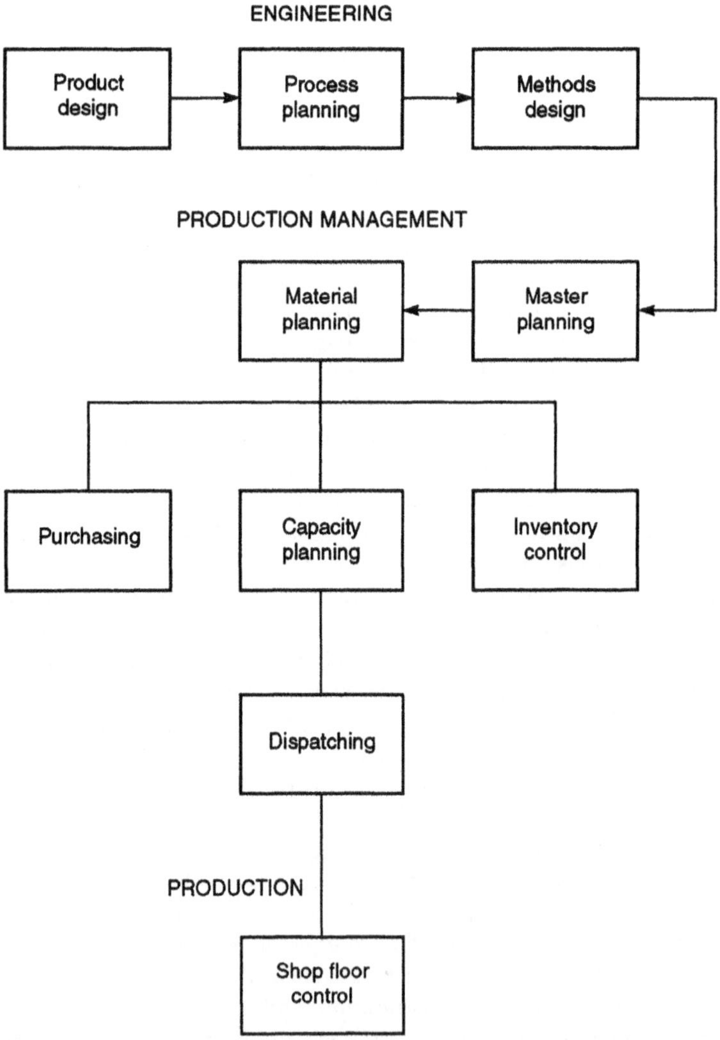

Fig. 1.4 Manufacturing chain of activities.

- Methods, jigs and fixtures design
- Production management:
  - master production planning
  - material requirement planning
  - capacity planning/purchasing/inventory control
  - dispatching
- Shop floor manufacturing and control
- Quality assurance
- Maintenance

However, now that the industrial products market is becoming more and more competitive, the emphasis is shifting from a suppliers' to a buyers' market. The market trend is towards decreasing product life cycle. These developments have led to a change in the goals and priorities in the industrial world. Rapid response to market demands in order to improve customer satisfaction is becoming a dominant factor, together with considerations of quality and cost. In practical terms, reduction of the manufacturing cycle time is becoming a primary goal in industrial activities.

To meet these new demands, a simultaneous engineering, or concurrent engineering, manufacturing methodology is proposed. The emphasis in these methodologies is on interaction between product design and process planning. Under the traditional chain of activities approach, interaction is achieved in an *ad hoc* way: the product designer, as and when deemed necessary, initiates interaction with the process planner. However, in order to reap the full benefits of concurrent engineering, such an approach will no longer be viable and it is necessary to develop a more systematic approach in order to allow close personal or computer-based interaction between product designer and process planner. The process planner must be involved in the design stage in order to ensure design for manufacturing (DFM).

DFM involves a streamlined design process using the skill of a multifunctional team to influence the design concept as it evolves into a product. In this sense, DFM is not a manufacturing process, but a product delivery process.

Similarly, DFA (design for assembly) is often the first step in the process, followed by DFQ (design for quality), DFR (design for reliability), serviceability, safety, user friendliness and time to market. All are complementary approaches supporting the introduction of concurrent engineering.

In CE, a strong emphasis is put on the role of process planning in the design stage in order to reduce lead times and costs. These two stages, product design and process planning, determine largely the minimum cost of the product. Of course, lead time and cost can also be increased at the production management stages. Therefore, process planning should also be matched with production planning.

Recently, substantial improvements in productivity have been mentioned in relation to the application of the CE concept. It can eliminate 30–90% of the lead time in introducing new products. In addition, DFM and DFA teams may reduce part costs and the drop in assembly cost may be as high as 65%.

FURTHER READING

Baur, A., Browne, J., Bowden, J. *et al.*, (1991) *Shop Floor Control Systems*, Chapman & Hall.
Halevi, G. (1980) *The Role of Computers in the Manufacturing Processes*, Wiley-Interscience.
Halevi, G. (1993) The magic matrix as a smart scheduler. *Computers in Industry*, **21**, 245–53.
Halevi, G. and Weill, R. (eds) (1992) *Manufacturing in the Era of Concurrent Engineering*. North-Holland.
Hunt, V.D. (1989) *Computer Integrated Manufacturing Handbook*, Chapman & Hall.
Kalpakjian, S. (1984) *Manufacturing Processes for Engineering Materials*, Addison-Wesley.
Parsaei, H.R. and Mital, A. (1992) *Economic Advanced Manufacturing Systems*, Chapman & Hall.
Parsaei, H.R. and Sullivan, W.G. (1993) *Concurrent Engineering*, Chapman & Hall.
Rolstadas, A. (ed.) (1988) *Computer Aided Production Management*, IFIP State of the Art Report. Springer-Verlag.
Smith, P.G. and Ghosh, A. (1991) *Developing Products in Half the Time*, Van Nostrand Reinhold Inc.
Stephanou, S.E. (1992) *The Manufacturing Challenge*, Van Nostrand Reinhold Inc.
Wu, B. (1991) *Manufacturing Systems Design and Analysis*, Chapman & Hall.

CHAPTER 2

Logical design of a process plan

To give a global idea of the preparation of a process plan, it seems appropriate to analyze a typical mechanical part and to define the different stages of decisions to be taken. The intention is not to present a complete set of methods of analysis – these are developed in the course of the book – but, rather, to give a general feeling of the nature of process planning as an introduction to the following chapters.

Although the process plan analyzed in this chapter relates to machining processes only, for reasons given earlier (a very high percentage of machining operations in manufacturing and the special difficulties of designing machining processes), it is clear that many of the logical steps taken in machining can be transferred easily to other processes and so the instructional value of the analysis applies generally. Process planning for other processes is discussed in Chapter 5.

2.1 PRELIMINARY ANALYSIS OF A MECHANICAL PART

Having correctly analyzed the part drawing from a geometric point of view (detailed in Chapter 3), it should be possible to define a feasible process plan on the basis of the information available on production facilities (machines, tooling, accuracy capabilities, etc.). Logical steps to reach this goal are described in the course of this chapter, but before going into detail on this, it is a good idea to study the outline of the part in order to think of ways of designing its final process plan. Of course, a good interpretation of the part drawing, as explained in Chapter 3, must come first.

To illustrate this approach, let us take as an example the part represented in Fig. 2.1 (Weill, Spur and Eversheim, 1982). This part comprises a number of features which need to be produced by basic primary manufacturing processes (presented in Chapter 5). The selection of a primary manufacturing process suitable for a given part depends heavily on the quantity of parts required. For the part in Fig. 2.1, in order to save raw material and machining time, and also because of the large quantity required, it was decided to select forming from

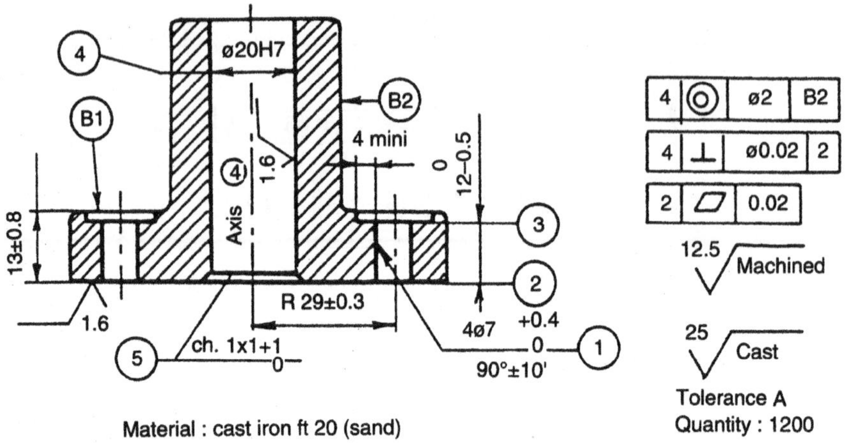

Fig. 2.1 Definition of a drawing of a mechanical part. Redrawn from Karr, J., *Methodes et Analyses de Fabrication Mecanique*, published by Dunod, Bordas, Paris, 1979.

liquid by sand casting as the initial job. Casting will produce a part with approximate dimensions which are defined according to a standard called, in this case, 'tolerance A', giving the accuracy of the raw part from casting. Afterwards, machining operations will finish the part to its final dimensions. However, it is important to note that subsequent tolerancing of the machining operations will relate to the accuracy of the raw part (Chapter 4).

For a better understanding of the steps in process planning, here are some general definitions. In machining, the combination of different processes to be executed on one machine tool is called a 'job'. By 'job', we mean a succession of 'operations', these being defined as elementary processes executed by a single type of tool, in several cuts or passes. For example, the cylindrical turning of a part on a lathe is called an operation. When more than one operation is executed in the same fixture, e.g. when producing a slot in the cylindrical part, the sequence of operations is called a 'subjob'. Changing fixturing on a machine will produce another subjob, and both subjobs are a 'job'. Figure 2.2 illustrates these concepts, showing how a subjob designs a subphase of a job when the part is turned over on the same fixture. The whole sequence of operations combined in jobs is the process plan.

In addition to these definitions, another concept is that of a group of surfaces in a part which have functional relations between them, e.g. tolerances of position or orientation. Such a grouping of surfaces is called 'associated surfaces' and is illustrated in Fig. 2.3. As shown in this figure, associated surfaces can be produced by form tools, by a system of tools, or more generally, by a unique set-up and processing of surfaces by a number of different tools. The highest degree of accuracy on a given machine tool is obtained through using associated surfaces grouping. In some other cases, as seen later, associated

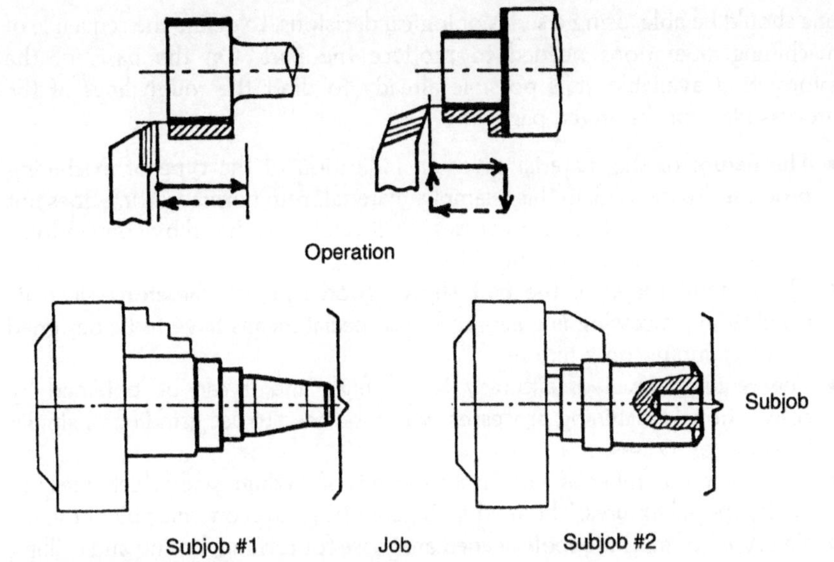

Fig. 2.2 Definition of: operation, job, subjob.

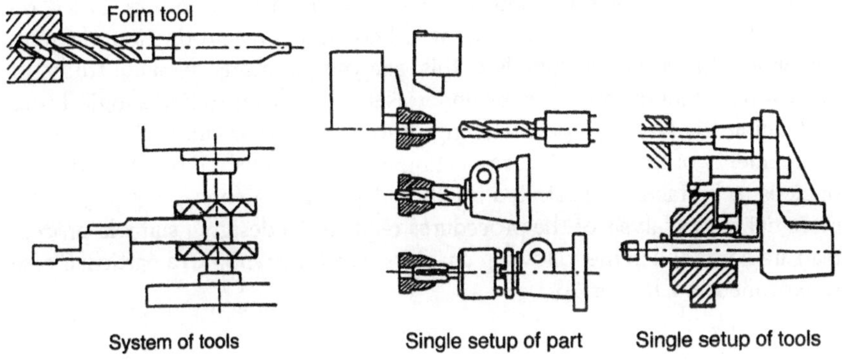

Fig. 2.3 Definition of associated surfaces.

surfaces are also used in order to gain time between operations, i.e. for reasons of economy.

Considering all the information in Fig. 2.1, i.e.:

- geometric shape
- dimensions and their tolerances
- geometric tolerances
- surface roughness
- material type and its hardness
- size of the raw material (given by its tolerance A)
- number of parts in the batch (1200),

one should be able, using a series of logical decisions, to define the sequence of machining operations needed to produce the part. On the basis of the information available, it is possible already to draft the rough lines of the process plan for the given part:

- The nature of the material gives an indication of the type of machining processes to be used. In this example, material from normal casting does not require special machining processes, i.e. it can be produced by conventional technologies, without the need for grinding.
- The general shape of the part shows good rigidity, therefore, since its stability in processing is guaranteed, no special means have to be designed for its clamping on a fixture.
- The required level of accuracy is not high and it can be obtained by conventional machining processes, without a need to use grinding or similar finishing processes.
- The required number of parts in the batch is of medium size, which does not justify special fixtures. This is advantageous from an economic point of view.
- The types of machine tools needed are those for turning, boring and drilling commonly used in manufacturing.

These general observations represent an initial estimation of production methods required for the manufacturing of a given part, and are based on technical and economic evaluations well known among production planning personnel. In order to complete this analysis, a more detailed study of processing methods must now be undertaken. The idea is to find a logical way of defining the sequence of machining processes in accordance with precision requirements of the part while respecting economic constraints. The sequence of steps to be taken is illustrated briefly in Fig. 1.2 in a conceptual way, but a more detailed analysis of the procedures required to design a suitable process plan are described in the following chapters. The adaptation to a particular case is explained in Chapter 16.

2.2 SELECTION OF MACHINING PROCESSES, TOOLS AND CUTTING PARAMETERS

A part is composed of a number of features which are produced by a succession of machining operations of different types. For example, in the case of the part in Fig. 2.1, Table 2.1 gives a description of the processes and tools needed to produce the different features with the precision and surface finish required.

For some features, such as feature (2) two possibilities are envisaged for the type of process and the number of passes (roughing and/or finishing). The final choice will depend on the capabilities of the machinery which is available. For feature (4), it appears that three passes (roughing, semi finishing, finishing) are necessary to comply with the accuracy requirements. The total number of tools for each operation is given in the last column of Table 2.1 (number of tools).

Table 2.1 Passes and tools selected for each operation

Feature to be machined	*Design specifications*	*Passes and tools*			*Number of tools*
		Roughing	*½ finishing*	*Finishing*	
(1)	Ø $7_{0}^{+0.4}$; R_a 12.5 29±0.3			Drill	1
(2)	13±0.8; R_a 1.6 planeity 0.02	Eventually miller or surfacing tool		Miller or surfacing tool	1 or 2
(3)	$12^{0}_{-0.5}$; R_a 12.5 4 mini.			Countersinking tool	1
(4)	Ø 20H7 – R_a 1.6 ◎ Ø2; ⊥ 0.02	Drill	Tool on bar	Reamer	3
(5)	Chamfer: $1 \times 1_{0}^{+1}$			Tool for chamfering	1

Source: Karr, J., *Methodes et Analyses de Fabrication Mecanique*, published by Dunod, Bordas, Paris, 1979.

The selection of the number of tools and passes is made according to tolerance considerations, e.g. for surface (2), the dimensional tolerance ±0.8, the planeity 0.02 and the surface roughness Ra 1.6 can be produced by a milling tool as well as a surfacing tool in turning. The need for a second pass depends on the values of the form and surface finish tolerances achievable by the available processes (milling or turning).

The same logic is applied to each feature by selecting processes and tools according to rules which are found in handbooks (Metcut Research Associates, 1972) or in computerized databanks (R. Weill *et al.*, 1978). Table 13.1 in this book gives an example of information on hole-producing processes with their boundaries in accuracy and roughness. More information concerning machining processes capabilities is given in Chapter 8.

The choice of process parameters at this stage is strictly tentative and may change during the development of the process plan. Of course, the selection of processes and tools has to be governed by the executing conditions, that is, the machining parameters (speed, feed, depth of cut, number of passes, tool angles, lubrication, etc.). This aspect of process planning is dealt with in detail in the appropriate chapters of this book; the overall objective is to attain optimum machining conditions according to economic considerations, explained in Chapter 6.

2.3 GROUPING OF PROCESSES INTO JOBS

Up to now, each feature has been considered as a separate item and no provision has been made to take into account requirements of position and orientation between features. In order to observe tolerances of location (position and orientation), it is, however, imperative to group operations in the same job so that the machine tool can be used at its best accuracy. A succession of set-ups for executing the sequence of operations would produce an undesirable stack-up of tolerances. For example, if surfaces (2) and (4) have to be perpendicular with a tolerance of 0.02, or, in other words, become associated surfaces, both surfaces should be machined in the same fixture on a lathe. Because grouping also has an economic impact, it is advisable in this case to associate surface (5) with the group (2–4) in order to be processed in the same job. A similar logic is applied to the group (1 + 3); this has no obvious tolerance relations, but it is more economical to produce the surfaces in the same job.

2.4 SELECTION OF MACHINE TOOLS

Following the grouping of surfaces in section 2.3, we can now proceed to the selection of machine tools appropriate for executing the jobs defined earlier. The main criteria of selection are:

- the number of tools necessary to carry out the different operations;
- the size of the batch which has to be considered in relation to the set-up time; and
- the accuracy of the selected operations.

Taking the part in Fig. 2.1, a lathe would be better for job (2 + 4 + 5) than a boring machine, where the number of tools to be used and the difficulty of adjusting the tools is much more critical. The type of lathe will depend on the size of the batch. The different types of lathes are:

- lathe with adjustable switches,
- turret lathe,
- semi-automatic and automatic lathe, and
- a CNC lathe,

all of which have specific characteristics in terms of accuracy and productivity.

Similar considerations apply to job (1 + 3) where a choice has to be made between drilling machines of different kinds, such as:

- a sensitive hand drilling machine,
- a rotating head drilling machine,
- a multispindle drilling machine,
- boring machines in line, etc.

Also, the use of guide bushings is possible. The final choice will depend on economic considerations.

The information available on the machining equipment will give details on the capabilities of the machines, and the process planner is guided by this when considering times of set-up operations, machine setting, cost of machine time, flexibility of machine types and so on. In modern enterprises, computerized machine files help in making fast and reasonable choices.

The choice of machine tools is basically between universal machines and special machines. In the case of a universal machine, the cost of machine time is high, the setting of tools is more time consuming, the clamping of the part in a single fixture requires the transfer of tolerances which lowers production tolerances and raises production costs and not all the capabilities of the machine are efficiently used. On the other hand, because the fixture is unique it is more economical and the machine is more flexible. Universal machines are usually used for single parts or small batches.

In the case of special machines, the dimensions are executed as direct dimensions, the machine cost is lower, the operators are cheaper and the setting is simpler, but the fixtures are special and more expensive. Also, a change from machine to machine is generally necessary and this influences the stack-up of tolerances. Special machines are usually used for mass production, or large series.

2.5 SEQUENCING THE OPERATIONS ACCORDING TO PRECEDENCE RELATIONSHIPS (ANTERIORITIES)

The operations defined in section 2.2 have to be put in a certain order according to precedence relationships or anteriorities based on technical or economical constraints. The different categories of anteriorities can be classified in the following way:

- dimensions with a datum as anteriority;
- geometric tolerances with data references as anteriorities;
- technological constraints in order to execute sequences of operations properly; and
- economic constraints which reduce production costs and wear or breakage of costly tools, etc.

Examples of these different types of anteriorities are explained below:

1. Figure 2.4 illustrates the case of a dimensional precedence: surface F1 being dimensioned in relation to the rough surface B1, it is logical to execute surface F1 before surface F6 (also because of the tighter tolerances ± 0.3). If for certain reasons, F6 has to be executed before F1, a transfer of dimensions would be necessary and a reduction of tolerances would be the result.
2. Figure 2.5 illustrates a geometric anteriority where the tolerance of coaxiality of hole (2) is referred to shaft (1), taken as a datum and therefore an anteriority.
3. Figure 2.6 shows an important practical case relating to the choice of a datum between a centering pin and a plane. The anteriority should go to the plane because when using the pin, the error of positioning is the sum of the

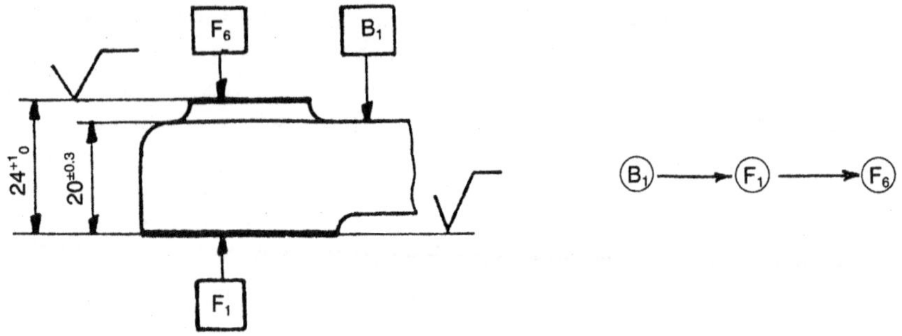

Fig. 2.4 Anteriorities: case of a dimensional precedence.

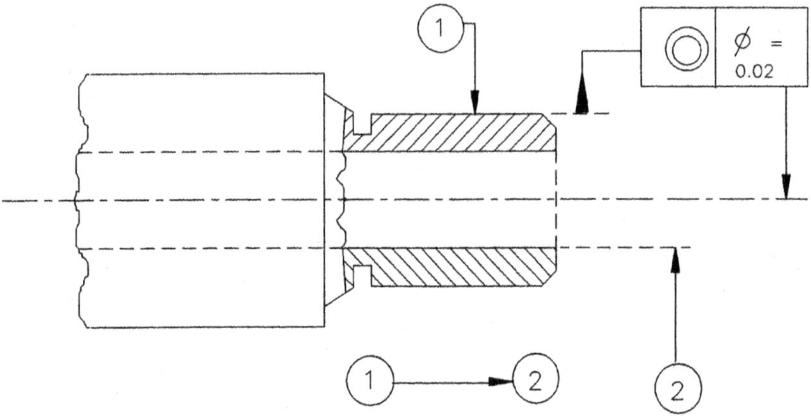

Fig. 2.5 Anteriorities: case of a geometric precedence.

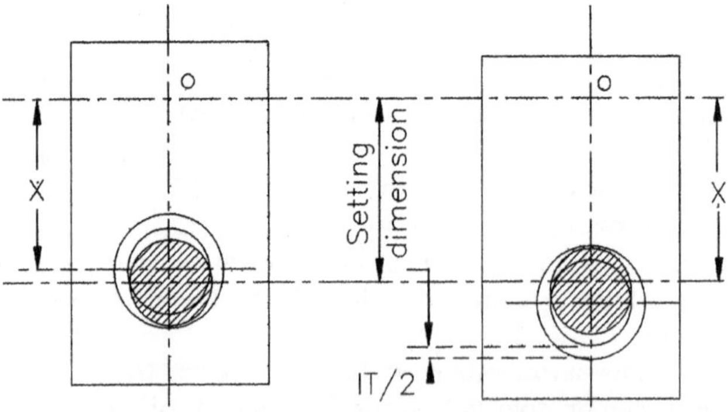

Fig. 2.6 Anteriorities: case relating to the choice of a datum.

tolerances of the hole (*IT*), plus the clearance between the minimal diameter of the hole and the diameter of the pin (*J*), plus the tolerance on the position of the plane.

On the other hand, using the plane as an anteriority, the only error comes from the error on the distance between the plane and the centre of the hole. Therefore, the error in the case of the pin taken as first datum is increased by $(IT+J)$.

4. Figure 2.7 gives an example of a technological constraint. This anteriority says that a hole having the smallest diameter or the longest depth has to be machined before another hole of a larger diameter or a lesser depth. This is because the straightness of a thinner hole or of a more precise hole can be impaired by the larger or less precise hole when they intersect (e.g. H7 has to be produced before H11).
5. Figure 2.8 shows two cases where burrs are not allowed in a precise hole. In one case, burrs are not accepted in hole (1) and therefore chamfer (3) has to be executed before hole (1). In the other case, hole (2) has to be executed before hole (1) because burrs are not allowed in hole (1).
6. Figure 2.9 illustrates economical constraints. In one case, drilling of two coaxial holes could be performed by drilling the long and small hole first.

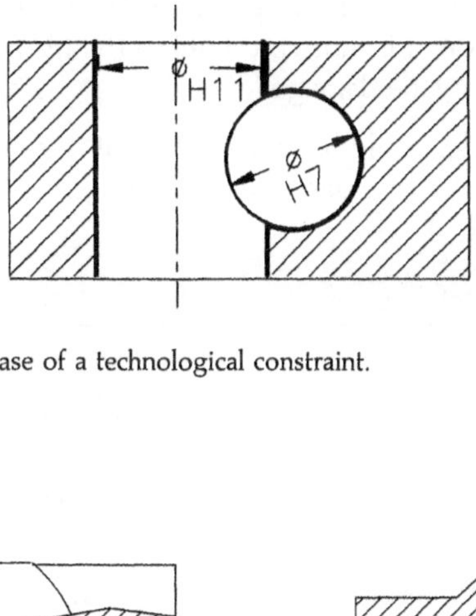

Fig. 2.7 Anteriorities: case of a technological constraint.

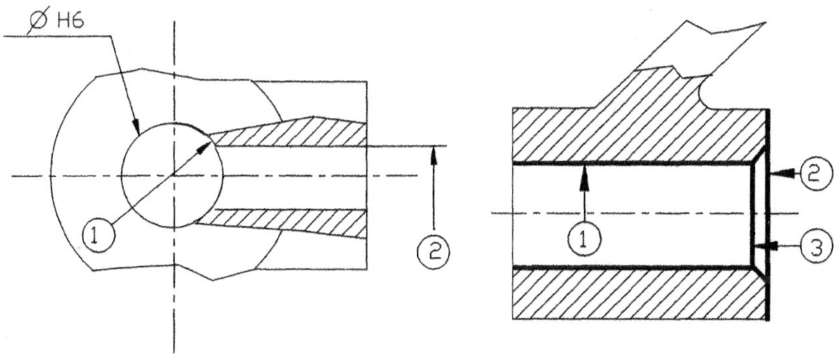

Fig. 2.8 Anteriorities: case of a burr.

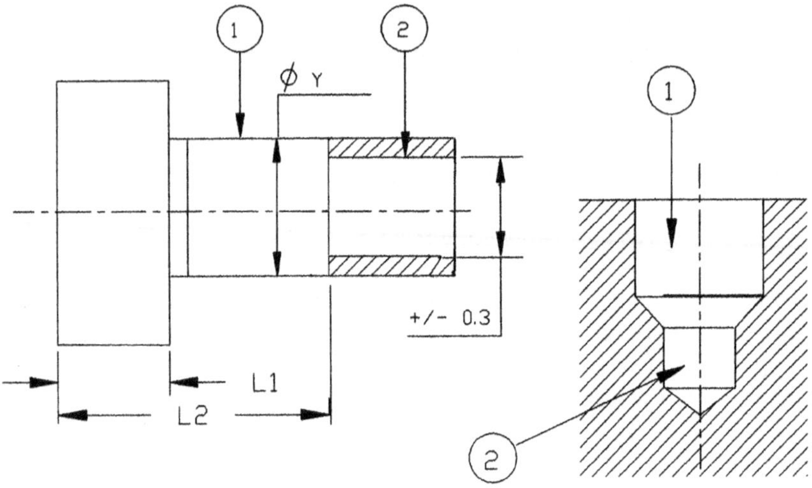

Fig. 2.9 Anteriorities: case of economical constraints.

However, by drilling the large hole first, the length and time of processing is reduced and time spent on the large hole is not influenced by the presence of the small hole. Therefore, it is more economical in fact to drill the large hole first. In a second case, it is more economical to finish surface (2) first so that surface (1) would not be damaged.

Many other examples could be mentioned, but it is more useful to consider Table 2.2, which relates to the part in Fig. 2.1 and gives the anteriorities in this

Table 2.2 Table of anteriorities

	Anteriorities			
Operations	*Dimensional*	*Geometr.*	*Technol.*	*Economical*
1F	4F	2F	orthogonal	
2R				$4\frac{1}{2}$F economy of machining
2F	B1		2R	
3F	2F	1F		
4R				
$4\frac{1}{2}$F			4R	
4F		B2 ◎ 2F ⊥	$4\frac{1}{2}$F 5F	2F, protection of tool 5F, no burrs
5F	2F			$4\frac{1}{2}$F economy of machining

F = finish. R = rough.
Source: Karr, J., *Methodes et Analyses de Fabrication Mecanique*, published by Dunod, Bordas, Paris, 1979.

particular case. Here, obviously, the anteriorities of B1 for 2F and of 2F for 3F and 5F are bound by dimensional tolerances, where the meaning of the symbols are: 2R for roughing of surface (2), 2F for finishing surface (2) and B1 raw surface as in Fig. 2.1. On the other hand, anteriorities B2 and 2F for 4F are governed by geometric tolerances. However, the anteriorities of 4F for 1F, as well as 2F for 1F are not clearly indicated in the drawing. In the first case, the anteriority is dimensional and is justified by the necessity to position the four holes (1) in relation to the center of hole (4). As for the anteriority of 2F to 1F, which is geometric (orthogonal) and the anteriority of 1F for 3F (coaxial), it is the decision of the process planner to add these anteriorities in order to improve the geometric quality of the workpiece. Such cases are very common in process planning and emphasize the importance of the knowhow of the process planner.

The technological anteriorities of 2R, 4R and $4\frac{1}{2}$F (semi finishing of surface 4) for 4F are obvious, but it was found necessary to add 5F because no burrs are permitted in hole (4) which is of a high geometric quality. The economic constraints for 2R and 5F are justified by the gain of time in machining surface 2R and 5F when hole (4) has already been semi finished because of a reduction in the quantity of material to be removed. Similarly, the anteriority 2F for 4F is justified by the fact that the reaming tool for 4F is costly and should be protected when it penetrates surface (2).

Having defined all the anteriorities, it is now possible to find the right sequence of operations. But it should be remembered that this sequence is the result of the definition of relevant anteriorities. This is not a very easy decision to arrive at, as illustrated earlier in a simple example, and it is explained in more detail in Chapter 16. The consistency of the anteriorities depends heavily on the experience of the process planner.

A well-known method of finding the order of precedence of the operations is based on the use of a matrix (Fig. 2.10) in which the left part records the anteriorities according to Table 2.2. For example, before executing 4F, it is necessary to complete operations 2F, $4\frac{1}{2}$F and 5F which are marked by an X in the matrix. On the right-hand-side of the figure (the levels), the number of anteriorities for every surface is recorded. In the first column, for example, for surface (3F), there are two anteriorities which have to be executed before surface (3). In the case of surface 4R, there is no anteriority and this surface can be executed. Its index is noted in the bottom of the matrix. Proceeding to level 2, it can be seen that surface $4\frac{1}{2}$F can be processed because the anteriority of 4R has already been removed in the matrix (as shown by putting (4R) with a ring in the first column). This surface can be put on the bottom line. Proceeding iteratively in the same manner, an order of precedence of operations will appear in the bottom as shown in Fig. 2.10. This means that no operation in the line can be carried out before the preceding operations are completed. The matrix selection can easily be implemented on a computer.

Similarly, another method for sequencing the operations, based on a graph theory approach, is also suitable for computerized treatment. The difficulty in both cases comes only from the assessment of the anteriorities in Table 2.2

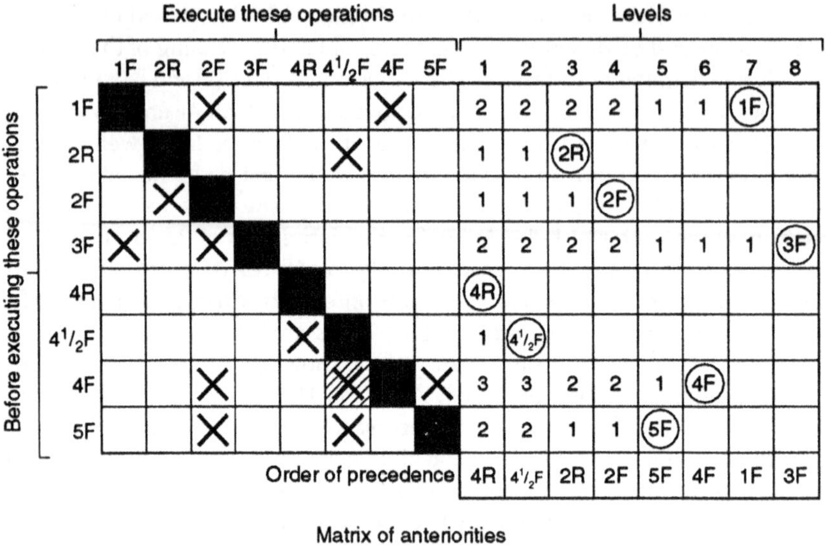

Fig. 2.10 Matrix of anteriorities.

which can result in contradictory conditions. It is then the task of the process planner to make decisions to solve the conflict by changing priorities respecting the most severe functional conditions.

After having established the order of execution of the individual operations, it is possible to make the necessary groupings as explained in section 2.3, in particular when position tolerances between features are to be respected.

As shown earlier, two groupings have been retained for accuracy reasons:

Job (1) : (2)+(4)+(5) (turning)
Job (2) : (1)+(3) (drilling)

They are also consistent with the order of precedence (Fig. 2.10). However, this is an ideal case; the conflicts have to be resolved by changes in the priorities decided by the process planner in relation to available machinery and functional requirements.

2.6 SELECTION OF WORKPIECE HOLDERS AND DIMENSIONAL DATA REFERENCES

Having chosen the operation groupings, it is now necessary to define positioning devices suitable for locating precisely the part in the operating coordinate system of the machine tool. In addition, the part has to be firmly clamped so that external forces caused by cutting or gravity cannot change the location of the part and destroy its stability. The first priority is, of course, given to the positioning function which influences directly the accuracy of the part. A more detailed analysis of the positioning and clamping functions is given in Chapter 7.

To simplify the locating problem and to have a proper starting point for the designing of fixtures, use is made of the principle of isostatism, which is explained in detail in Chapter 7. In the meantime, it is sufficient to explain that the isostatism principle in manufacturing is applied practically by using the 'six points principle'. This consists of defining six points of contact on the part which define its location uniquely, i.e. if the part is removed from its jig, it can be replaced exactly in its former position. For parts in a batch, it means that they are all located in an identical manner in relation to the machine reference system.

For example, in the case of the part in Fig. 2.1, to carry out the first job, the part can be located as shown in Fig. 2.11. Three points (1, 2 and 3) are placed on the plane B1 as shown by the locating symbols, and two points (4, 5) on the cylinder B2. The last degree of freedom is cancelled by point 6 (rotation around axis 4) which, in this special case, also has a clamping function.

This distribution of the six points has been chosen because of the way surface (2) relates to surface B1: it is sufficient to place three points on B1 to define a plane and to cancel three degrees of freedom (two rotations around x and y, one translation along z). The accuracy achieved by such a locating arrangement is 1.6 in translation (± 0.8) and 0.02 (i.e. 1.6/80) in rotation, assuming that the

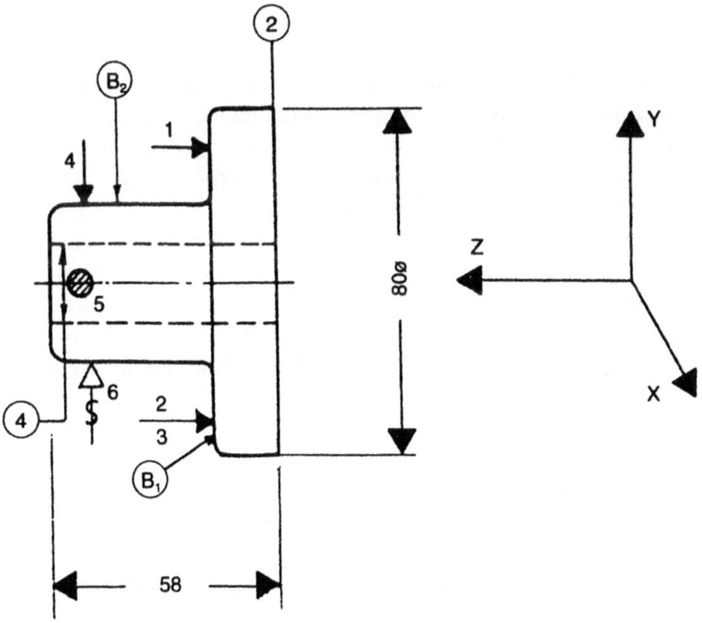

Fig. 2.11 Positioning and clamping of workpiece.

Table 2.3 Selection of locating points on isostatic jig

Surf. mach.	*Dist. to raw material*	*Surf. of contact*	*Dimens./angul. tolerance*	*Max. no. of contacts and d.o.f. cancelled*	*No. of contacts chosen and d.o.f. cancelled*	*Type of jig*
(2)	13 ± 0.8	B_1	Dim.: 1.6 Ang. for R_x: 1.6/80: 0.02 Ang. for R_y: 1.6/80=0.02	3 T_z R_y R_x	3 T_z R_y R_x	Plane on B_1
(4)	Center. Ø 2	B_2	Dim. for T_x, T_y 2 Ang. for R_x, R_y: 2/58=0.036	4 T_x T_y R_x R_y	2 T_x T_y	Center on B_2

Source: Karr, J., *Methodes et Analyses de Fabrication Mecanique*, published by Dunod, Bordas, Paris, 1979

diameter of face (2) is 80. Similarly, surface (4) is located in relation to its datum B2 by four points cancelling the four degrees of freedom of a cylinder (two translations along x and y, two rotations around x and y). The translation tolerances here are 2 and the rotational ones 0.053, assuming the height of the cylinder (4) to be 58. Obviously, a conflict arises in choosing six points for an isostatic positioning because, in principle, there are seven points involved instead of a possible five. The choice is finally made by placing three points on B1 and only two points on B2 because the precision is higher for surface (2). The sixth point in this case has an auxiliary role only (to cancel the rotation around axis 4) and does not influence the preceding analysis.

The choice of locating points are summarized in Table 2.3.

Concerning the order of execution of job (1) before job (2), already justified by anteriority reasons, it could be argued that a better accuracy of the operations is necessary for job (1) in comparison with job (2). This is a general principle in defining machining data references in process planning. It is more efficient from a tolerance point of view to begin processing more precise features so that the stack-up of tolerances for these surfaces if processed later will not create problems of tolerance reductions which are not welcome. More details on the datum surface selection are given in Chapter 7.

2.7 FINAL PREPARATION OF THE PROCESS PLANNING FILE

Having defined the sequence of operations and the positioning devices, it is now possible to establish a process planning file for the part. It describes in detail every job as shown in Fig. 2.12 for job (1) of the part already analyzed. Such a process sheet summarizes the sequence of operations defined earlier

Job	Subjob	Op.	Description	M/T	Positioning clamping	Tooling inspection
10			Turning	Semi Automatic Lathe	short center	chuck #3
		a	Drilling ø18±0.3		on (B2) (points 4,5) plane	jaws drill standard
		b	Face rough 14.5+0.5		on (B1) points 1,2,3	tool for facing
		c	Face finish 13±0.8		clamping #6	tool for chamfer
		d	Internal turning ø19.4–0.2			insert on bar borer ø20H7
		e	Chamfer			plug gauge 20H7
		f	Boring ø20H7			Standard surface finish (visual)

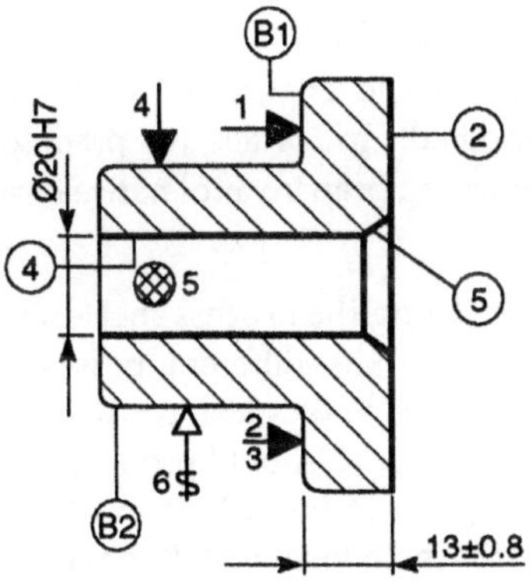

Fig. 2.12 Example of process sheet.

(operations 10a–f, see Fig. 2.1) and indicates the machine tools used, the method of positioning and clamping, the tooling for every operation and the inspection instrument to be used. A simplified sketch shows the part in relation to the surfaces to be processed (identified by thick lines).

However, in order to deliver a complete information package to the production department, it is necessary to add to the process file a certain amount of complementary data:

1. exact sizes of the cuts to be executed, e.g. in Fig. 2.12, for operation 10b, facing rough $14.5 + 0.5$;
2. details on the machining conditions for each operation (depth of cut, feed, lubrication, type of tool, cutting speed, length of machined segment, machine times);
3. description of the inspection instruments necessary to check the different features and to guarantee their conformity with the definition drawing; and
4. editing of the process plan file as a comprehensive document.

Detailed explanations of these different items are given in the following sections.

2.7.1 Transfer of dimensions and tolerances in design to those in manufacturing

The dimensions and tolerances in manufacturing are not necessarily the same as in design. According to the process plan chosen, one of two situations can result:

1. The dimensions and tolerances on the drawing can be executed directly during the processing of the part. This is the case in Fig. 2.12 for the size 13 ± 0.8 during facing of surface (2).
2. The dimensions and tolerances on the drawing cannot be translated directly in production. They are the result of a chain of dimensions obtained during the production processes and eventually, both their tolerances and their dimensions, are modified. In order to predetermine the production tolerances, it is necessary to prepare a tolerance charting table (explained in detail in Chapter 4).

For the purpose of completing the process sheet in Fig. 2.12, it is sufficient to calculate the depth of cut of the different passes to obtain the final size according to accuracy and technological considerations. For example, the operations 10b and 10c in Fig. 2.12 should produce a final size of 13 ± 0.8. It is assumed that the roughing operation 10b from the raw material can have a tolerance of ± 0.5. If the finishing facing depth has to be between 0.7 and 2.8, it is possible to calculate the dimensions of the rough cut as follows:

Minimum depth cut in roughing
$= $ minimum size for roughing (R_{min}) $-$ maximum final size

or

$$0.7 = R_{min} - 13.8$$
$$R_{min} = 13.8 + 0.7 = 14.5$$

Maximum depth of cut in roughing
$=$ maximum size of roughing (R_{max}) $-$ minimum final size

or

$$2.8 = R_{max} - (13.0 - 0.8)$$
$$R_{max} = 12.2 + 2.8 = 15.0$$

Therefore, the size of the rough cut is:

14.5 + 0.5 with a tolerance of 0.5 as requested.

For other cases, when R_{max} or R_{min} are not imposed, the tolerance of the rough cut can be imposed. The sizes of the operations 10a, 10d and 10f have been calculated according to the preceding assumptions of minimum/maximum depth of cut and tolerances of the intermediary passes.

In the following chapters, the values of depth of cut will be looked at from an economic rather than technical point of view. Both have to be considered in the final optimization of working conditions.

2.7.2 Determination of machining conditions, times and costs

As well as determining the cutting depth as carried out in section 2.7.1, it is necessary to give to the production personnel some indication of the machining conditions to be used (using symbols according to ISO Standard 3002, 1977):

1. the cutting speed (v_c) and the speed of rotation of the part or of the tool (n) which is the practical input in machine tools;
2. the feed per revolution (f) or the feed speed in translation of the machine elements (v_f);
3. the depth of cut (a) (or engagement) determining the width of the removed chip;
4. the length (l), width (b) and diameter (d) of the part; and
5. the type of tool chosen and its angles, as well as the type of lubrication.

Figure 2.13 gives an example of an instruction sheet in the case of a roughing operation in turning. This illustrates the choice of these parameters of machining using this method.

The machining parameter values for each operation are calculated according to practical values found in handbooks (Metcut Research Associates, 1972) or from experience. They can be worked out more comprehensively by using calculations giving optimum values of the parameters described in Chapters 6 to 11. Both methods are used in the development of technical data banks (Weill *et al.*, 1978).

The instruction sheet is also used to evaluate the production times and, consequently, the production costs. In Fig. 2.13, the different time components are indicated for the example used here, mainly manual times, machining times, preparation times and the total time for a batch. If the latter is found to be too high, some of the time components will have to be changed. Repeating this evaluation for every operation on a part will produce an estimated production

Job 20 S/J B	Turning ø48±0.5	batch size = 100 pcs.

L = Machining length, mm
Tc = Machining time, min
Tm = Manual time, min
Tz = Hidden time, min
Ts = Preparation time, min
v_c = Cutting speed, m/min
f = Feed rate, mm/rev
n = Revolution per minute
a = Depth of cut, mm
v_f = Feed speed, mm/min

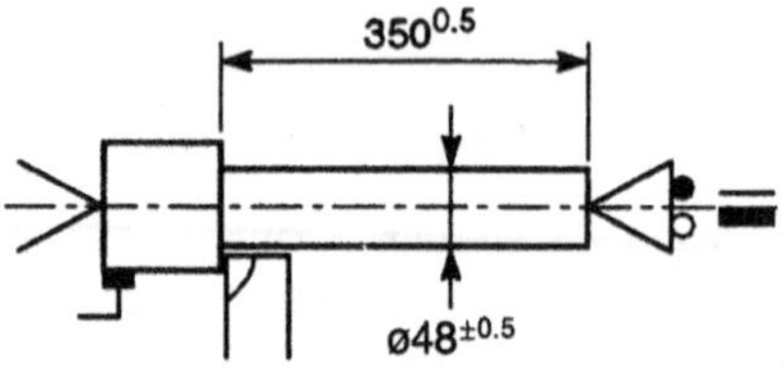

	v_c	f	n	a	v_f	L	Tc	Tm	Tz	Ts
Preparations										
Set up tool										15.0
Adjust speed										0.4
Setting 1st part										
Clamping between centers										
Turn	125	0.3	723	3.5	217	353				3.5
Unclamp part										0.2
Machining parts										
Clamping between centers								0.4		
Turn	125	0.3	723	3.5	217	353	1.63			
Unclamp part								0.2		
Inspection								0.3		
Tear down set up										6.0
Other paperwork										12.0
Total							1.63	0.9		37.1
Increase for fatigue							8%	15%		10%
Total							1.76	1.0		40.8

Total batch time = 40.8 + (1.76+1.04) x 100 = 320.8 min.

Fig. 2.13 Example of operation sheet.

time which can be converted into a production cost, to be used for an economic estimation of the production plan.

It should be mentioned that the optimization of a single operation cannot give a good optimum for the processing of the whole part. In Chapter 10, the problem of optimizing the production of a whole part is looked at and procedures proposed for the optimization of the whole sequence of operations executed on a part.

2.7.3 Selection of inspection instruments to check the final part

Because the finished part can never be guaranteed without checking its dimensions, quality control methods must be used at the end of the production process. The inspection process is a final judgment on the dimensional quality of a part and should be executed independently of the production department. It is sufficient here to mention the conventional metrology instruments used in simple quality checking operations, such as slide gauges, micrometers, plug and snap gauges, dial gauges, etc. The plug gauge 20H7 in Fig. 2.12 is an example. More advanced equipment is available for checking complex features. They take the form of three-dimensional coordinate measuring machines (CMM) which can be programmed to execute a large number of controls automatically. This equipment checks tolerances in the coordinate system of the part itself, i.e. in a functional way.

2.7.4 Editing the complete process plan

After preparation of the set of information to be transferred to the production department, a file should be established. It can of course, be created using a computer and processed accordingly. Figures 2.12 and 2.13 indicate how the information contained in the file should be presented. Using a computer, it is also possible to compare different process plans in terms of time and/or costs.

2.8 APPLICATION TO A REAL INDUSTRIAL PART

A fully developed process plan for a real industrial part is presented in Chapter 16. It reviews all the phases of a process plan from inspection of the raw material to the finished part, going through the choices of manufacturing technologies used in the production of the part. We show how important it is for a process planner to possess enough expertise to make decisions which do not necessarily result from the specifications in the part drawing. The example demonstrates how the methodologies developed in this book can be applied to a real case.

2.9 CONCLUSION

The purpose of this chapter has been to present to the non-expert a logical method of designing a process plan for a given part. However, it left out a number of aspects which are important in practical terms but which are of a more empirical nature and therefore did not fit into the proposed logic.

To complete, at least partially, the description of a process plan, it seems reasonable to review briefly the aspects which were left out, and which are dealt with in specialized documents:

1. Size of the series influences very critically the choice of machine tools and associated equipment. For example, special fixtures are only justified for large series. In general, the choice of equipment is not obvious and economic calculations have to be carried out in order to compare the different options.
2. Qualifications of personnel also have a bearing on the number of parts to be produced. For large series, the machines are automated and do not need highly qualified operators, although very detailed instruction sheets must be prepared for the setting and operating of the machines. For smaller series, these functions can be left to the initiative of the better-qualified operators who are employed for this purpose.
3. Any hazards and security factors concerning the operators have to be taken into consideration when planning production equipment, as they will, of course, influence manufacturing productivity.
4. When preparing the raw blank, it is important to allow for excess material so that external damaged layers resulting from casting and forging can be removed. It is also important to make sure that any jointed surfaces on the blank are in a non-functioning area of the part.
5. Primary processes, thermal treatments, or the removal of large volumes of material can cause internal stresses in the part. It is necessary therefore to implement machining passes during manufacture to remove any stresses before finish machining. Alternatively, stabilization treatments should be applied before the final operations and after thermal treatments such as quenching and annealing. All the deformations have to be removed before final grinding or honing in order to achieve a high standard of accuracy.
6. To improve the quality of mechanical surfaces, a great variety of surface treatments are applied during the manufacturing process. The most important of these are thermal treatments such as case hardening and nitriding which improve the hardness of the external surfaces, or mechanical treatments such as sand blasting, shot peening, polishing and anticorrosion spraying. These treatments change the surface properties and influence the process plan, i.e. they change the hardness of the part which then needs the application of non-conventional processes such as precision grinding, electrodischarge machining, electrolytical machining and so on.

This short review of the very important factors influencing a process plan has probably convinced the reader that a good process planner must acquire much expertise in many technological fields. The purpose of the logical approach to process planning discussed in this chapter merely offers the process planner a draft plan which he or she must build on progressively through qualifications and experience in manufacturing technologies.

REFERENCES

ISO Standard 3002 (1977) *Geometry of the active part of cutting tools.*

Metcut Research Associates (1972) *Machining Data Handbook,* Cincinnati, Ohio, USA.

Weill, R., Lemaitre, F. and Agaise, C. (1978) The development of technological data banks for small manufacturing systems, *Proc. of the CIRP Seminar on Manufacturing Systems,* **7**(1), 15–26.

Weill, R., Spur, G. and Eversheim, W. (1982) Survey of computer-aided process planning systems. *CIRP Annals,* **31**(2), 539–51.

CHAPTER 3

Geometric interpretation of technical drawings

3.1 FUNCTIONAL ANALYSIS OF TECHNICAL DRAWINGS

When a technical drawing of a mechanical part is handed over to the production planning department, this drawing must be considered as a contract, to be honored in all its details. This drawing, prepared by the design department, expresses certain functional requirements which have been defined in relation to the functionality of the part in the framework of the complete product.

The part is defined in such a way that, when assembled with the whole mechanism, it will fulfill its technical functions and be dimensioned and toleranced so that it can be mounted in a subset of parts in a completely unterchangeable manner. Of course, the designer must also take into consideration the feasibility of manufacturing the part and should try to find the most suitable technology for producing it (manufacturability). For completeness, the design should incorporate ways and means of inspecting the finished part in an economical way so that its quality assurance can be guaranteed. Obviously, respecting all these conditions, some of which can be contradictory, is a difficult task for the industrial designer. The recently developed 'concurrent engineering' approach tries to provide a solution to this problem.

These requirements are translated into a technical 'language' recognized by the production department and depicted in the technical drawing, as shown in the example in Fig. 3.1. Apart from general information such as the nature of the material of the component, its designation, its coding, the number of parts to be produced, its weight and ultimately, the primary process used to obtain, an accurate rough blank (Chapter 2, Fig. 2.1), the basic information on the drawing indicates all the dimensions and tolerances (dimensional and geometric) of the different features in the part. These parameters, defining size and accuracy of the different features, and their relative locations, have been determined by detailed studies of the functionality of the part through 'geometric analysis'. It is not the intention of this book to discuss the different aspects of this approach, as details can be found in more specialized documents (Gladman, 1967). The following analysis will be limited to a basic interpretation of the symbols and quantities found on a drawing and how they are referenced in international standards.

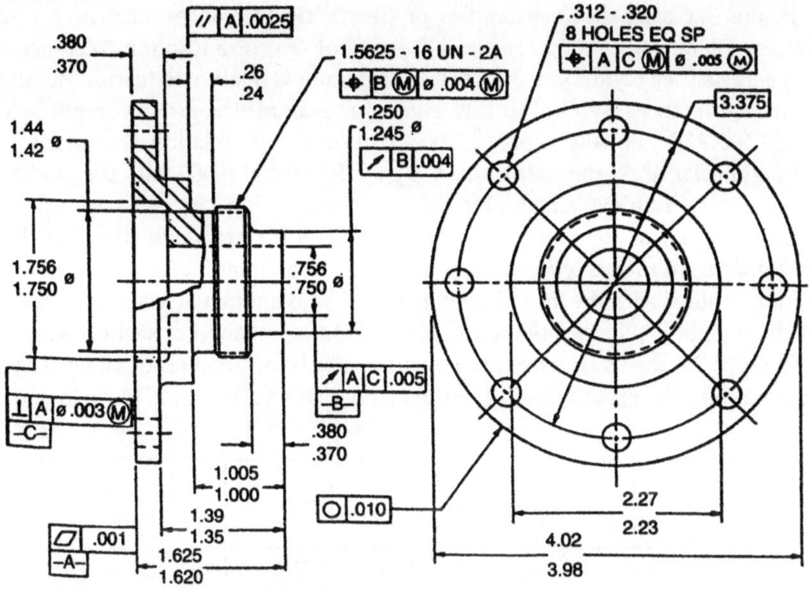

Fig. 3.1 Drawing of a mechanical part.

3.2 TOLERANCING AND DIMENSIONING STANDARDS ISO AND ANSI

Over the last decade, the standards organizations have built up a consistent range of dimensioning and tolerancing rules which represent an international language for all technical drawings. The International Standard Organization (ISO) is a worldwide organization which has issued a number of international standards in the field of dimensioning, tolerancing, inspection gauges, etc., which guarantee the interchangeability of industrial parts.

National standards organizations, such as the American National Standard Institute (ANSI, 1982) have completed these international standards by interpretation rules and, sometimes, additions of concepts which are needed for a better application of the ISO standards. But, basically, the ISO standards, which are also constantly updated, represent a well-accepted body of conventions and rules which applies to all industrial products in the world and permit a good understanding for successful dimensioning and tolerancing of industrial products.

As shown in Fig. 3.1, the drawing defines the sizes and dimensional tolerances for all the features contained in the drawing, e.g. for the diameter Ø 1.756/1.750, it is understood that this diameter should be between 1.750 and 1.756 inches. It can also be written as 1.753Ø $\pm$0.003. The exact meaning of the dimensional quantities is detailed in the *ISO Standard System of Fits and Limits* (ISO Standard 286, 1988).

Figure 3.1 also shows examples of geometric tolerances such as a form flatness tolerance of 0.001 for plane A (symbol -A-) or a location tolerance of perpendicularity (symbol ⊥) of Ø0.003 of feature C relative to feature A. Also, position tolerances (symbol ⌖) are given; for example the group of eight holes Ø0.312–0.320, equally spaced, is positioned in relation to plane A (perpendicular to A) and to the hole C by a tolerance of 0.005 with the modifier Ⓜ, which will be explained later.

The geometric tolerances of form and position are defined in the *ISO Standard for Tolerances of Form and Positions* (ISO Standard 1101, 1983), with a section dealing with the important concept of 'maximum material condition' or MMC (symbol Ⓜ) (ISO Standard 2692, 1988), explained in section 3.3.

In Fig. 3.2a, the usual terminology and symbols for geometric tolerances are represented. The exact meaning of the symbols is explained in Figs 3.2b and c,

Category	Characteristic	Symbol	Datum references
Form	Flatness	⏥	Never uses a datum reference
	Straightness	—	
	Circularity	○	
	Cylindricity	⌭	
Orientation	Perpendicularity	⊥	Always uses a datum reference
	Angularity	∠	
	Parallelism	//	
Location	Position	⌖	
	Concentricity	◎	
Runout	Circular runout	↗	
	Total runout	⌰	
Profile	Profile of a line	⌒	May use a datum reference
	Profile of a surface	⌓	

Fig. 3.2a Geometric symbols. Redrawn from Krulikowski, A. (1990), *Geometric Dimensioning and Tolerancing, Self Study Workbook*, published by Effective Training Inc., 20968, Wayne Road, Westland, MI 48185.

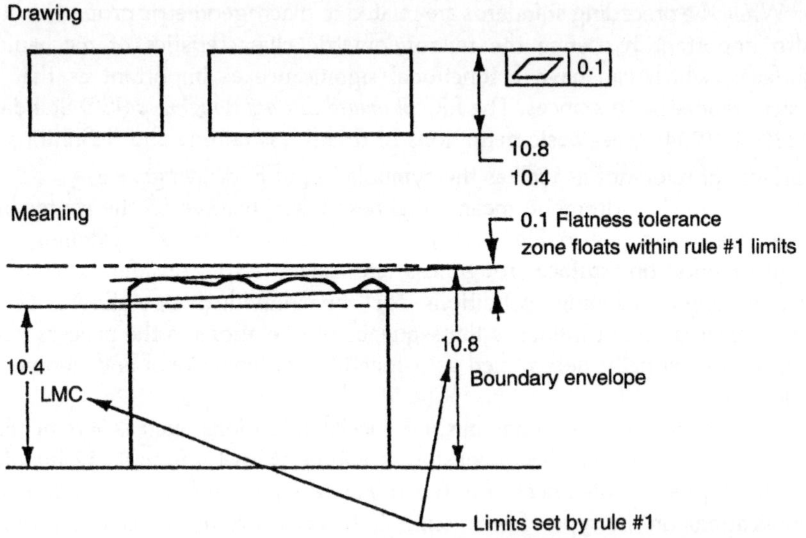

Fig. 3.2b Geometric symbols: flatness. Redrawn from Krulikowski, A. (1990), *Geometric Dimensioning and Tolerancing, Self Study Workbook*, published by Effective Training Inc., 20968, Wayne Road, Westland, MI 48185.

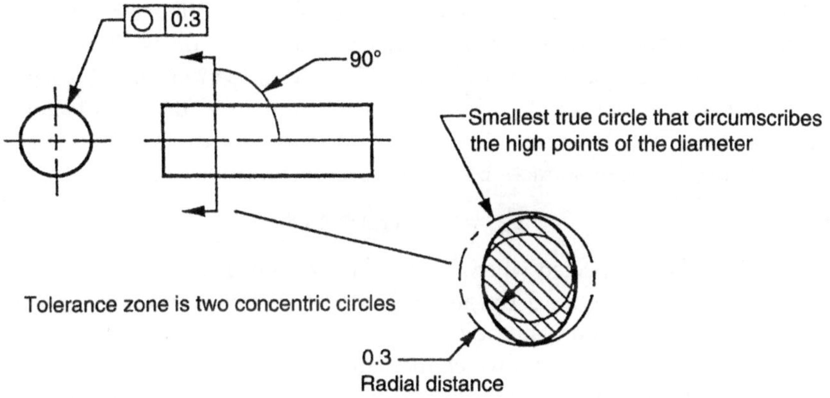

Fig. 3.2c Geometric symbols: circularity. Redrawn from Krulikowski, A. (1990), *Geometric Dimensioning and Tolerancing, Self Study Workbook*, published by Effective Training Inc., 20968, Wayne Road, Westland, MI 48185.

giving examples of a tolerance of flatness and a tolerance of circularity. For location tolerances a datum is indicated, e.g. hole C for the group of eight holes in Fig. 3.1. The exact meaning of datum surfaces is given in ISO Standard 5459 on datum systems for geometric tolerances (1981). Indications on practical ways of simulating theoretical data by real surfaces of high geometric quality are also given in the standard.

While the preceding standards are related to macrogeometric properties, it is also important to define the microgeometric characteristics of mechanical surfaces, which can have a functional significance as important as that of macrogeometric tolerances. The *ISO Standard Surface Roughness* (ISO Standard 4287–1, 1984) gives basic definitions of roughness criteria and definitions of surfaces of reference as well as the symbols to put in drawings, e.g. $\sqrt{}^{R_a 3.2}$ to characterize the arithmetic mean roughness taken relative to the center line reference (mean line in Fig. 3.3). For the purpose of process planning, the requirements on surface roughness are taken into consideration when determining machining conditions for the individual operations. These requirements do not influence the sequence of operations in the process plan, which is essentially determined as a function of dimensional and geometric tolerances.

The drawings can also contain some special indications, such as '4 mini' (Fig. 2.1), which means that this dimension should be at least 4 mm (0.157 in) wide (without precise tolerances) for functional reasons. Such an indication has implications on the type of processing, in this case on the choice of a tool for countersinking and respecting this requirement. Indications of the same type relate to the concept of 'liaison to the raw blank' which expresses the need to retain a minimal distance between a feature in the part and the surface of the raw blank, so that conditions of material strength or of collision are obeyed. On the other hand, the rough surface is commonly used as a starting surface for the processing of the part, as explained later in Chapter 7, on positioning and clamping. For example, in Fig. 3.4, which relates to the part in Fig. 2.1, a liaison of 58 mini to the rough surface B3 from plane 2 is imposed. Because of another already toleranced dimension (13 ± 0.8), it is necessary in this case to tolerance the distance between surfaces B1 and B3 by a minimal dimension:

$$b_{min} = 58 - 12.2 = 45.8$$

which has to be respected when casting the rough blank.

In some other cases, a size can also be limited by a maximum value, e.g. when defining the depth of a blind hole.

Despite the accuracy with which the dimensional and geometric tolerances have been defined by the standardization bodies, it is well known that the indications on a drawing have, in many cases, to be interpreted to ensure the interchangeability of parts. New standards or complements to existing

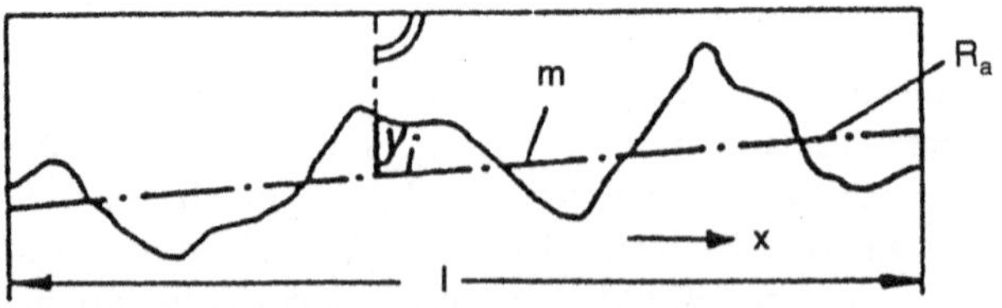

Fig. 3.3 Surface roughness: center line average (CLA). R_a is the mean value of y_i.

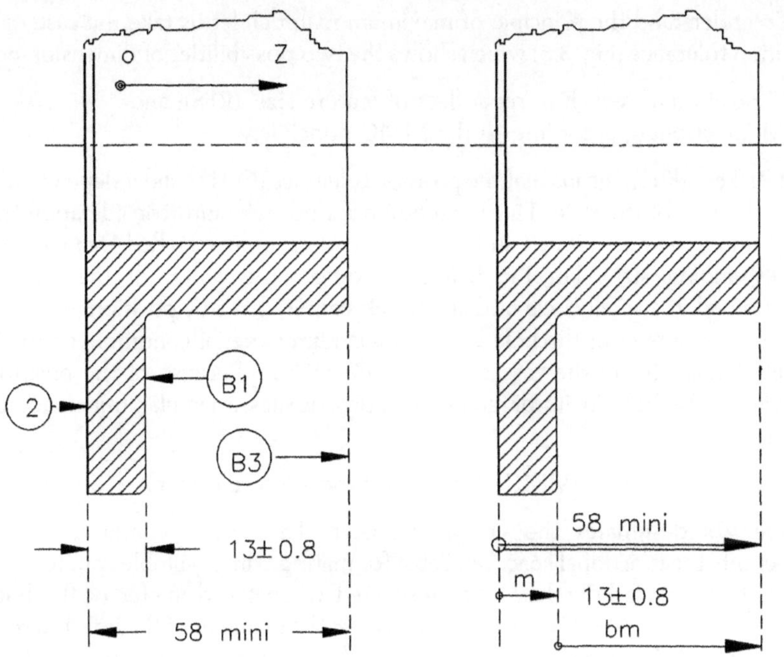

Fig. 3.4 Special indication 'mini'.

standards have therefore been issued and will be addressed later on. A part of the improvements brought to the old standards are inspired by the will to permit easier and more economical manufacturing. Of course, the requirements of functionality remain a priority. A very important example of this way of thinking is illustrated in the concept of MMC (maximum material condition).

3.3 TOLERANCING FOR MANUFACTURABILITY

3.3.1 Application of the maximum material principle

As a good example of tolerancing for manufacturability, let us take the 'maximum material principle' defined in ISO 2692 (1988) and which states:

> There is a mutual dependence of tolerances of size and tolerances of form, position or orientation which permits an increase to the geometric tolerance as the tolerance-related features depart from their maximum material size provided the part does not violate the virtual condition.

The MMC principle therefore gives a bonus to the geometric tolerance in the case when a feature is not at the limit of maximum material size (the normal case) and, as a consequence, allows an enlargement of manufacturing tolerances which are certainly welcome from the point of view of economy.

To understand the principle of maximum material, let us take the case of a position tolerance (Fig. 3.5) which shows the two possibilities of dimensioning:

- 'S' positioning which is 'regardless of feature size' (RFS); and
- 'M' positioning according to the MMC principle.

The RFS condition means that the position tolerance ($\varnothing t$) is valid independently from the size of the hole. This condition does not take into consideration the function of the hole (which, in the majority of cases, is matched to a shaft of corresponding size). For the MMC condition, the center of the hole is constrained in a small circle of diameter $\varnothing t$ centered at true position, but only when the diameter of the hole is at its maximum material condition, i.e. at its smallest diameter in the case of a hole ($\varnothing - t\varnothing/2$). Because of the position tolerance, the hole in its different positions defines a circular free space of diameter:

$$\varnothing v = \varnothing \text{MMC of hole} - \varnothing t \text{ (position tolerance)}$$

where $\varnothing v$ designates the virtual diameter. The virtual condition which represents the functional space available for mating with a suitable counterpart, i.e. a shaft with a hole, has to be respected when the diameter of the hole deviates from the MMC diameter. In this case, the diameter of the hole is larger than the MMC diameter and therefore, the tolerance of position will be enlarged while respecting the 'free volume' defined by the virtual condition. The tolerance of position $\varnothing t$ will become

$$\varnothing t \text{ (actual)} = \varnothing t \text{ (nominal)} + (\varnothing \text{actual} - \varnothing \text{MMC})$$

which shows the bonus of tolerance in position which has been gained. The maximum bonus which can be gained is obtained when the hole is at its maximum diameter or at the 'least material conditon' (LMC), i.e.:

$$\begin{aligned}\text{maximum deviation of position} &= \varnothing t + [\varnothing t + t\varnothing/2 - (\varnothing t - t\varnothing/2)] \\ &= \varnothing t + t\varnothing\end{aligned}$$

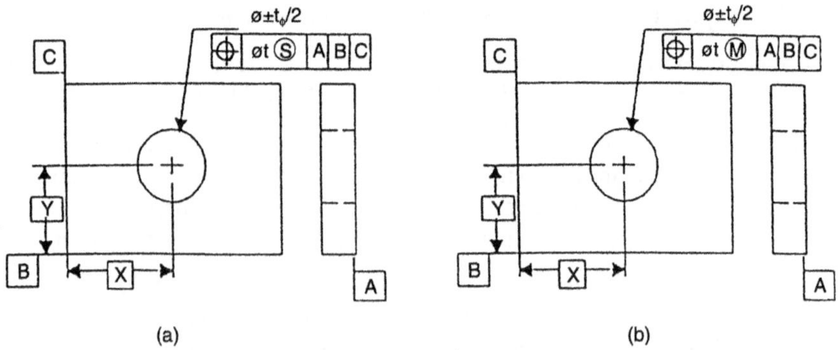

Fig. 3.5 Position tolerance specification: (a) RFS; (b) MMC.

To illustrate the gain of tolerance in a practical situation, Fig. 3.6 illustrates the case of two holes with a nominal distance of 50.8 (called basic) and position tolerances of 0.14 at MMC and four holes of diameters Ø6.4 + 0.1/0 (only two are presented in the figure). In the upper picture, the holes at MMC are represented at their theoretical position. Also, a mating part is represented,

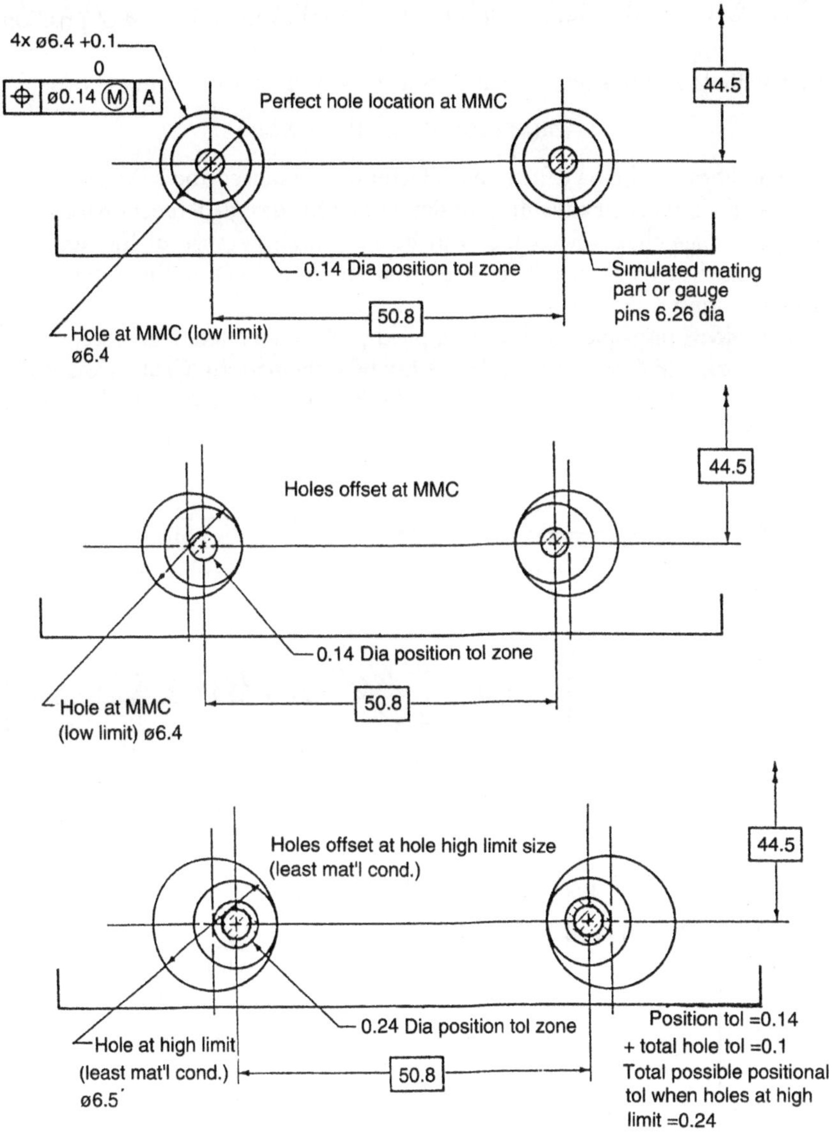

Fig. 3.6 Dimensioning holes positions. Redrawn from Krulikowski, A. (1990), *Geometric Dimensioning and Tolerancing, Self Study Workbook*, published by Effective Training Inc., 20968, Wayne Road, Westland, MI 48185.

i.e. a gauge pin, simulating the virtual condition with a diameter of (Ø6.4 − 0.14 = 6.26) as explained earlier. In the middle picture, the situation represented relates to a deviation of 0.14 of the holes at the MMC limit, but leaves a virtual space of diameter Ø6.26 in the worst situation represented. In the lower part of the figure, the diameter of the holes has increased to Ø6.5, the high limit or the least material condition (LMC). As seen earlier, this circumstance gives a bonus of (0.14 + 0.1 = 0.24) to the tolerance of position without impairing the virtual space condition in the worst situation represented. The position tolerance is increased up to its maximum value:

$$\varnothing t(\text{actual}) = 0.14 + 0.1 = 0.24$$

In addition to the benefits in manufacturing tolerances, the MMC principle has also the important advantage of simplifying the design of functional gauges and to reduce their cost, which can be very high because of the accuracy required for gauges. Thus it represents an important gain in the economy of manufacturing.

The MMC principle can also be applied to the data references of geometric features, e.g. C(M) as shown in Fig. 3.7, where the pattern (C) at MMC is the reference for the pattern of three holes. In this case, the gain by application of the MMC condition to the sizes of the three holes in the pattern is again

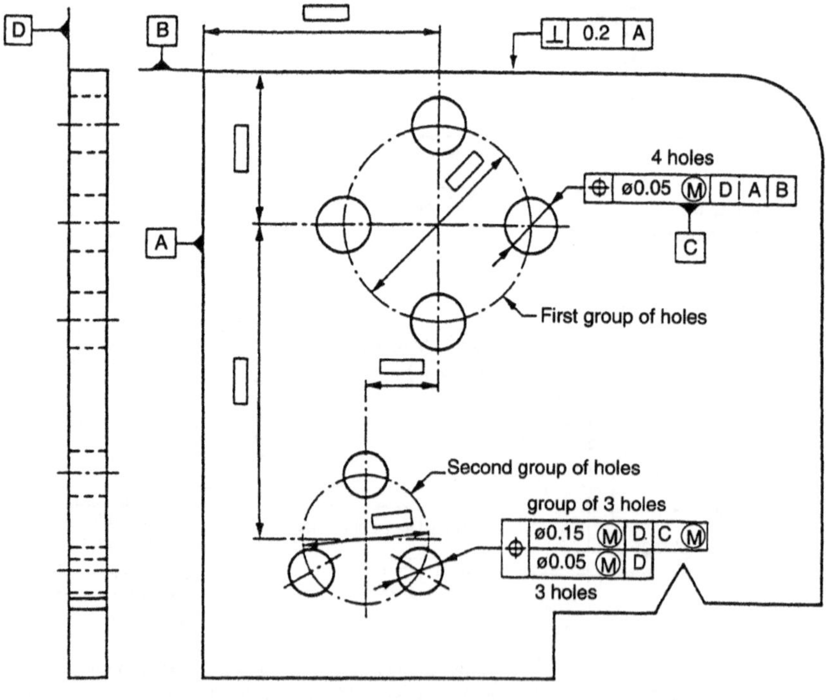

Fig. 3.7 Group of features nominated as datum.

increased by the MMC condition applied to the pattern (C) (composite tolerances). In other words, the positions of the three holes equally spaced benefit from two bonuses, one coming from the feature-related tolerance Ø0.05 and the other from the pattern-related tolerance of C(M), i.e. Ø0.15. Concerning the tolerancing of the position of pattern (C), there are tolerances of position Ø0.05 relative to the absolute data references A and B.

Without applying the principle of MMC, the design of the inspection gauge for such groups of patterns would have been very complex.

3.3.2 Application of the 'principle of independence'

Another interesting example of tolerance definitions which benefit manufacturing is represented by the 'principle of independence' defined in ISO 8015 (1985). It states:

> There is no dependence between dimensional tolerances and geometric tolerances, each being defined and checked independently of the other.

This statement means that if a part is dimensioned according to the independence principle, its size is checked by two point measurements (e.g. a number of diameters along the generator of a cylinder) without taking into account any influence of form errors. This interpretation is, of course, not compatible with the concept of 'mating', which ensures the assembly of part and counterpart, or male and female parts. Figure 3.8 shows the meaning of 'mating

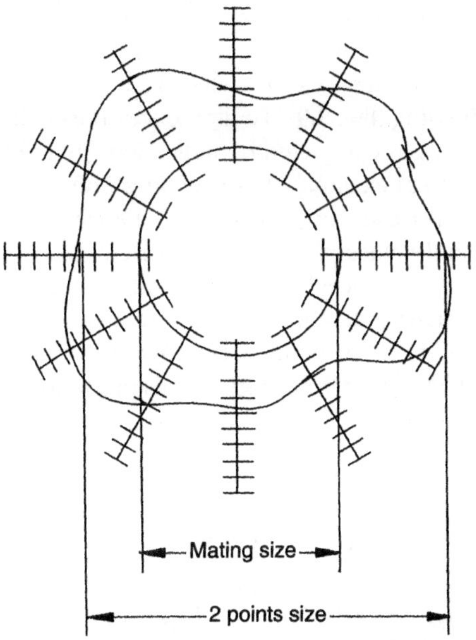

Fig. 3.8 The meaning of 'mating size'.

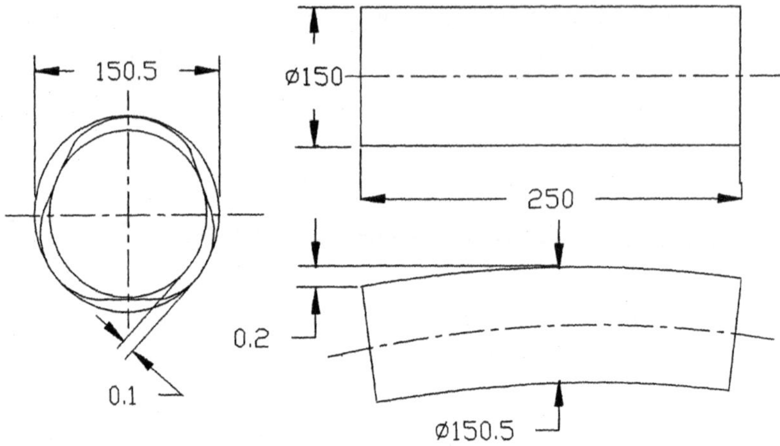

Fig. 3.9 Tolerancing according to the independence principle.

size' compared to '2 point' sizes, these latter being of course in infinite number and scattered within a given range. The difference between the two concepts is very important in the case of parts which have to be assembled with counterparts.

In many other cases, the principle of independence provides great simplification in tolerancing and economies of production for parts which have not to be toleranced in form or position. Figure 3.9 shows the dimensioning and tolerancing of a cylindrical part according to the independence principle. Obviously, the dimensioning has become very simple. The dimensions are defined by a class of accuracy – *mH* – which is defined in ISO 2768 (1989) and indicated in the drawing (Fig. 3.9). To limit excessive deviations of dimensions and form, the drawing is completed by indications giving the tolerances to be respected. In the present example, class *m* defines a dimensional tolerance of 0.5 and class *H* defines a class of geometric tolerances which define values of circularity of 0.1 and values of straightness of 0.2.

The use of this new concept of independence should greatly simplify the interpretation of drawings and execution of processes in manufacturing and its use is recommended as much as possible. However, because the old standards are still in common use, there is a danger of misinterpreting the indications on a drawing. This problem, and other, similar ones, are examined in the following section.

3.3.3 Problems of interpretations of existing standards

As a first example, let us take the problems raised by the application of Taylor's principle or the 'envelope principle'. As shown in Fig. 3.10, which relates to a plain cylindrical part, Taylor's principle indicates that the part should be without error forms when its dimension is at the maximum material limit – in

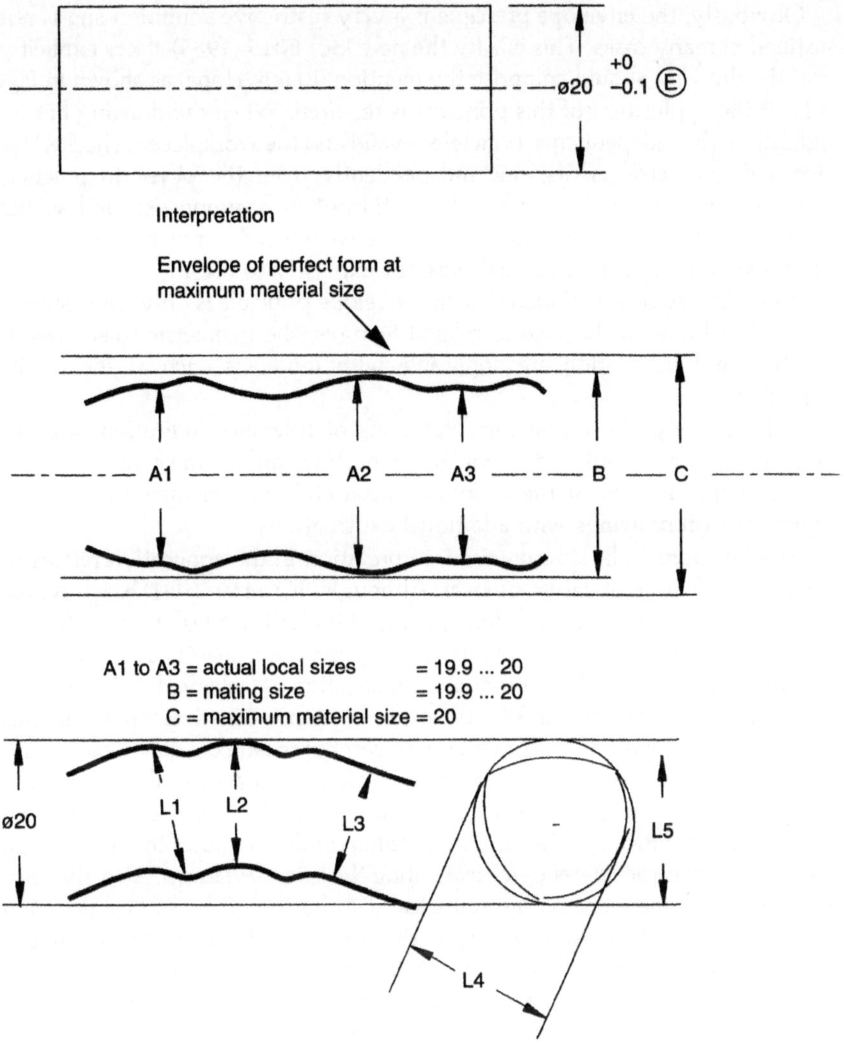

Fig. 3.10 Tolerancing according to the envelope principle.

this case at the maximal diameter – in every section along the length of the part. This condition is necessary when one is thinking of mating the part with a cylindrical hole of ideal form which has the same size as that shown in the figure. When, however, the part is not at its maximum material size, form errors could be allowed which will not prevent the shaft being introduced into the cylinder of the maximum size if they are limited in number. Concerning the minimum material size (LMC), on the other hand, there is no reason to prescribe form limitations and it is sufficient to check its value by two point measurements in different points along the part.

Obviously, the envelope principle is a very restrictive condition and is not justified in many cases. This is why the new ISO 8015 (1985) states explicitly that the drawing should comprise the mention (E) (envelope) as shown in Fig. 3.10, if the application of this principle is required. When the drawing has no indication, the independence principle is valid and the workpiece is checked for size and geometric errors, one independently from the other. In practice, however, the use of the symbol (E) is still not in common use, and so the interpretation of a drawing can be too restrictive when the envelope principle is observed, although the intention was not explicitly to use it.

It should also be mentioned that the envelope principle is only applicable to isolated features. In the case of related features, the geometric tolerances of position and orientation are applicable with modifiers such as the MMC condition.

Unfortunately, the recent interpretations of tolerance indications are not commonly understood and it will be some time before their application is correctly implemented. In the meantime, it is useful to supplement the standard indications on drawings with additional explanations.

Another domain lacking precise interpretation in the application relates to the exact definition of datum surfaces. Although ISO 5459 (1981) has provided significant clarification of this domain, it is still a difficult task to manufacture physical datum surfaces to the required geometric quality and for many different dimensions. A substantial simplification is achieved by using the MMC principle, but not all situations can be dimensioned according to this principle. In the case of measurements of computerised coordinate measuring machines (CMM), the problem of defining datum surfaces is also of primary importance and can give rise to errors in finding the real inaccuracies. The reason for this difficulty originates in the different definitions of the substitution surface, which is the reference for evaluating the form deviations. Also, the form error itself is dependent on the statistical distribution of the surface points in relation to the substitution surface, so that the exact datum as defined in ISO 5459 (i.e. the external surface of nominal form enveloping the real surface as closely as possible) is not precisely known in CMM measurements. Of course, practical datum surfaces (referred to as simulated in the standard) are successfully used in industrial applications for most current situations.

More common limitations in the use of dimensioning standards are due to ambiguities in the indications found on the drawings. In Fig. 3.11a, for example, the axis of the cylinder Ø20H7 (C) is positioned in relation to planes P_1 and P_2. However, there is an ambiguity because one can define its location in different ways:

1. the axis of C has to be inside two planes parallel to P_2, situated at a distance between 14.9 and 15.1 from P_2 and at a distance between 18.05 and 17.95 from P_1;
2. if P is a plane perpendicular to P_1 and containing axis C, plane P_2 should be comprised of two planes parallel to P and situated at 14.9 and 15.1 from it; and/or

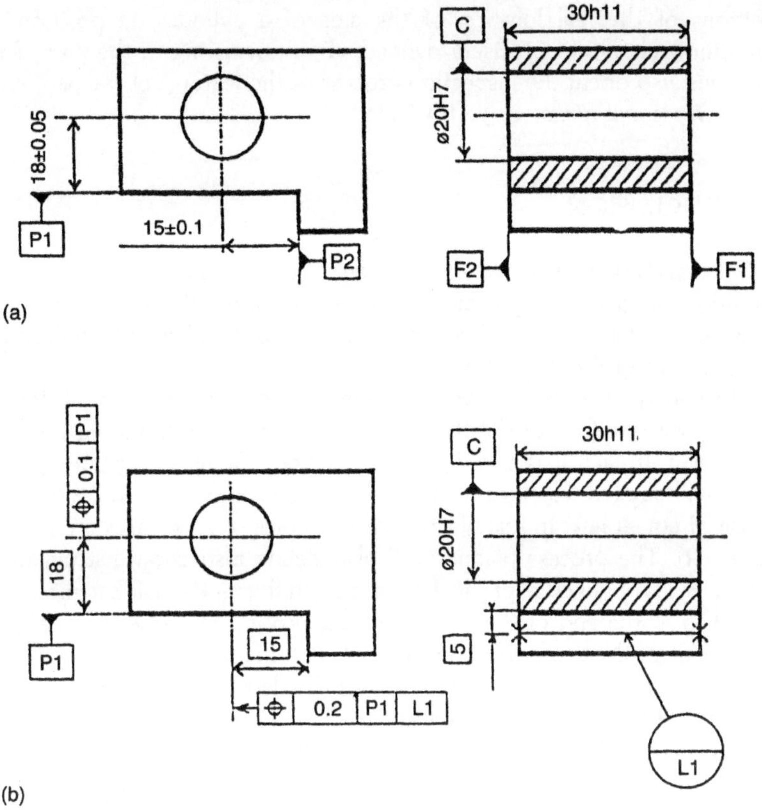

Fig. 3.11 (a) and (b) Resolve tolerancing ambiguity by geometric position tolerances.

3. point O being the intersection of the axis of C with plane F_1 and P a plane perpendicular to F_1 and P_1 passing through O, P_2 should be inside two parallel planes to P situated at distances of 14.9 and 15.1 from P.

This ambiguity can be resolved by geometric position tolerances (Fig. 3.11b). It means that axis C has to be between two planes perpendicular to P_1 and distant by 0.1 on each side from an ideal plane at a basic distance of '15' from the reference L_1. Also, in the other direction, axis C should be contained between two planes parallel to P_1, distant by 0.05 on each side from the basic plane at a distance of '18' from P_1. Obviously, the first interpretation of tolerancing conformed to this geometric positon tolerancing, but it must be stated that the solution implies a choice of a datum (P_2 in this case) which could be erroneous. If, for example, according to its function, plane P_1 were a primary plane and a line L_1 were the secondary datum (Fig. 3.11b), the tolerance indications of Fig. 3.11b would conform to the functional constraint.

In many other instances, to solve such cases of ambiguity, it is advisable to add to the drawing information regarding the real meaning of a tolerance, e.g.

definitions of the real location of the axis of a cylinder (in position and orientation) relative to precisely defined surfaces existing in the part. These indications also orient the inspection process of the features of the part, either with conventional means or with CMM.

3.4 CONCLUSION

A detailed and comprehensive examination of the drawing of a part is not only a condition *sine qua non* to produce the part so that it is functionally correct, but it is also the best approach for finding a suitable process plan for manufacturing and inspection of the desired part.

At this stage, it is also important to emphasize that the technical drawing does not limit the freedom of the process planner when designing a suitable process plan. In fact, it is possible that, in certain circumstances, the process planner will suggest changes in the design, for example a better tolerancing because of constraints in manufacturing. Examples of such cases are given in Chapter 16. The process planner will also define inspection instruments to check the sizes obtained in production, although the functional dimensions will be checked at the end of the manufacturing process in a special department independent of the production department. If conflicts arise between the two departments, it is the duty of the heads of the departments to solve them in favor of the functional requirements.

In conclusion, the process planner has plenty of freedom in designing the process plan, fulfilling first of all the functional conditions defined by the design office. A careful examination of the drawing will guide him/her in finding suitable manufacturing technologies; in addition, the process planner's experience in manufacturing will be a big determining factor in generating an optimal and robust process plan.

REFERENCES

ANSI – American National Standard Institute (1982) Y14.5, Standard on Dimensioning and Tolerancing.

Gladman, C.A. (1967) The role of geometric analysis in design for production, *Proceedings of the 8th International MTDR Conference*, The University of Manchester, Institute of Science and Technology, Sept., Pergamon Press, Oxford and New York, 1249–66.

ISO Standard 286 (1988) System of fits and limits.

ISO Standard 1101 (1983) Geometrical tolerancing.

ISO Standard 2692 (1988) Geometrical tolerancing, Maximum Material Principle, MMC.

ISO Standard 2768 (1989) General tolerances.

ISO Standard 4287-1 (1984) Surface roughness.

ISO Standard 5459 (1981) Geometrical tolerancing, Datum and datum systems for geometrical tolerances.

ISO Standard 8015 (1985) Fundamental tolerancing principle.

CHAPTER 4

Dimensioning and tolerancing for production

4.1 THE ACCURACY PROBLEM IN PRODUCTION

A part is defined by an engineering drawing which gives complete information on its geometry and associated data. It is the fundamental means of communication between the designer, the process planner and all other functions in the enterprise; it is the medium by which the designer's intentions and wishes are expressed in an unambiguous manner.

Dimensioning and tolerancing convert the drawing into a powerful language where minute variations of the designer's intentions can be expressed. Because it is impossible to execute exactly prescribed dimensions, variations in the dimensions are acceptable; this variation is called tolerance. Real parts can have errors, but these are acceptable as long as the dimensions are included in the tolerance range, called from now on the 'interval of tolerance' (or IT). Furthermore, by proper dimensioning of major functional tolerances, less important tolerances can be relaxed while ensuring that no difficulties will arise with the part when it is assembled with other parts. The designer achieves this by being able to use as many datum references as desired and convenient for the tolerancing procedure.

The process planner's task is to translate the requirements expressed by the rich and powerful language (the drawing) into a machinery language (the machine, the fixture, the tool) with a much more limited vocabulary. It is imperative in production to observe strictly the dimensions and tolerances which have been defined according to functional criteria. These quantities are to be found in the definition drawing of the part to be manufactured.

However, for various reasons related to the selected process plan, the mode of clamping of the part on its fixture and economic considerations, it very often happens that the functional dimensions are not executed directly in manufacturing. In this case, the functional dimensions are obtained as indirect dimensions, rather than direct dimensions or, in other words, as resultant dimensions of a chain of direct dimensions (also called component dimensions). The tolerance of a resultant dimension is then the sum of the tolerances of the component dimensions which are given by the process used in manufacturing.

Obviously, the result of this is that the tolerances of the component dimensions have to be small enough for their sum to comply with the tolerance of the resultant dimension given on the drawing. This can raise problems of tolerancing in production when production equipment is not able to produce parts to the small tolerances required. In this case, the only solution is to increase the tolerance of the resultant dimension, which can contradict design requirements, or to change the process plan and to use more precise equipment, which means increasing the cost of manufacturing.

This situation can be considered as the fundamental accuracy problem in manufacturing. It constitutes the background for this chapter, which proposes solutions for resolving this difficulty.

Before analyzing the proposed tolerancing strategies, let us review briefly the main factors contributing to inaccuracies in production. There are various causes of geometric inaccuracy. For instance, flatness, angularity and perpendicularity errors in milling can have one of several causes: machine tool geometric errors, workpiece deflection, cutting tool deflection, tool eccentricity, tool flatness and, in the case of producing the part in more than one subphase, refixturing of the part on separate surfaces.

Concentricity, run out and true position inaccuracies will occur when separate features are being machined on separate fixtures in more than one subphase. As can be seen in Table 7.2, each refixturing of the part introduces a large error. Machine tool errors, tool deflections and part deflections contribute to inaccuracies as well. In order to devise a process that meets geometric tolerance specifications, the following precautions should be observed:

- *Fixturing* When a geometric tolerance is specified, the only way to meet the specification is to machine the relevant surfaces in a single subphase, i.e. in one fixturing. Any refixturing or changing of fixture, will produce non-conforming parts (unless lengthy manual operations are executed).
- *Machine accuracy* Parts can only be as accurate as the machine on which they are produced.
- *Tool accuracy* Similarly, parts can only be as accurate as the tools to produce them.
- *Tool deflection* Tools deflect under the load generated by the cutting forces, so these forces have to be controlled by appropriate cutting conditions.

There are many other factors such as temperature influences, vibrations, material heterogeneities, kinematic and control errors and so on. In spite of the accumulation of all these errors, it is possible to produce accurate parts by careful choice of machine tools, machining conditions, appropriate tooling and accurate fixtures, and last, but not least, an optimal choice of tolerancing strategies as discussed later in this chapter. Nevertheless, errors will subsist in production because of uncontrollable factors. They can be compensated by proper resetting strategies (section 4.8). However, to minimize the in-process interventions, it is valuable to design tolerancing procedures which predict a good degree of accuracy in tolerancing on the basis of realistic accuracy values obtainable in production. The available strategies are detailed below.

However, before going into the details of tolerancing, it is useful to define the different types of dimensioning encountered in manufacturing:

- *workpiece drawing* – defined by the designer to ensure correct functioning of the workpiece;
- *machined or manufacturing drawings* – specified by the process plan to instruct the machine operator or NC programmer to ensure that the workpiece will conform to the drawing; and
- *setting drawing* – defined by the process planner, defining the tools and fixture positioning in the machine system of reference.

The process of transferring dimensions from the workpiece drawing to the machining and setting drawings is analyzed and discussed in this chapter.

4.2 THE ELEMENTARY MODEL OF TOLERANCING FOR PRODUCTION

In order to illustrate the problem just described, let us consider a simple milling operation to obtain a step as represented in Fig. 4.1. The functional dimensions to be produced are defined as D_1 and D_2, but the chosen process plan will not execute them as direct dimensions. For reasons of machining efficiency, the step is obtained by a set of milling tools (Fig. 4.1) which execute dimension D_1 as a direct dimension, called now C_{m1} because it is a manufactured dimension distinct from the design dimension D_1. Dimension D_2 is obtained as a resultant dimension depending on the dimensions C_{m1} and the tool dimension C_t. Obviously, the mean dimension of D_2, now called C_{m2} because it is a manufacturing dimension, is given by:

$$\text{mean}(C_{m2}) = \text{mean}(C_{m1}) - \text{mean}(C_t) \qquad (4.1)$$

because mean values can be added algebraically. This relation is easy to express when the dimensions are toleranced symmetrically, e.g. 29 ± 0.3. Concerning the tolerance of C_{m2}, called by its interval of tolerance IT(C_{m2}), we obtain the

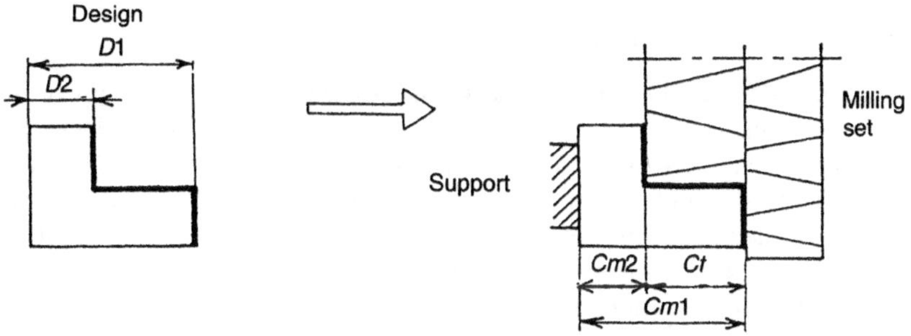

Fig. 4.1 Milling of a step. Redrawn from Karr, J., *Methodes et Analyses de Fabrication Mecanique*, published by Dunod, Bordas, Paris, 1979.

following equation (4.2) which expresses the well-known result that the resultant tolerance of a sum or a difference of component tolerances is the sum of the individual tolerances:

$$\mathrm{IT}(C_{m2}) = \mathrm{IT}(C_{m1}) + \mathrm{IT}(C_t) \tag{4.2}$$

The fundamental condition to respect for C_{m2} is given by

$$\mathrm{IT}(C_{m2}) \leqslant \mathrm{IT}(D_2) \tag{4.3}$$

Of course, one will try to increase $\mathrm{IT}(C_{m2})$ as far as possible in order to approach the maximum possible, i.e. $\mathrm{IT}(D_2)$, in order to achieve the highest economic gain possible. In this case, as the upper limit of $\mathrm{IT}(C_{m2})$ is given, one has to choose $\mathrm{IT}(C_{m1})$ correctly because generally $\mathrm{IT}(C_t)$ has some fixed value. From (4.2), we obtain a limit value of $\mathrm{IT}(C_{m1})$:

$$\mathrm{IT}(C_{m1}) \leqslant \mathrm{IT}(D_2) - \mathrm{IT}(C_t) \tag{4.4}$$

To be feasible, (4.4) implies that

$$\mathrm{IT}(D_2) > \mathrm{IT}(C_t) \tag{4.5}$$

which means also that $\mathrm{IT}(C_{m1})$ given by this relation should not be too small to be executable on a machine. If this would not be the case, another process plan would have to be chosen or more precise equipment would become necessary.

In a similar way, it is possible to define the extrema (minimum and maximum) of the resultant dimension C_{m2}:

$$\min(C_{m2}) = \min(C_{m1}) - \max(C_t) \tag{4.6}$$

$$\max(C_{m2}) = \max(C_{m1}) - \min(C_t) \tag{4.7}$$

The equations (4.6) and (4.7) are equivalent to (4.2), but it is important to notice that they are arithmetic relations resulting from the fact that they express the severest condition for limiting C_{m2}, i.e. they consider the combination of the extrema of C_{m1} and C_t. The equations (4.6) and (4.7) have to be written in the given form and cannot be replaced by apparently similar relations as permitted by algebra. For example, writing:

$$\min(C_{m1}) = \min(C_{m2}) + \min(C_t) \tag{4.8}$$

is not equivalent to (4.6) or (4.7) and would result in wrong limitations of the component dimensions. If C_{m1} is the controlling variable because it is set on the machine, its values are taken from (4.6) and (4.7) and not from (4.8). Returning to conditions (4.6) and (4.7), it is important to emphasize that these conditions are only justified when the possibility of combining these extrema values of the component dimensions is envisaged. As the probability of this event is very low, the preceding relations are, in practice, replaced by less severe conditions based on statistical considerations as seen in section 4.9. The importance of the loss of good parts, functionally speaking, due to the use of relations (4.6) and (4.7) will now be explained.

Coming back to (4.1), one can consider C_{m2} as a linear function of C_t as follows

$$C_{m2} = -C_t + C_{m1} \tag{4.9}$$

which is represented in Fig. 4.2 as straight lines, although for only two values of C_{m1}: max(C_{m1}) and min(C_{m1}). The total permissible area for C_{m2} and C_t should be the area of the rectangle ABCD. However, because of the limitations imposed on C_{m1} by (4.6) and (4.7), the only part of this rectangle which can be used is represented by the parallelogram BFDE which corresponds to the extremes of C_{m1}, i.e. min(C_{m1}) and max(C_{m1}). All the values of C_{m1} checked in production, i.e. between max(C_{m1}) and min(C_{m1}), are good parts. However, the parts falling in the triangles CDF and ABE are rejected in production although they are functionally correct. For those parts, a supplementary dimension control after production should be carried out, but this means increasing the cost of manufacturing.

The arithmetic method leads therefore to a loss of good parts evaluated by a comparison of areas of BEDF and ABCD. Such a loss is only justified for special situations where a 100% assurance for correct dimensions of the parts is required, e.g. very precise assemblies or security parts. For common production

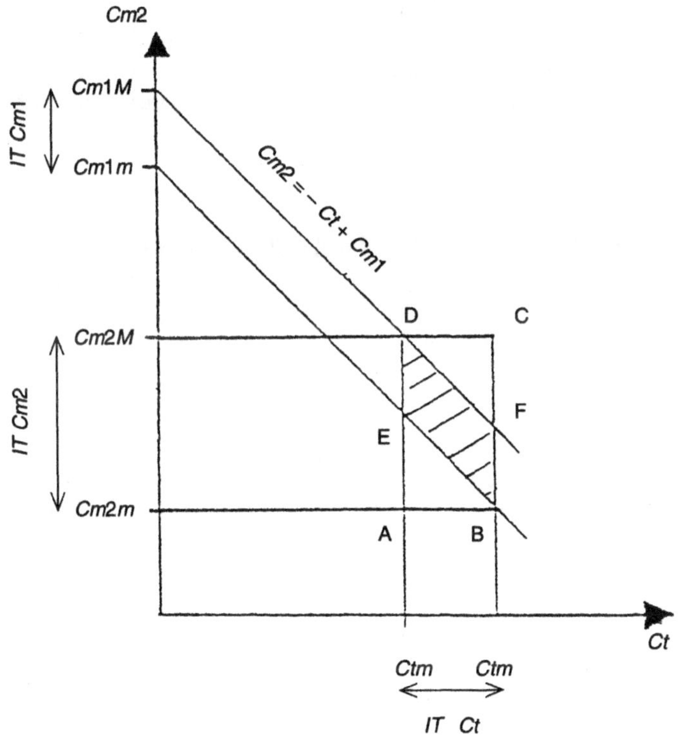

Fig. 4.2 Permitted tolerance zones.

conditions, it is more efficient to use the statistical method for tolerancing, which enables an increase of the tolerances of the components at the expense of the possibility of bad parts. This possibility is however very limited and can be accepted in practice. The statistical approach is discussed in section 4.9.

In order to illustrate better the problems of tolerancing, different examples of linear chains of dimensions are described below.

4.3 EXAMPLES OF TOLERANCING LINEAR CHAINS OF DIMENSIONS IN PRODUCTION

Before examining specific examples, we should remember that tolerances may be given in a unilateral or bilateral way. The term unilateral means that the total value of the tolerance in relation to the nominal dimension is given in one direction (e.g. 20 1/0). The term bilateral tolerance means that the tolerance value is divided in two equal parts, having sign '+' and '−' (e.g. 20.5 ±0.5). Both definitions are used in practice and one can be converted to the other without difficulty.

4.3.1 Stack-up of tolerances by the arithmetic method

The basics of tolerance arithmetics are explained in the following examples: Figure 4.3a shows a chain of four dimensions with their tolerances. One task is to define the overall length of the part. The nominal length will be obviously:

$$L=A+B+C+D$$

The maximum length will be:

$$A+a+B+b+C+c+D+d=A+B+C+D+(a+b+c+d)$$

The minimum length will be:

$$A+B+C+D-(a+b+c+d)$$

and the tolerance will be:

$$l=a+b+c+d$$

Figure 4.3b shows the total length with its tolerance ($L \pm l$) as well as the tolerances of dimensions A,E,D. The problem is to define the tolerance of C.

The nominal dimension of C is:

$$C=L-(A+B+D)$$

The maximum length will be:

$$C=L-(A+B+D)+l+a+b+d$$

The minimum length will be:

$$C=L-(A+B+D)-(l+a+b+d)$$

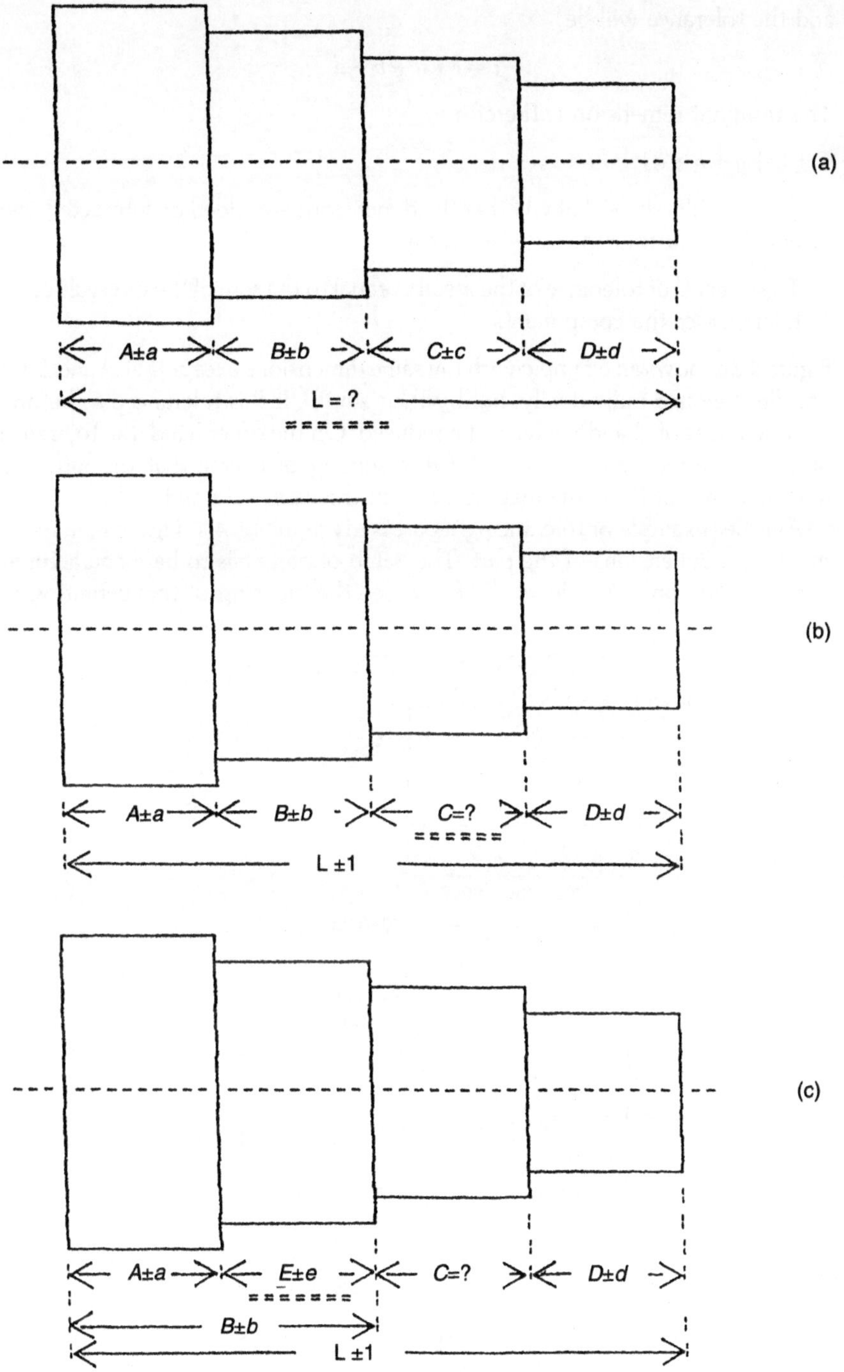

Fig. 4.3 Dimensioning method effect on tolerance stackups.

and the tolerance will be:

$$c = l + a + b + d$$

The resultant dimension is therefore:

$C \pm (l + a + b + d)$

These results show that whether the dimensions are added or subtracted, the resultant law of tolerancing is as follows:

> The interval of tolerance of the result is equal to the sum of the intervals of tolerance of the components.

Figure 4.3c shows an example with the same dimensions except that *A* and *E* are not dimensioned individually, but by their sum *B*. If *B* is toleranced as before, the tolerances of *A* and *E* have to be reduced. On the other hand, the tolerance of *C* will be reduced to: $c = l + b + d$, assuming of course that the different tolerances are of the same magnitude as in the cases 4.3a and 4.3b.

Another example of tolerancing a part is given in Fig. 4.4. Figure 4.4a gives the design dimensions of the part. The batch of parts has to be manufactured with a $\pm 6\sigma$ confidence level, i.e. $k = 2$ (see the meaning of this definition in

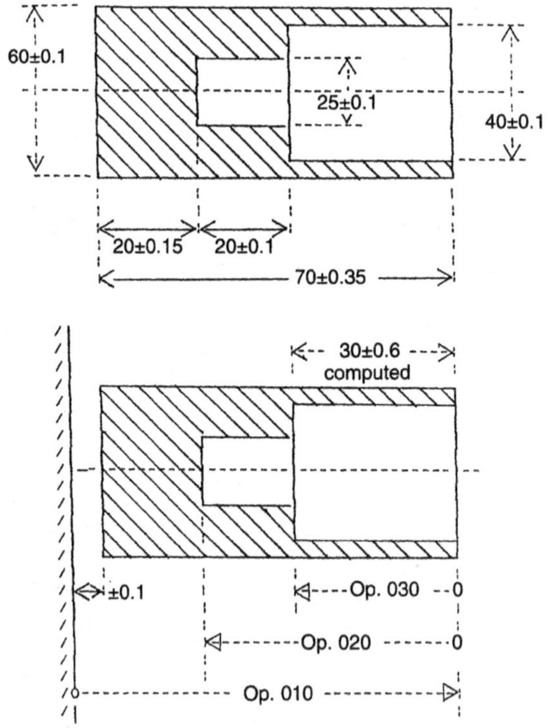

Fig. 4.4a Example: part to machine.

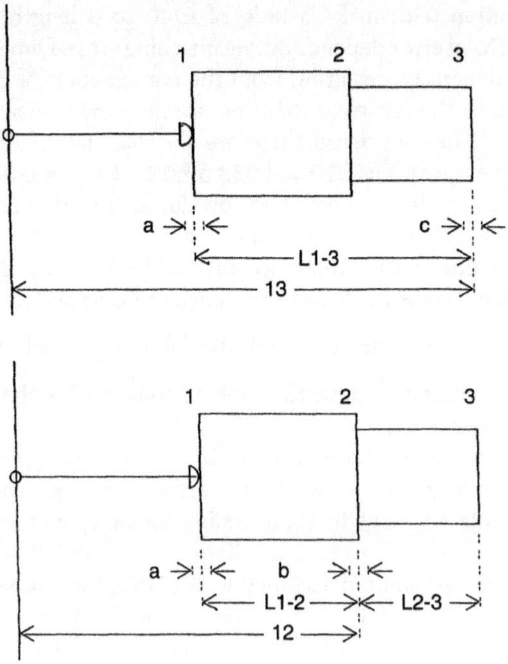

Fig. 4.4b Machining with one and two sub-phases.

Chapter 12, section 12.4). The raw material consists of blanks of Ø60 ± 0.1 and 72.0 length. The problem is to determine the tolerances for the machining operations, knowing that the fixturing repeatability is ± 0.1. The dimensions and tolerances for the length of hole Ø40 will be according to the law of tolerancing:

$$\text{length} = 70 - 20 - 20 = 30$$

$$\text{tolerance: } 0.35 + 0.1 + 0.15 = 0.6$$

and the length of the segment will be: 30 ± 0.6. The process planning steps are shown in Fig. 4.4b.

1. Chuck the blank in the lathe.
2. Execute operation 010: machine the face to a length of 70 ± 0.35. The total accuracy of this dimension should consider one machine error and one fixture error. Therefore, the accuracy for operation 010 will be:

$$\frac{0.35 - 0.1}{(1 \times 2)} = 0.125$$

3. Execute operation 020: make a hole of Ø25 with a length (70 − 20 =)50 from the datum. If the tolerance of fixturing is neglected because the datum surface is machined, the total tolerance will depend on one machine error.

4. Execute operation 030: make a hole of Ø40 to a length of 30. In this operation, the total error depends on one machine error. Dimension 20 ± 0.1 is an indirect dimension resulting from the combination of operations 020 and 030. Because the errors of fixturing are neglected, there are only two machine errors to be considered, therefore, the tolerance of 20 ± 0.1 requires tolerances in the operations 020 and 030 of: $0.1/(2 \times 2) = 0.025$. Dimension 20 ± 0.15 is a resultant dimension produced by the combination of operations 010 and 020. Its tolerance is determined by two machine tolerances and one fixture tolerance or: $(0.15 - 0.1)/(2 \times 2) = 0.0125$ for both operations. This is the lowest tolerance to be respected in production.

From the previous derivations, we reach the following conclusions.

- The machine accuracy is normally not established by the smallest part tolerance in design.
- The length of hole Ø40 is computed to be 30 ± 0.6, but its real dimension will probably be 30 ± 0.0125, i.e., the actual tolerance is given not according to the design tolerance but to the machine tolerance and the sequence of operations.
- If no machine of the required accuracy is available, then either improve the accuracy of the fixture to 0.05 or machine the part from a bar instead of from cut pieces. The minimum machine tool accuracy is then only 0.025. If a confidence level $\pm 3\sigma$ is acceptable, then the minimum machine accuracy is 0.05.

4.3.2 Tolerancing in forming operations

Parts made from forming from liquid or from solid by deformation employ dies. The die designer will consider shrinkage, deflections, spring back, part ejection and many other parameters to design a die.

The accuracy of the formed part depends on the accuracies of the die, the accuracy of the machine and of the position of the die in the machine. Each one of these parameters affects different dimensions of the part. Figure 4.5 shows the three types of dimensions:

1. tool dependent dimensions;
2. non-tool-dependent dimensions; and
3. non-tool-dependent dimensions in the direction of closing the tool.

Type (1) dimensions should be handled by a method similar to those mentioned in previous sections. Type (2) dimensions are a result of the set-up accuracy of the dies in the machine, especially the concentricity and the parallelism of the separate portions of the dies. Also, the machine accuracy and the freedom in the guideways may affect these dimensions. Type (3) dimensions are mostly dependent on the machine accuracy, the deflections and power stability. The application of manufacturing tolerancing in process planning, as described in the previous sections, holds true for this method of

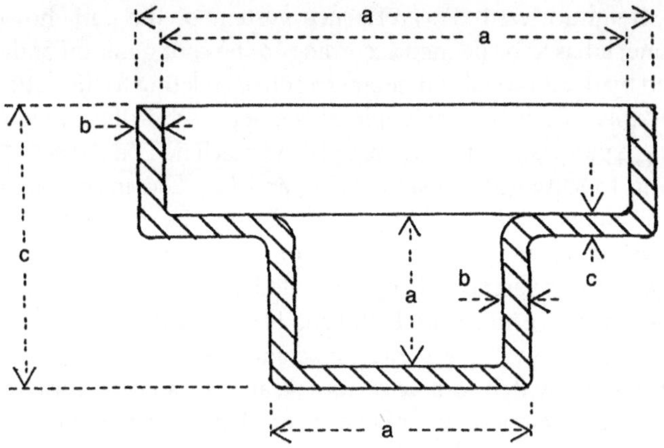

Fig. 4.5 Type of dimensions in forming.

forming. However, the process planner has to isolate each type of dimension and to design it separately. Type (1) dimensions will probably affect forming by metal deformation in producing each section of the die. Type (2) dimensions will depend on machine jigs and fixtures, or the datum surfaces of the individual sections of the die. These surfaces have no direct bearing on the shape of the produced part, but on the accuracy of the die assembly and set-up. Type (3) dimensions are used to determine the machine accuracy for the job.

4.4 A GENERALIZED MODEL FOR TOLERANCING LINEAR CHAINS OF DIMENSIONS

A more comprehensive model of tolerancing, superior to the elementary model seen earlier, has been developed and is described in detail in Fainguelernt *et al.* (1986). It takes into account all types of errors such as fixturing, setting and tool wear errors and has been computerized and optimized as explained in section 4.3. One of its superiorities is to make a clear distinction between dependent and independent dimensions in a linear chain, and by so doing, to avoid wrong determinations of tolerances or, in some cases, to avoid purely fictive tolerancing.

The problem facing a process planner is to use coordinate datum elements of the different reference systems used in machining, as follows:

- the machine datum
- the fixture datum
- the tool datum
- the part datum.

The first three elements can be handled without special difficulty and their setting is straightforward. The reference system of the part, however, is a changing one; it has to be defined according to the configuration of the part and is influenced by dimensional and geometric errors relating to the datum surfaces on the workpiece itself. The principle of the approach is illustrated in Fig. 4.6 representing a part clamped on surface (1) and machined on surfaces (2), (3) and (4) according to setting dimensions L_2, L_3 and $L_{3/4}$. Dimensions L_2 and L_3 are independent dimensions resulting from setting the tools in the machine reference system, i.e. the coordinate system XOY (hereafter considered as the absolute system of reference). They are called setting or machine dimensions. Their proper tolerance interval – IT(L_2) and IT(L_3) – is the result of a number of error factors such as setting accuracy, machining accuracy, kinematic accuracy of the control loop, switch accuracies, tool wear, mechanical deformations in the machine structure, temperature influences on dimensional accuracy and many other factors. It is assumed that all these errors are summed up in the tolerance interval IT(L_i) of the L dimensions.

Two other types of dimensions with their interval of tolerance are also represented in Fig. 4.6. One relates to the positioning tolerance of the part in the machine tool system and is called L'_1 with an interval of tolerance IT(L'_1). It concerns the position of the clamping face of the part (surface 1) in its fixture. Its tolerance IT(L'_1) is a function of the clamping error due to the clamping force and to geometric errors in the reference face of the fixture or in surface (1) which are

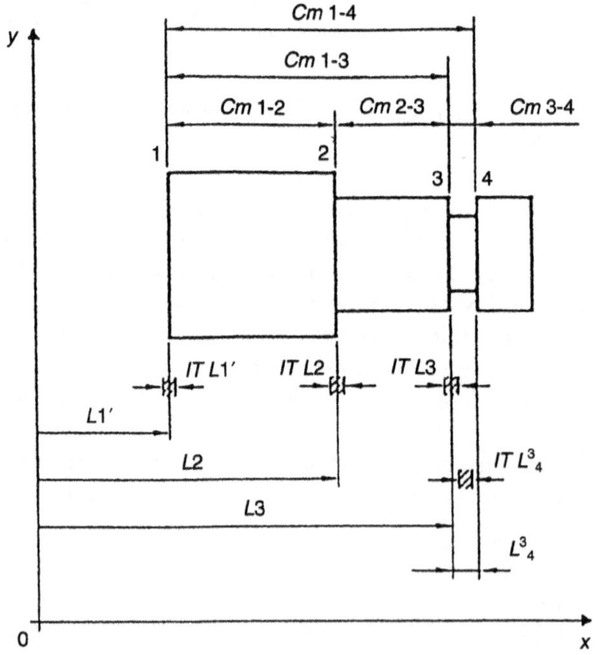

Fig. 4.6 Generalized model of dimensioning and tolerancing.

the result of previous machining operations. For example, Fig. 4.7 shows the case of form errors in the clamping face which are defined as $\Delta f'$. The resulting error of positioning $\mathrm{IT}(L_1')$ can be equal to $\Delta f'$ or to $2\Delta f'$ depending on the type of positioning, by a plane contact or by a punctual contact. These positioning errors $\mathrm{IT}(L_1')$ have been neglected in the elementary model although their value is of the same order of magnitude as the other production errors, as can be seen in Table 7.2. As will be shown later, this simplification, if it is not justified by a very small value of $\mathrm{IT}(L_1')$, is causing errors in tolerancing.

Another type of dimension is the tool dimension $L_{3/4}$ which has no relationship with the machine dimensions and which can be considered as independent, with its interval of tolerance $\mathrm{IT}(L_{3/4})$.

Other dimensions appearing in Fig. 4.6 are the sizes of the segments between two machined surfaces (C_{mi-j}) or between the clamping surface and a machined surface. These C_{mi-j} sizes are the result of the processing of the part according to the setting dimensions L_i. They are called manufactured dimensions and can be measured on the finished part. Of course, these manufactured dimensions have to be compatible with the design dimensions as expressed in (4.3) or

$$C_{i-j} \subset D_{i-j} \tag{4.10}$$

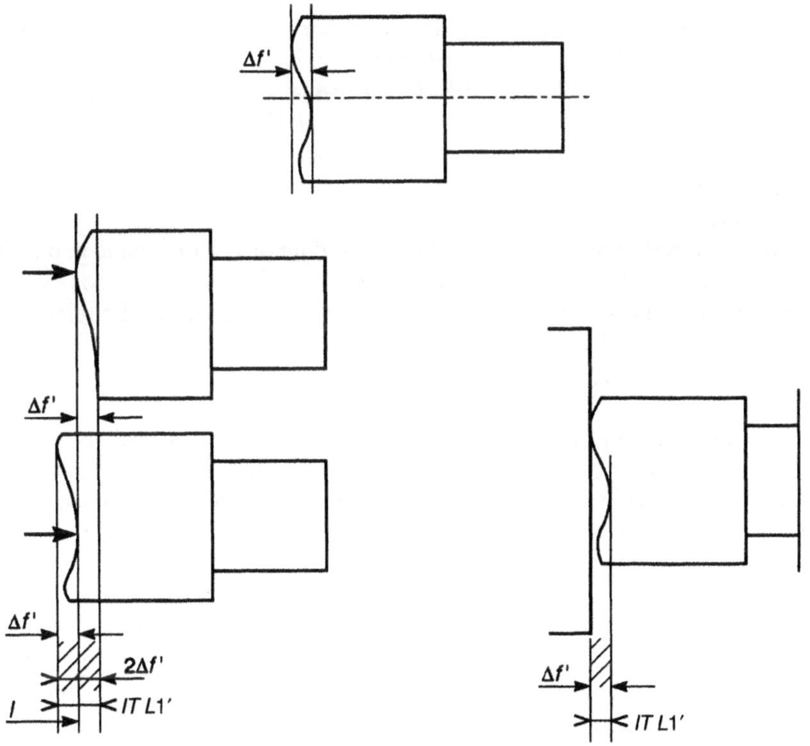

Fig. 4.7 Influence of form errors of datum.

meaning C_{i-j} is included in D_{i-j} (design dimension) in a mathematical sense. Because, in this model, dimensions are expressed as centered dimensions, it is equivalent to state that

$$\text{mean}(C_{i-j}) = \text{mean}(D_{i-j}) \tag{4.11}$$

and

$$\text{IT}(C_{i-j}) \leqslant \text{IT}(D_{i-j}) \tag{4.12}$$

Because manufactured dimensions are generally resultant dimensions dependent on the setting tolerances and not identical with design dimensions D_{i-j}, their tolerance has to be determined by an optimization procedure, explained below.

Having assumed that the setting dimensions L_i are independent, it is possible to write the exact relations for the resultant dimensions C_{mi-j} for the part in Fig. 4.6:

$$\begin{aligned} \text{IT}(C_{m1-2}) &= \text{IT}(L_1') + \text{IT}(L_2) \\ \text{IT}(C_{m1-3}) &= \text{IT}(L_1') + \text{IT}(L_3) \\ \text{IT}(C_{m2-3}) &= \text{IT}(L_2) + \text{IT}(L_3) \end{aligned} \tag{4.13}$$

Knowing the values of each $\text{IT}(L_i)$, the equations in (4.13) give the values of the manufactured dimensions C_{mi-j}, or, inversely, having fixed the values of the resultant C_{mi-j}, the values of the $\text{IT}(L_i)$ necessary are a result of these same relations. As seen later, the determination of the values of each $\text{IT}(L_i)$ is determined by an optimization procedure which takes into account the design requirements as well as the realistic values of the $\text{IT}(L_i)$s in production (machine accuracy).

Before developing the optimization procedure, it is worthwhile explaining the problem raised by combining dependent or independent components. For example, calculating the expression of $\text{IT}(C_{m2-3})$ as a result of the relations in (4.13), it comes out as

$$\text{IT}(C_{m2-3}) = \text{IT}(C_{m1-2}) + \text{IT}(C_{m1-3}) - 2\text{IT}(L_1') \tag{4.14}$$

This expression shows that $\text{IT}(C_{m2-3})$ is not only the sum of $\text{IT}(C_{m1-2})$ and $\text{IT}(C_{m1-3})$ as would be expected if the C_{mi-j}s would be independent dimensions, but has a correction term $\text{IT}(L_1')$. In the elementary model of tolerancing, the relation would have been

$$\text{IT}(C_{m2-3}) = \text{IT}(C_{m1-2}) + \text{IT}(C_{m1-3}) \tag{4.15}$$

However, this relation is incorrect in the general case and gives erroneous values of the tolerances which may lead to manufacturing impossibilities. For example, if C_{m1-2} and C_{m1-3} would have high values of IT, it could happen that C_{m2-3} will have a very high value of its IT – according to (4.15) – which is no more compatible with design requirements. In reality, because of the correction factor $\text{IT}(L_1')$ in (4.14), it could happen that the $\text{IT}(C_{m2-3})$ is perfectly acceptable and that no changes in the process plan will be necessary. This results from (4.13) which gives $IT(C_{m2-3})$ as the sum of $IT(L_2)$ and $IT(L_3)$ which is smaller than $IT(C_{m1-2}) + IT(C_{m2-3})$. Of course, if for a given process plan, the values of the C_{mi-j}s appear to be independent, then (4.15) can be applied. Also, if it is assumed that the positioning error $IT(L'_1)$ can be ignored, this relation is applicable.

To illustrate this situation, Fig. 4.8 shows a part processed in two subphases 10 and 20. The relation between the tolerances are as follows:

$$\Delta C_{m1} = \Delta l'_1 + \Delta l_3 \ \Delta C_{m2} = \Delta l'_4 + l_2$$
$$\Delta C_{m3} = \Delta l_2 + \Delta l'_4 + \Delta C_{m1} = \Delta l_2 + \Delta l'_4 + \Delta l'_1 + \Delta l_3 \qquad (4.16)$$

or in a simplified form:

$$\Delta C_{m3} = \Delta C_{m1} + \Delta C_{m2}$$

This result shows that independent manufactured dimensions can be combined simply. However, to avoid ambiguities, it is recommended to use the equations in (4.13) as in the optimization procedure later on. As for the values of the ITs for tool dimensions, such as $IT(L_{3/4})$, or for fixture dimensions, (for example the distance between two tools mounted in the same jig and machined simultaneously), they can be taken as independent dimensions.

To complete the generalized model, it is necessary to add the calculations of the average values of the setting dimensions L_i and to take into account other limitations which are required for correct manufacturing results. For example, the thickness of the first chip to be removed from the raw material should have a minimal thickness so that the remaining layers will be without disturbances caused by, for example, casting or forging. Inversely, the chip thickness should not be too large so as not to generate too high a cutting force. Another factor to be considered is the wear of the tools, which will influence the calculation of the production tolerances. All these additional aspects are explained in the next section, as well as the procedure used to determine the tolerances and average dimensions in production.

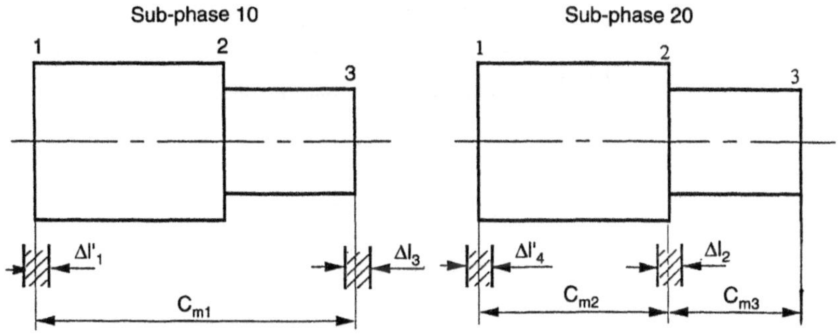

Fig. 4.8 Tolerancing: case of two separate jobs.

4.5 APPLICATION OF THE GENERALIZED MODEL TO A SIMPLE PART

An example of a part is shown in Fig. 4.9. The design dimensions are given as D_{i-j}. To produce the part, a process plan has been designed as follows.

- Subphase 10: fixture on surface 4 and machining of surface 1.
- Subphase 20: fixture on surface 1 and machining of surface 2.
- Subphase 30: fixture on surface 1 and machining of surface 3.

The setting dimensions are indicated on the drawing as (l_i, Δl_i) and the positioning dimensions as l'_i or l''_i. To determine the setting dimensions while respecting the design dimensions D_{i-j}, one has to write the following equations:

$$\Delta D_{1-4} = \Delta l_1 + \Delta l'_4$$
$$\Delta D_{1-3} = \Delta l''_1 + \Delta l_3$$
$$\Delta D_{2-4} = \Delta l'_1 + \Delta l_{1-2} + \Delta C_{1-4} \text{ (execution of } D_{1-4}\text{)}$$
$$= \Delta l'_1 + \Delta l_2 + \Delta l_1 + \Delta l'_4 \qquad (4.17)$$

These equations can be summarized in Table 4.1 (tolerance table) which in the upper part gives values to l_is and l'_is corresponding to the lowest tolerance feasible in production for the given process plan. If these tolerances are not compatible with the design constraints, as shown by comparing ΔD_{i-j} to ΔC,

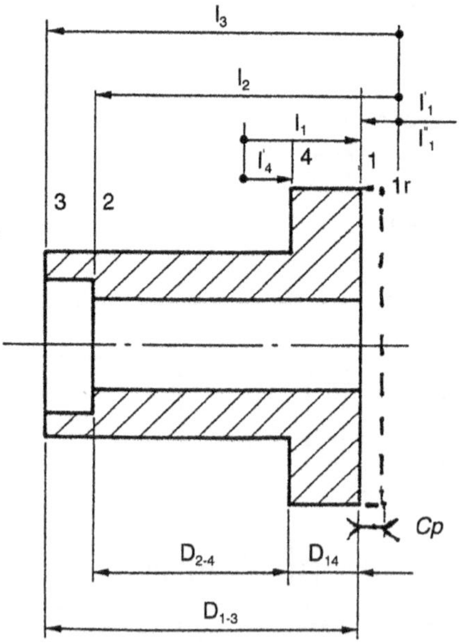

Fig. 4.9 Part to be machined.

Table 4.1 Optimization of tolerancing

	Δl_1	$\Delta l'_1$	$\Delta l''_1$	Δl_2	Δl_3	$\Delta l'_4$	ΔC	ε	n	ε/n
$\Delta D_{1-4}=0.8$	0.1					0.5	0.6	0.22	2	0.1
$\Delta D_{2-4}=1.0$	0.1	0.05		0.1		0.5	0.75	0.25	4	0.0625
$\Delta D_{1-3}=0.6$			0.05		0.1		0.15	0.45	2	0.225
$\Delta D_{1-4}=0.8$	0.163					0.562	0.72	0.08	—	—
$\Delta D_{2-4}=1.0$	0.163	0.112		0.163		0.562	1.0	0	0	0
$\Delta D_{1-3}=0.6$			0.05		0.1		0.15	0.45	2	0.225
$\Delta D_{1-4}=0.8$	0.163					0.562	0.72	0.08	0	—
$\Delta D_{2-4}=1.0$	0.163	0.112		0.163		0.562	1.0	0	0	0
$\Delta D_{1-3}=0.6$			0.27		0.33		0.6	0	0	0

there is no solution for the proposed process plan. In general, these values can be increased as will be shown in the optimization procedure. In Table 4.1, in column ΔC, the sum of the tolerances is given and its difference with ΔD is expressed by ε. If ε is positive, the proposed process plan is feasible. If not the process plan has to be changed.

In Table 4.1, the εs are positive which means that a bonus of tolerance exists. The optimization consists then in sharing this bonus among the different Δl_is and is explained in the middle part of Table 4.1.

Before, for the purpose of simplification, the sharing was made equally on every active dimension. This is expressed by n, the number of active dimensions for a line and ε/n which gives the possible bonus to be added to each dimension. To be realistic, the sharing begins with the lowest value of ε/n which creates a limit for increasing the tolerances. In Table 4.1, this minimum value is 0.0625 in line D_{2-4}. Adding it to the previous values of the Δl_is gives the new values of the Δl_is for the whole table. Normally, it is advantageous to continue to increase the Δl_is as shown in the lower part of Table 4.1, but line D_{1-4} is already saturated. Only line D_{1-3} can be optimized by adding a bonus of 0.225 to the Δl_is.

This optimization procedure, although empirical, is efficient. Other refinements can be added to it as seen in section 4.5. But it should also be remembered that further increase of tolerances in production may not give a higher benefit because the cost of production will not change with larger tolerances above a certain limit.

Having found the optimal tolerances in production, it is now the task of the process planner to check how to prescribe exactly the tolerance in production. As has been seen already in the example in Fig. 4.4, there are rules based on statistics which can guarantee the tolerance but with a certain degree of risk. This point is explained fully in section 4.9, but to simplify, it can be said here that a common rule to attain a given interval of tolerance is to equal this interval to the $\pm 3\sigma$ range which ensures a rejection rate of 0.25% with a well-centered production. If this latter condition cannot be guaranteed, it is customary to use a range of ± 6 or a range of $\pm 9\sigma$ to achieve a production in tolerance. (In this latter case, up to 99.998% of the parts will be in the drawing specifications.)

To complete the analysis of tolerancing for production, the average values of the L_is have to be determined. This is very straightforward because no restrictions of statistical dependency are applicable. As the values of the averages of the C_{i-j}s are made equal to those of the D_{i-j}s, it comes out as

$$\begin{aligned} D_{1\text{-}4} &= l_1 - l'_4 \\ D_{2\text{-}4} &= l_2 - l'_1 - (l_1 - l'_4) \\ D_{1\text{-}3} &= l_3 - l''_1 \end{aligned} \tag{4.18}$$

These three equations can give the values of the l_is provided that the number of variables is also 3. This is not the case here because there are six variables. However, it can be admitted that the mean values of the l'_i are 0 because the dimensions l_i can be referred to the mean position of the fixture. The system of linear equations in (4.18) is then resolvable.

Other limitations, already mentioned, are related to manufacturing constraints. A common example is concerning the calculation of the raw material dimension lB (dimension $4 - 1r$) where $1r$ is the raw surface. In order to satisfy the restriction of a minimum thickness for the first chip to be removed, and if C_{pmin} is the minimum thickness required and ΔC_{p} the tolerance of the thickness, one can write:

$$\text{mean}(C_{\text{p}}) = C_{\text{pmin}} + \tfrac{1}{2}\Delta C_{\text{p}} \tag{4.19}$$

If we take as an example the determination of the raw material dimension lB for dimension D_{1-4}, one can write an equation for the chip thickness $(1 - 1r)$ with a tolerance of ΔC_{1-1r} between the machined surface 1 and the raw surface $1r$:

$$\Delta C_{1-1r} = \Delta lB_{4-1r} + \Delta C_{4-1} = \Delta lB + \Delta l'_4 + \Delta l_1 \tag{4.20}$$

Knowing the raw material tolerance lB in the primary process, (4.20) can give the chip thickness tolerance. It is then possible to calculate the chip thickness average as follows:

$$\text{mean}(C_{\text{p}(1-1r)}) = C_{(1-1r)\text{min}} + \tfrac{1}{2}\Delta C_{1-1r} \tag{4.21}$$

and the average of the raw material dimension:

$$\text{mean}(lB) = \text{mean}(l_1) + \text{mean}(C_{\text{p}(1-1r)}) \tag{4.22}$$

which gives the dimension of the raw material to prepare. Its tolerance ΔlB is known for every specific primary process.

Another manufacturing constraint is related to the wear of tools in fabrication. If the wear value is not significant, it can be included in the overall value of l_i. But in order to be more precise, its influence should be analysed.

To explain the influence of wear, let us take the simple example of Fig. 4.10 showing how wear is a component of the total value of Δl_2. The systematic part

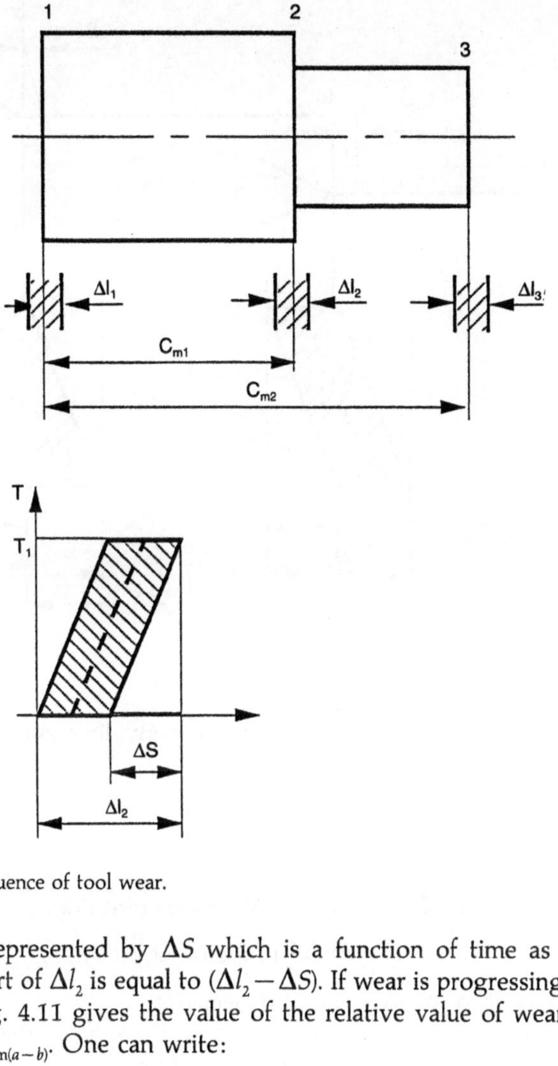

Fig. 4.10 Influence of tool wear.

of wear is represented by ΔS which is a function of time as shown. The stochastic part of Δl_2 is equal to $(\Delta l_2 - \Delta S)$. If wear is progressing in the same direction, Fig. 4.11 gives the value of the relative value of wear $|S_a - S_b|$ on dimension $C_{m(a-b)}$. One can write:

$$0 < |S_a - S_b| < \max(\Delta S_a, \Delta S_b) \tag{4.23}$$

The tolerance of the segment $C_{m(a-b)}$ can be expressed as:

$$\Delta C_{m(a-b)} = (\Delta l_a - \Delta S_a) + (\Delta l_b - \Delta S_b) + \max(\Delta S_a, \Delta S_b) \tag{4.24}$$

where $(\Delta l_i - \Delta S_i)$ represents the stochastic part of tolerance Δl_i. This expression can also be written:

$$\Delta C_{m(a-b)} = \Sigma \Delta l_i - \Sigma \Delta S_i + \max(\Delta S_a, \Delta S_b)$$

or

$$\Delta C_{m(a-b)} = \Sigma \Delta l_i - \Delta U \quad \text{with} \quad \Delta U = \Sigma \Delta S_i - \max(\Delta S_i). \tag{4.25}$$

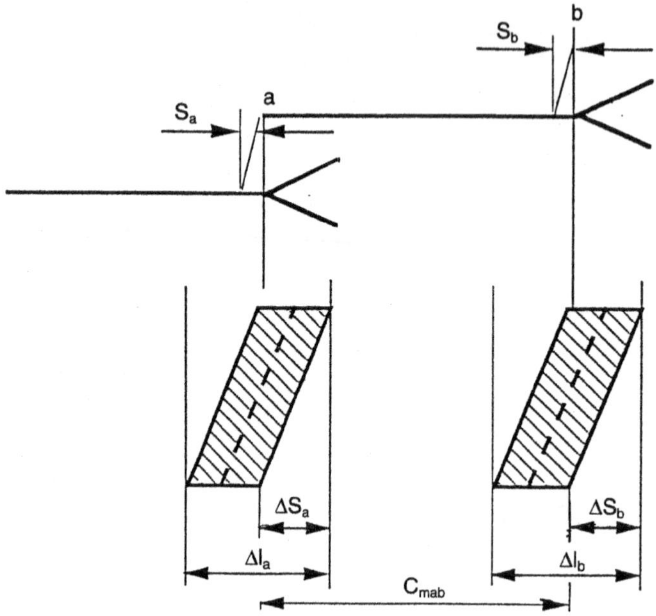

Fig. 4.11 Case of wear in two identical directions.

This expression can be generalized for wear in any direction (either the same direction or opposite):

$$\Delta C_{m(a-b)} = \Sigma\, \Delta l_i - \Delta U$$

with

$$\Delta U = \Sigma \Delta S_i - \Sigma\, \max(\Delta S_i \text{ in one direction}) - \Sigma\, \max(\Delta S_i \text{ in the other direction}) \qquad (4.26)$$

which can be applied to any chain of dimensions.

The correction due to tool wear should be reported in the optimization procedure as shown in Table 4.1. Instead of sharing the excess of tolerance on the total Δl_is, it is necessary to subtract the value of ΔU of the wear effect from Δl_i, and proceed to the usual optimization as shown before. In other words, the value of ΔC will become $(\Sigma \Delta l_i - \Delta U)$ and the value of ε will become $(\Delta D - \Delta C)$. A complete analysis of the wear correction can be found in Fainguelernt *et al.* (1986).

4.6 SIMULATION AND COMPUTERIZATION OF TOLERANCING

The principles explained above can be summarized in a table called 'charting' (Table 4.2), also called 'simulation'. In this table, the part and its design

Table 4.2 Charting table

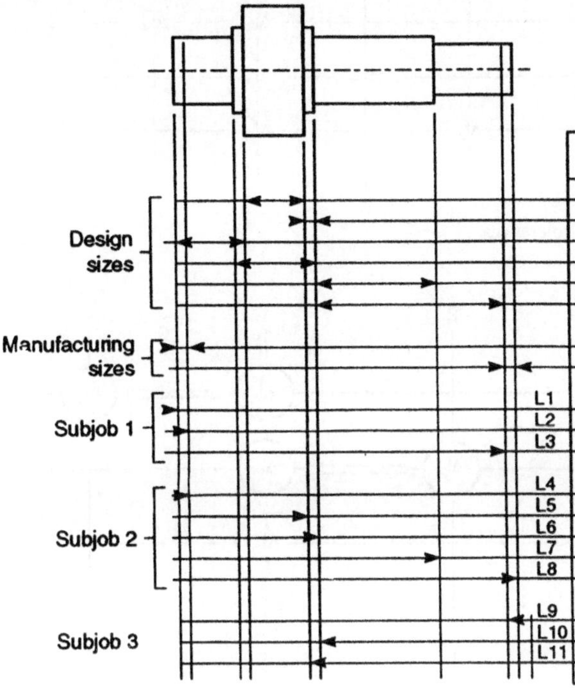

	Mean sizes	Tolerance	Max	Min
	20	0.8		
	2.5	0.5		
	21	0.5		
	25	0.8		
	40	0.5		
	65	1		
				2
				2
L1	0	0.115		
L2	2.14	0.15		
L3	108.36	0.15		
L4	0	0.12		
L5	41.36	0.13		
L6	43.86	0.145		
L7	83.86	0.355		
L8	108.86	0.855		
L9	0	0.12		
L10	84.86	0.13		
L11	87.36	0.115		

dimensions are noted, together with the production limitations (e.g. minimal chip thickness equals 2 mm for the first cut). The setting dimensions L_i, as calculated earlier, are also noted with their average dimensions and their tolerance intervals. The resulting manufactured dimensions C_{i-j} can be calculated from the values of the L_is and should be compatible with the design dimensions D_{i-j}.

As the preparation of tables such as Table 4.2 can become very cumbersome when the number of dimensions becomes large, it was obvious that the whole process of tolerancing should be computerized (Fainguelernt *et al.*, 1986). Some computerized procedures are now briefly described.

Initially, the adopted process plan is represented in a matrix as shown in Fig. 4.12, corresponding to the part in Fig. 4.9. The surfaces are indicated in the columns and the subphases in the rows. The production tolerances (Δl_is) are reported at the intersection of columns and rows.

To write the relation for a given tolerance, e.g. ΔD_{2-4}, the following algorithm is used (Fig. 4.13):

1. Except for the tolerances belonging to the extremities of the considered dimension (here Δl_2 and Δl_4), the tolerances which are single in a column can

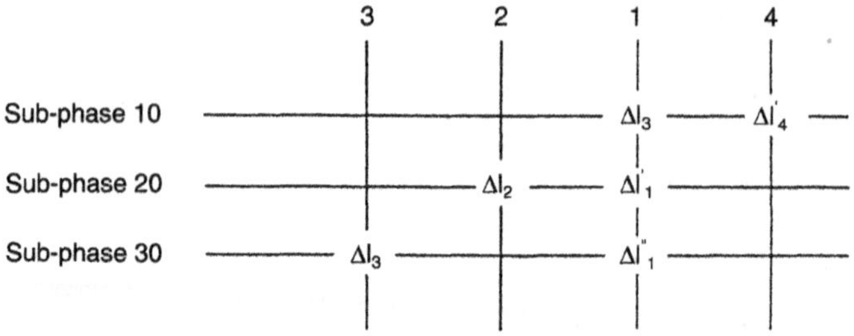

Fig. 4.12 Matrix of setting tolerances.

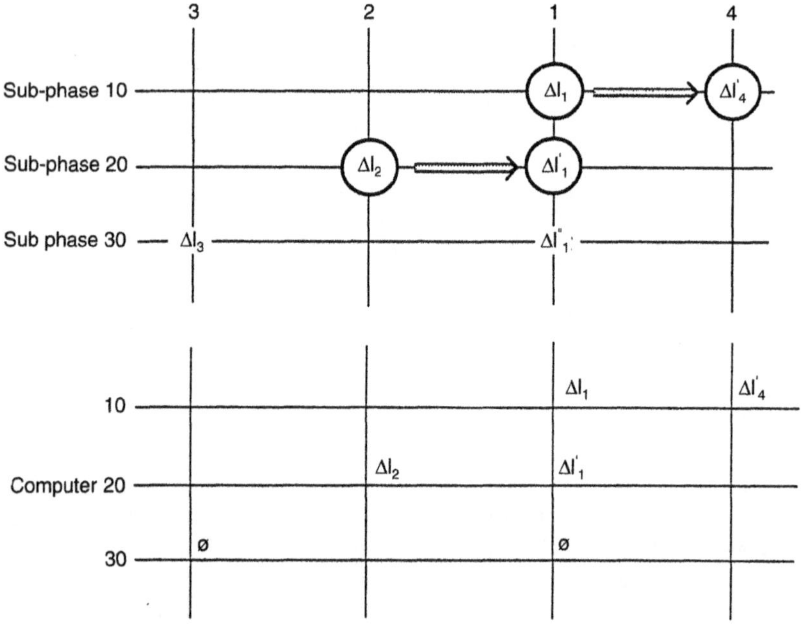

Fig. 4.13 Tolerancing by computers.

be eliminated (because they cannot serve as a link to another line). Therefore Δl_3 is cancelled.

2. Similarly, a tolerance isolated in a line can be cancelled because a segment in a line cannot have a single tolerance. Therefore, $\Delta l_1''$ is eliminated.

Proceeding in this manner across the whole table, a number of tolerances which cannot be eliminated will remain and constitute the components of the tolerance stack-up for the considered dimension, in this case:

$$\Delta D_{2\text{-}4} = \Delta l_1 + \Delta l_1' + \Delta l_2 + \Delta l_4'$$

which is, of course, identical to (4.17).

Having completed the first part of Table 4.1 by this procedure, it is then straightforward to computerize the optimization procedure as explained earlier.

A more sophisticated optimization can be achieved when the difficulty of setting the dimensions on the machine are taken into account (Weill, 1991). The principle of this approach is explained in Figs 4.14a and 4.14b which describe a simple process plan to machine the proposed part in one subphase. The setting dimensions are R_1 and R_2 to obtain dimensions C_1 and C_2, taking into account

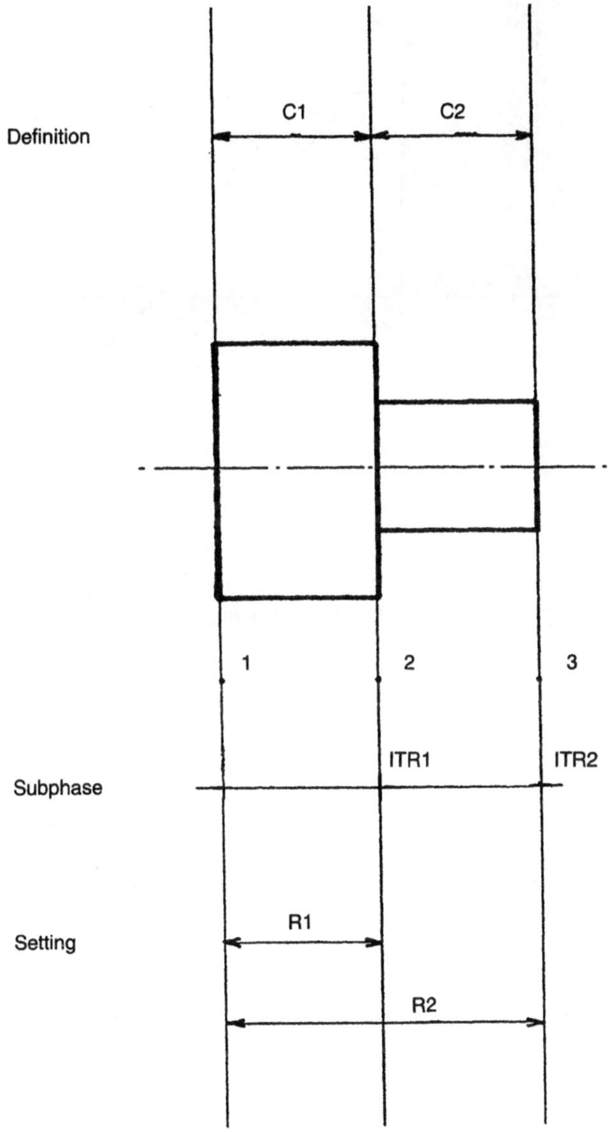

Fig. 4.14a Definition of a sub-phase.

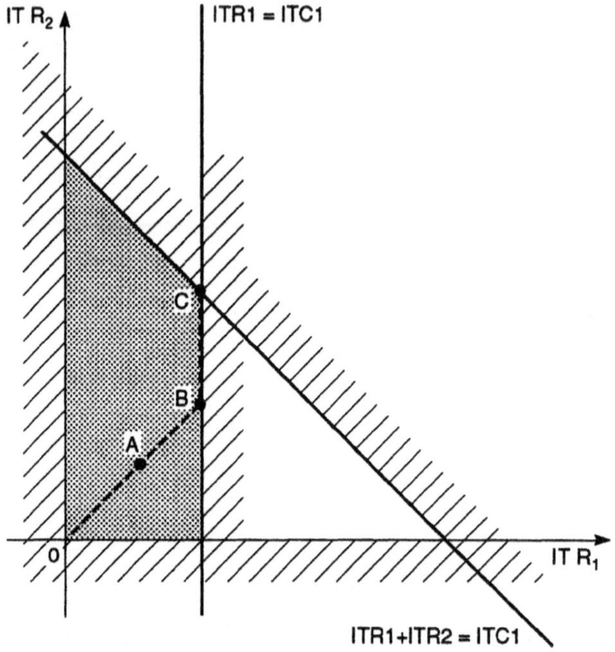

Fig. 4.14b Optimization of the setting tolerances.

their tolerances IT(R_1) and IT(R_2). In this case, the positioning tolerance, which is constant, is not influencing the optimization and is therefore not included in the derivations. Representing the real setting tolerances IT(R_1) and IT(R_2) in a coordinate graph, the tolerancing relations here are:

$$\Delta C_1 = \Delta R_1$$
$$\Delta C_2 = \Delta R_1 + \Delta R_2 \tag{4.27}$$

They are represented by two straight lines in Fig. 4.14b.

If the setting difficulties are the same for R_1 and R_2, the controlling point will follow a straight line such as AB, inclined by 45°, up to the limiting constraint in point B. The tolerance IT(R_1) cannot be enlarged more whereas tolerance IT(R_2) can be increased up to point C which is the absolute highest limit for both tolerances. Based on the same approach, i.e. considering the difficulty of setting per operation, algorithms have been developed which give the highest values of the IT(R_i), respecting the design constraint and the setting difficulties.

Concerning the determination of the average values of the setting dimensions by computers, a very useful method is a procedure based on graph theory. Details have been omitted here because of its complexity. It is explained in Fainguelernt (1986).

As a summary of the above, Fig. 4.15 gives the flowchart of a computer program for dimensioning and tolerancing setting dimensions l_is. It begins by

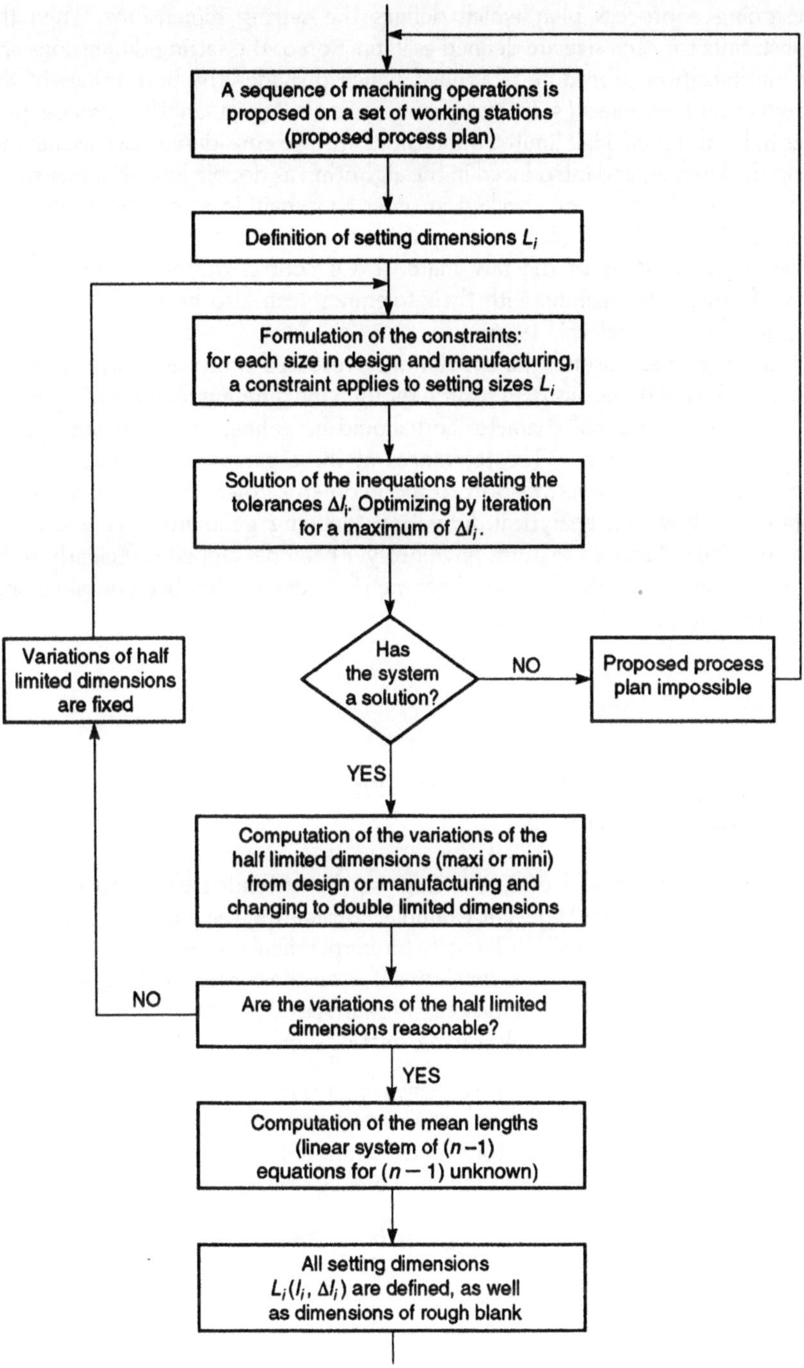

Fig. 4.15 Flow chart of dimensioning and tolerancing optimization.

designing a process plan which defines the setting dimensions. Then the constraints for each size are defined as a function of the setting dimensions and an optimization procedure is applied which provides the best values of the production tolerances l_is. If this system has no solution, another process plan has to be designed. Half limited dimensions are also considered, such as minimal chip thicknesses, and introduced in the algorithm as double limited dimensions. Their optimal values are checked in order to remain in a realistic production domain. Finally, the average setting dimensions are determined, as well as the average dimensions of the raw material. Of course, the new values of the manufactured dimensions with their tolerances can also be provided by this simulation (Fainguelernt, 1986).

Although very useful, the simulations presented up to now are limited to linear chains of dimensions. In reality, though, the different features of a part are of a three-dimensional character and should be defined by six parameters as explained in Chapter 7. The tolerances of these parameters of position and orientation are in direct relation to the geometric tolerances required by the drawing. However, analytical derivations defining geometric tolerances in a chain of manufactured features have not yet been developed sufficiently to be detailed here. The introduction of geometric tolerances has been possible only in simple cases as shown below.

4.7 INTRODUCTION OF GEOMETRIC TOLERANCES IN DIMENSIONAL CHAINS

The elementary model of tolerancing can be extended easily to chains of tolerances of a similar type. For example, considering the part in Fig. 4.16, the problem there is to transfer tolerances for perpendicularity between surfaces (3) and (2) (equal to 0.05). This constraint is a resultant of executing surface (3) perpendicular to surface (1) and executing surface (2) parallel to surface (1) in the chosen process plan. The tolerancing relationships are

$$\mathrm{IT}(3/2) = \mathrm{IT}(3/1) + \mathrm{IT}(2/1) \qquad (4.28)$$

where the ITs are angular tolerances. Translating the given tolerances of perpendicularity relative to a similar reference basis, gives

$$\mathrm{IT}(3/2) = \frac{0.05}{17} = \frac{0.3}{100}$$

and taking

$$\mathrm{IT}(2/1) = \frac{0.15}{100}$$

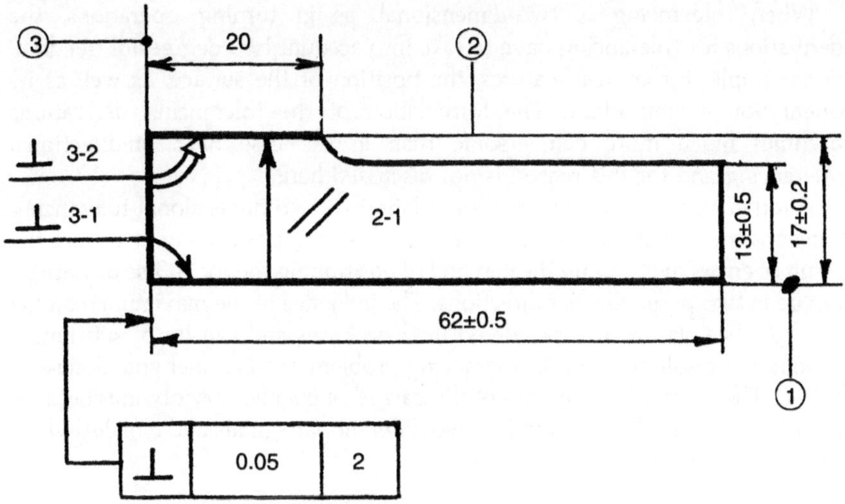

Fig. 4.16 Example of a geometric chain of dimensions. Redrawn from Karr, J., *Methodes et Analyses de Fabrication Mecanique*, published by Dunod, Bordas, Paris, 1979.

as a realistic production tolerance as well as

$$\mathrm{IT}(3/1) = \frac{0.15}{100}$$

it is proven that for these values of ITs, (4.28) is verified.

Another example of introducing geometric tolerances involves chains of linear dimensions integrating components of geometric errors as a part of the global setting error $\Delta l'_i$. If the geometric errors are small and included in the total error Δl_i, there is no additional influence from the geometric error in the optimization procedure of section 4.4. This circumstance is very general because a practical rule admits that only one third of the tolerance interval is devoted to geometric errors and should not therefore, be taken into account. However, for other cases, such as that illustrated in Fig. 4.7 concerning form errors, the influence of geometric errors has to be taken into consideration. As mentioned earlier, the form error $\Delta f'$ in Fig. 4.7 can generate errors of positioning of $\Delta f'$ or $2\Delta f$ according to the type of positioning jig. These errors will be integrated in the total error Δl_is for the optimization procedure.

In another important case where geometric tolerances influence a tolerance chain, the situation should be analyzed in detail. This example concerns the use of the maximum material principle in tolerancing positions of features, such as a linear chain of holes or shafts. As explained in Chapter 3, a bonus for the position tolerance can be gained when the diameter of a hole is larger than its maximum material size. In this case, the geometric tolerance influences the tolerancing of the chain of features.

When tolerancing is two-dimensional, as in turning operations, the derivations for tolerancing have to take into account two degrees of freedom. For example, for conical features, the position of the surface, as well as its orientation, is important. The formulation of the tolerancing derivations becomes much more cumbersome than in the case of a unidirectional tolerancing and for this reason is not discussed here.

Another very important situation relating to two-dimensional tolerancing concerns the boring of holes in a plate. In this case, positioning errors are the result of errors of clamping the part and of positioning the tools. The deviations can be in two perpendicular directions. The influence of the maximum material principle has also to be considered in these cases and can be of substantial benefit. For a solution to this tolerancing problem, see Lehtihet and Gunasena (1991). The practical importance of this case is, of course, very obvious because it relates to the problem of mating two different parts (male and female), which have to be toleranced accordingly.

4.8 THE FUNCTION OF SETTING IN MANUFACTURING

4.8.1 The basic setting problem

In section 4.4, a model was developed in order to tolerance setting dimensions (the Δl_i) on the machine tool. One of the difficulties of this procedure is to find reliable values for the tolerances in production (the Δl_i). This task is relatively difficult because the information in this field is vague, and to find more precise values, experimentation would be necessary. Since this is generally not done, the values of the Δl_i are not very precise. Nevertheless, the optimization model is helpful in checking the feasibility of a process plan and in providing a first approximation in production tolerancing and dimensioning.

To overcome this difficulty, it is common practice to observe what happens in production and to adjust the setting dimensions accordingly. But before describing some well-known setting strategies in production, it is useful to clarify the concept of 'setting' in manufacturing.

Up to now, the setting tolerance Δl_i was considered as a global result of many factors. Among these factors, it is useful to distinguish two categories. The first one, later called D_a, is the accumulation of all the errors occurring in manufacturing and having a completely stochastic character. These errors are the result of inaccuracies in the kinematic control system or stopping switch, in tool wear or in deformations due to cutting forces, etc. The second results from inaccuracies in the tool setting procedure *per se*. These errors are of a systematic character and do not change during manufacture of the part as long as the tool is not changed and reset. An advantage of resetting the tools in production is that it is in the hands of the operator who can adjust the manufacturing process so that it is more productive, using statistical strategies based on sampling methods (Chapter 14).

To illustrate this situation simply, Fig. 4.17 shows setting tools for the machining of a design dimension (D) with its given tolerance IT(D). The manufactured dimension (C) with its tolerance IT(C), here equal to IT(D), can only be obtained when the machine dimension (called L) will stay in a tolerance interval of IT(C–e) where e is the positioning tolerance of the part in its jig.

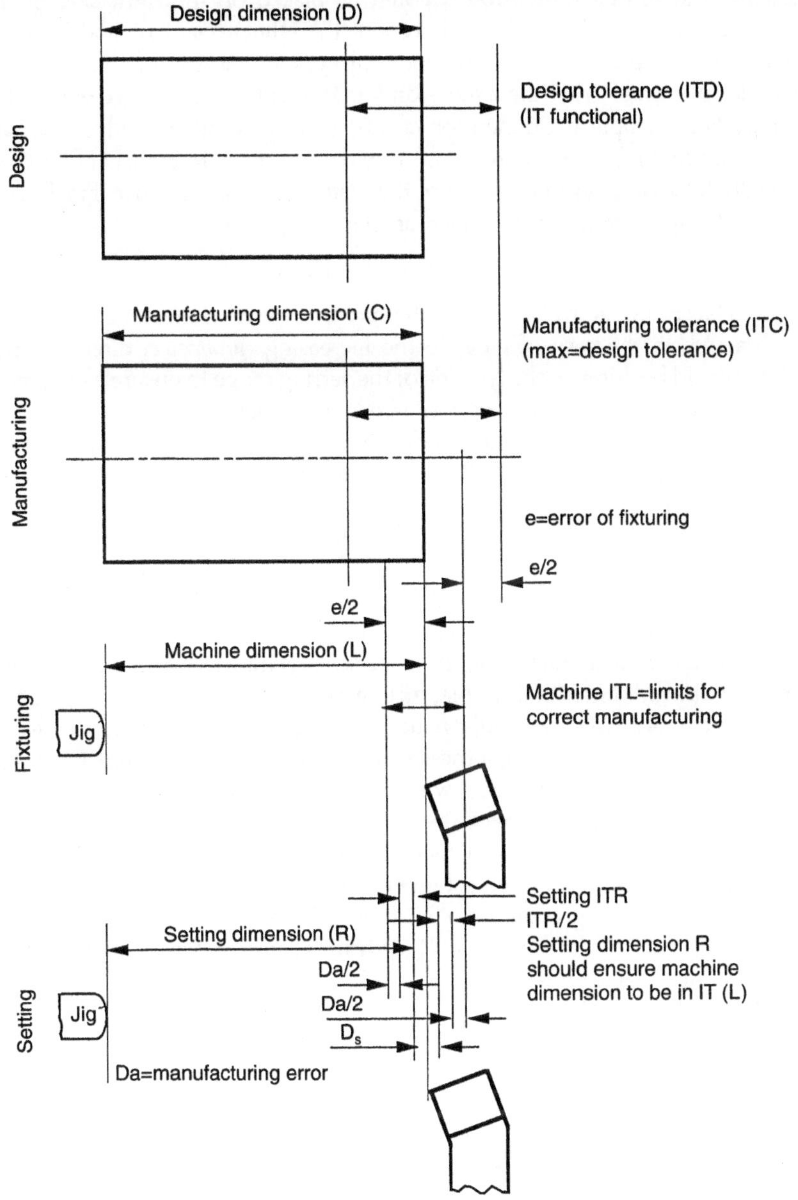

Fig. 4.17 Model of tool setting.

However, the setting dimension (called now R) with its tolerance IT(R) has to be adjusted so that the influence of the stochastic component of IT(L) (now called D_a) and the setting tolerance IT(R) are not producing manufactured dimensions out of tolerance. This condition is illustrated in Fig. 4.17 where the setting dimension R is at its minimum possible size, taking into account D_a and IT(R). Similarly, the setting dimension R could be placed on the right side of the tolerance interval IT(L) taking into account the influences of D_a and IT(R). In practice, ignoring how the production will evolve (because of tool wear for example), it is usual to set the dimension R in the center of the range of possible settings. For example, if a dimension of $20 \div 0.2/0$ is to be realized, one would set the machining dimension to 20.1 with a setting tolerance of ± 0.01, generally taken as a tenth or a fifth of the manufacturing tolerance. Finally, the whole setting range for R is defined as D_s:

$$D_s = \mathrm{IT}(L) - D_a - \mathrm{IT}(R) \tag{4.29}$$

The tool has to be set in this range to be sure that the machine dimension L will be obtained. In the case of wear in the increasing dimension direction, this setting would be done on the left side of the setting range to ensure a maximum number of good parts when the tool is being reground.

In practice, the setting strategies do not use the formulae set out in (4.28) because the values of D_a and IT(R) are not well known. Their general principles are reviewed below.

4.8.2 Review of setting strategies

As already explained in normal production, when no information is available about trends in manufacturing, it is good practice to define the setting dimension in the center of the tolerance interval.

When several passes are needed to achieve better precision, the strategies are different according to the size of the series and the precision required. For single parts or parts in very small series, without accuracy requirements, it is usual to take several passes and to check the result of the machining operation every time. This is a time-consuming, inefficient procedure and is not used frequently.

When the precision required is higher, about 0.1 to 0.05, it is customary to take two passes, a rough one and a finish one. The strategy is explained in Fig. 4.18 where a part of raw dimension L_b (measured) has to be machined. A rough pass is defined in such a way that the finish pass has a given chip thickness p_f. The depth of cut of the rough pass will therefore be:

$$p_e = L_b - \mathrm{mean}(L_f) - p_f \tag{4.30}$$

with mean(L_f) being the average size of the finished part. After roughing the part, the size of the part is measured and L_e is found. Then a finish pass is performed with a chip thickness of p_f:

$$p_f = L_e - \mathrm{mean}(L_f) \tag{4.31}$$

and finally, L_f is checked directly.

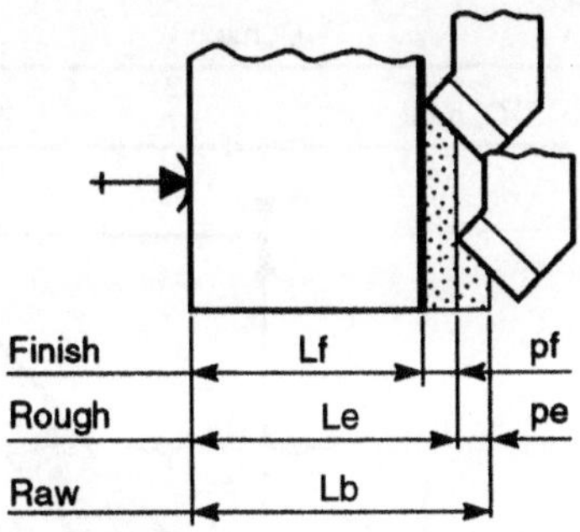

Fig. 4.18 Example of setting two passes.

In the case of a high precision part, for unitary or very small series, the procedure is refined to produce a more precise part.

The principle is to take equal passes for the semi finish and the finish passes so that the deformations will be identical. For example, if it is necessary to machine a part with a dimension $D = 27 + 0.02/0$, two final passes are recommended, with a chip thickness of 0.3, as follows:

1. roughing up to a dimension of $(27 + 0.3 + 0.3 =)27.6$
2. semi finishing by taking a pass of 0.3
3. measuring the size obtained, e.g. 27.33
4. finishing by taking a pass of

$$p_f = D_{\frac{1}{2}\text{finish}} - \text{mean}(D_{\text{final}})$$
$$= 27.33 - 27.01 = 0.32.$$

5. checking the final dimension.

This procedure ensures a precision of 0.02 and a final dimension near to mean(D_{final}). But as mentioned earlier it is time-consuming and only justified for precision manufacturing.

In the more common cases of medium and large series of parts, it is customary to check the first part machined in the batch, or more commonly, to check a small sample of parts at the beginning of the series. The average dimension of the batch is determined as shown in Fig. 4.19 (Xe) and the setting is corrected by adjusting the mean measure Xe to the ideal setting as defined in section 4.8.1, i.e. between R_{min} and R_{max} in Fig. 4.19. In this figure, the ideal setting is at R_{min} because the dimension C_m is increasing steadily during production and a maximum number of parts is required before a tool change. The scatter observed for the sample can be used to determine D_a which is used in (4.28). Application of statistical methodologies will provide a more reliable correction of setting. Details on these procedures are given in section 4.9.

Up to now, only unidirectional dimensions have been considered. In industrial practice, this procedure of setting is adopted for simplicity and efficiency.

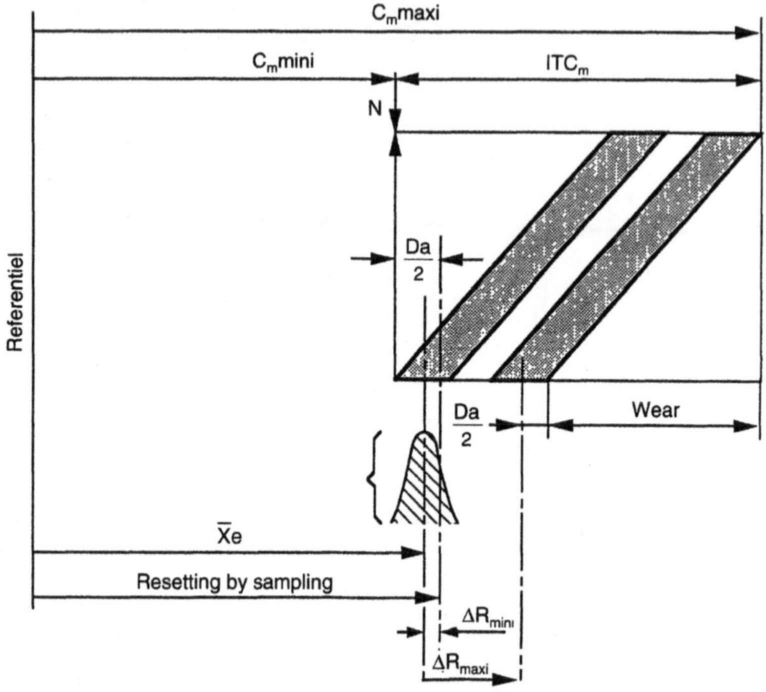

Fig. 4.19 Setting by a sampling method.

However, for situations where a number of features have to be produced with any type of spatial relationships, it becomes necessary to take into account the position and orientation of the different features and to correct them when parts are not in tolerance. For example, when machining a matrix of holes in a plate as represented in Fig. 4.20, it is necessary to correct the orientation as well as the position when errors in fixturing occur (Fig. 4.20(a)). The result of the corrective action can be seen in Fig. 4.20b when an average of the orientation and positioning errors is compensated. If only positioning errors were considered, the correction would not be optimal and might even amplify the errors.

When three-dimensional errors are analyzed, it is usually necessary to execute compensation in six degrees of freedom. The corrections are determined by an analysis in matrix form as explained in Chapter 7 (equation 7.3). The compensation can be executed with the carriages of the machine tool or by readjusting the location of the part in its fixture by small motions of the contact supports. Figure 4.21 shows conceptually the principle of compensation for deviations occurring only in a plane. The three locators (M_4, M_5 and M_6) will be readjusted in this case to compensate the location errors of the considered feature.

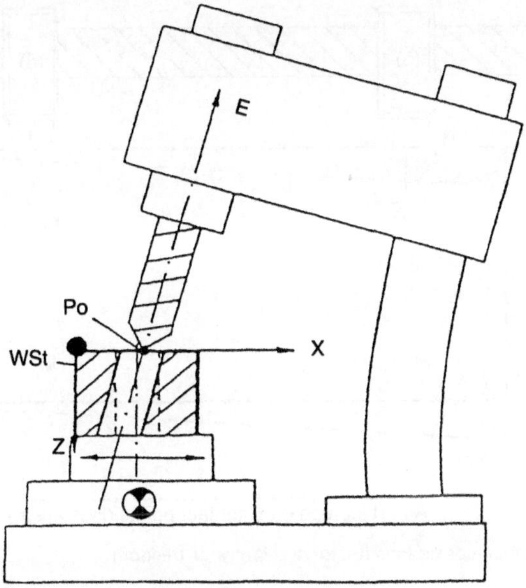

Fig. 4.20a Example of error in orientation.

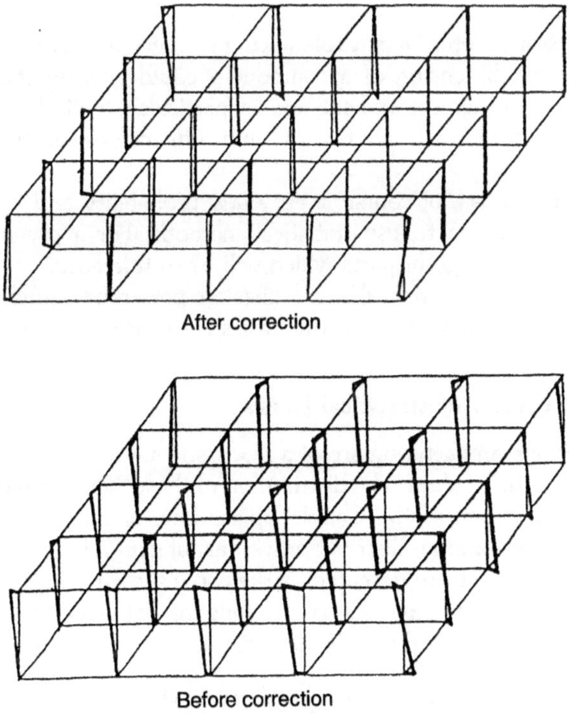

Fig. 4.20b Effect of corrections in position and orientation.

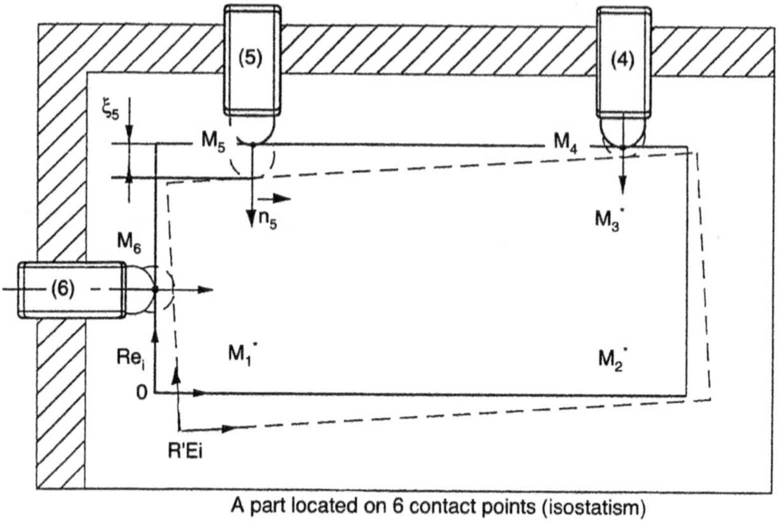

Fig. 4.21 Conceptual corrections for three degrees of freedom.

4.9 STATISTICAL INTERPRETATION OF TOLERANCING AND SETTING PROCEDURES

As explained earlier, deterministic tolerancing is a very restrictive procedure. It assumes that any dimension of a component could be combined with any dimension of the other component and in particular, limit dimensions of a component could be associated with limit dimensions of the conjugate component.

This assumption has, of course, a very low probability and it is therefore much more appropriate to use statistical methods. For a given risk, these determine the percentage of parts which will be in tolerance.

Two applications of statistical approaches are presented below, but to help the reader, a short review of statistical basics is given in the next section.

4.9.1 Short review of statistical basics

The quantities in manufacturing are of a stochastic nature because many error factors influence their values. The distribution of these values can be described by statistical laws such as the Gauss–Laplace law, or normal law, which is applicable to many random distributions in manufacturing.

The normal law can be represented by the curve shown in Fig. 4.22 or by an equation of the probability density $f(x)$ as a function of the random variable x. If the mean of the distribution is $\bar{x}$ ($\bar{x}=0$ in Fig. 4.22), the equation is:

$$f(x)=\frac{1}{\sqrt{2\pi}}\exp\left(\frac{-(x-\bar{x})^2}{2\sigma^2}\right) \tag{4.32}$$

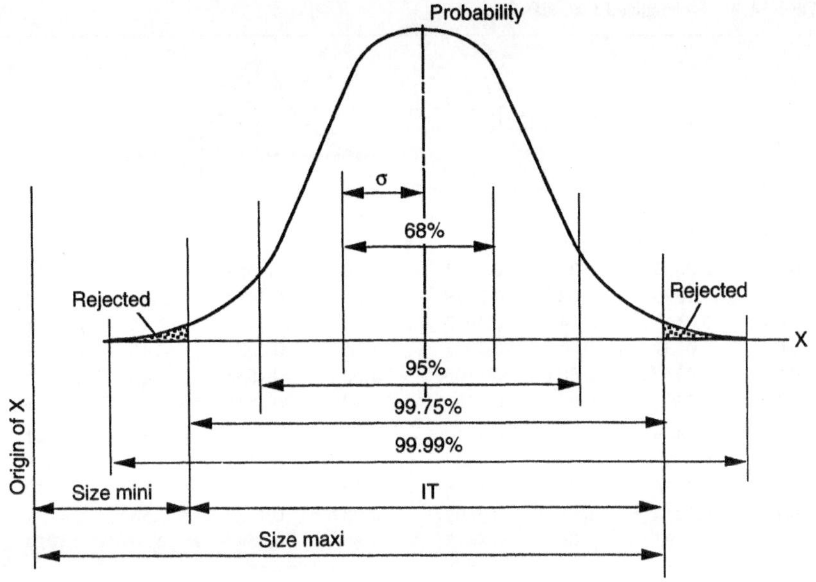

Fig. 4.22 Normal law and interval of tolerance IT.

or, expressed in normalized coordinates with $X=(x-\bar{x})/\sigma$

$$f(\chi)=\frac{1}{\sqrt{2\pi}}\exp\left(\frac{-\chi^2}{2}\right) \tag{4.33}$$

Figure 4.22 shows that 68% of the values are in the range $\pm\sigma$ and that 99.75% are in the range $\pm 3\sigma$ which is taken as the tolerance interval because it rejects only a percentage of 0.25% of the parts when the production is well centered in the tolerance interval. More generally, if a dimension is given by its tolerances, e.g. $25.92+0.10/-0.13$, and when the production of the parts has a mean of $m=25.92$ and a standard deviation of $s=0.04$, it is possible to determine the proportion of parts in tolerance. In this example, normalizing the experimental values comes out as

$$X_{\text{inferior}}=X_i=\frac{-0.13}{0.04}=-3.25$$

$$X_{\text{superior}}=X_s=\frac{0.10}{0.04}=2.5$$

Using these values in normalized tables such as Table 4.3, it is possible to calculate the proportion of rejected parts as follows:

$$\begin{aligned}\text{Prob}(X>2.5)&=0.5-\text{Prob}(0<X<2.5)\\&=0.5-0.4938=0.0062\end{aligned}$$

Table 4.3 Normalized Gaussian law

$$s = \frac{1}{\sqrt{2\pi}} \int_0^t e^{-x^2/2} \, dx$$

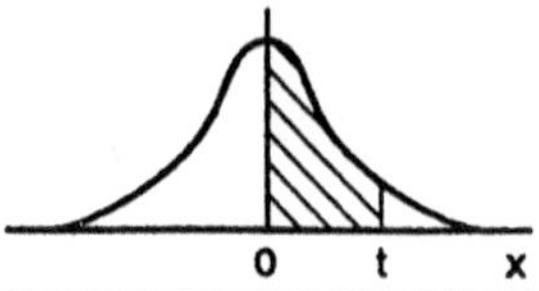

t	s	t	s	t	s	t	s
0.00	0.0000	0.80	0.2881	1.60	0.4452	2.40	0.4918
0.02	0.0080	0.82	0.2939	1.62	0.4474	2.42	0.4922
0.04	0.0160	0.84	0.2995	1.64	0.4495	2.44	0.4927
0.06	0.0239	0.86	0.3051	1.66	0.4515	2.46	0.4931
0.08	0.0319	0.88	0.3106	1.68	0.4535	2.48	0.4934
0.10	0.0398	0.90	0.3159	1.70	0.4554	2.50	0.4938
0.12	0.0478	0.92	0.3212	1.72	0.4573	2.52	0.4941
0.14	0.0557	0.94	0.3264	1.74	0.4591	2.54	0.4945
0.16	0.0636	0.96	0.3315	1.76	0.4608	2.56	0.4948
0.18	0.0714	0.98	0.3365	1.78	0.4625	2.58	0.4951
0.20	0.0793	1.00	0.3413	1.80	0.4641	2.60	0.4953
0.22	0.0871	1.02	0.3461	1.82	0.4656	2.62	0.4956
0.24	0.0948	1.04	0.3508	1.84	0.4671	2.64	0.4959
0.26	0.1026	1.06	0.3554	1.86	0.4686	2.66	0.4961
0.28	0.1103	1.08	0.3599	1.88	0.4699	2.68	0.4963
0.30	0.1179	1.10	0.3643	1.90	0.4713	2.70	0.4965
0.32	0.1255	1.12	0.3686	1.92	0.4726	2.72	0.4967
0.34	0.1331	1.14	0.3729	1.94	0.4738	2.74	0.4969
0.36	0.1406	1.16	0.3770	1.96	0.4750	2.76	0.4971
0.38	0.1480	1.18	0.3810	1.98	0.4761	2.78	0.4973
0.40	0.1554	1.20	0.3849	2.00	0.4772	2.80	0.4974
0.42	0.1638	1.22	0.3888	2.02	0.4783	2.82	0.4976
0.44	0.1700	1.24	0.3925	2.04	0.4793	2.84	0.4977
0.46	0.1772	1.26	0.3962	2.06	0.4803	2.86	0.4979
0.48	0.1844	1.28	0.3997	2.08	0.4812	2.88	0.4980
0.50	0.1915	1.30	0.4032	2.10	0.4821	2.90	0.4981
0.52	0.1985	1.32	0.4066	2.12	0.4830	2.92	0.4983
0.54	0.2054	1.34	0.4099	2.14	0.4838	2.94	0.4984
0.56	0.2123	1.36	0.4131	2.16	0.4846	2.96	0.4985
0.58	0.2190	1.38	0.4162	2.18	0.4854	2.98	0.4986
0.60	0.2257	1.40	0.4192	2.20	0.4861	3.00	0.49865
0.62	0.2324	1.42	0.4222	2.22	0.4868	3.10	0.49904
0.64	0.2389	1.44	0.4251	2.24	0.4875	3.20	0.49931
0.66	0.2454	1.46	0.4279	2.26	0.4881	3.30	0.49952
0.68	0.2517	1.48	0.4306	2.28	0.4887	3.40	0.49966
0.70	0.2580	1.50	0.4332	2.30	0.4893	3.50	0.49976
0.72	0.2642	1.52	0.4357	2.32	0.4899	3.60	0.49984
0.74	0.2703	1.54	0.4382	2.34	0.4904	3.80	0.49993
0.76	0.2764	1.56	0.4406	2.36	0.4909	4.00	0.49997
0.78	0.2823	1.58	0.4429	2.38	0.4913	4.00	0.49997

and

$$\text{Prob}(X < -3.25) = 0.5 - \text{Prob}(-3.25 < X < 0)$$
$$= 0.5 - 0.49941 = 0.00059$$

Altogether, the total probability of rejected parts will be about 0.7%, which is an acceptable proportion of parts to lose in practice.

Another important law in statistics concerns the distribution of the sum of a number of random variables whose means and standard deviations are known. The mean value of the sum is the sum of the mean values of the components, and the variance (or the square of the standard deviation σ) of the sum is the sum of the variances of the components. When the distributions are given in interval tolerances IT and it is admitted that the normal law is applicable, the same relation can be written for the intervals IT, i.e.:

$$(\text{IT resultant})^2 = \Sigma(\text{IT components})^2 \tag{4.34}$$

As mentioned before, a proportion of about 0.25% of the parts will be rejected. To apply (4.34) correctly, some precautions have to be observed.

- The standard deviation σ should be less than or equal to IT/6.
- The individual σ should be independent statistically.
- Every mean dimension in production has to be adjusted to the center of the tolerance interval.
- When the number of components is greater than 3, it is possible to accept non-normal distributions of the components because the resultant will be of normal distribution.

Of course, the advantage of a statistical evaluation will increase with the number of components.

The application of the preceding equations to determine optimized values of setting dimensions, as shown in Table 4.1, is straightforward by using (4.34) which becomes, in the case of l_is and C_{ij}s:

$$\text{IT}(C_{i-j}) \geqslant \sqrt{(\Sigma(\text{IT}(l_i))^2)} \tag{4.35}$$

Equation 4.35 can be applied easily by using Fig. 4.23 which considers the ratio:

$$t = \frac{\text{ITdesign}}{\sqrt{\Sigma \text{IT}(i)^2}}$$

and gives accordingly the proportion of rejected parts. Equation 4.35 has to be introduced in Table 4.1 instead of the arithmetic equation:

$$\text{IT}(C_{i-j}) \geqslant \Sigma(\text{IT}(l_i)) \tag{4.36}$$

The application of (4.35) presupposes however that the components $\text{IT}(l_i)$ are all of a stochastic nature. This condition is only true when considering the stochastic part of l_i, known as D_a, and not the systematic part IT(R) concerning

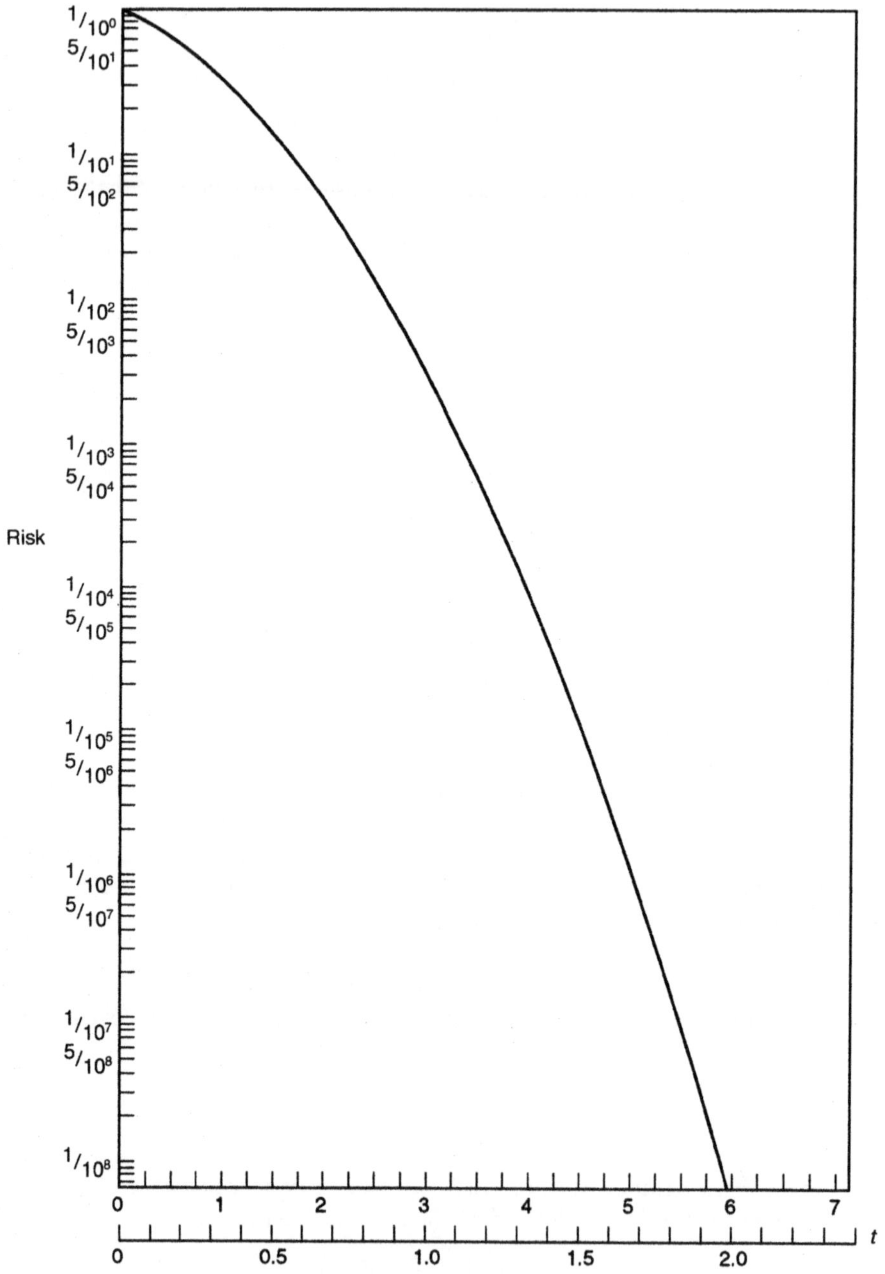

Fig. 4.23 Risk of rejects as a function of t.

the setting of the tool. It is therefore necessary in the optimization procedure to make this distinction so that (4.35) can be written:

$$\mathrm{IT}(C_{i-j}) \geqslant \sum(\mathrm{IT}(R)) + \sqrt{\Sigma(D_{ai}^2)} \tag{4.37}$$

To demonstrate the advantage of statistical interpretations, two examples taken from production situations are now presented.

4.9.2 Application of statistical methods to production problems

Example 1

Assuming that four independent segments are combined to give a sum which should have a tolerance interval of $T = 0.24$ and assuming that the standard deviation of the production process is:

$$\sigma_k = 0.02 \quad k = 1,\, 2,\, 3 \text{ and } 4$$

we need to know which tolerances to give to each component.

In the arithmetic method, the tolerance would be shared among the components as follows:

$$\mathrm{T}_k = \frac{\mathrm{T}}{4} = 0.06$$

which means that the ratio to the standard deviation in production would be:

$$r = \frac{\mathrm{T}_k}{2\sigma_k} = \frac{0.06}{0.04} = 1.5$$

i.e. according to the statistical tables, 13.2% of the parts would have to be rejected by sorting the parts in production.

On the contrary, using the statistical method, (4.35) gives the value of the standard deviation of the resultant dimension:

$$\sigma_{\text{resultant}} = 0.02 \times \sqrt{4} = 0.04$$

The new ratio to the standard deviation becomes:

$$r = \frac{\mathrm{IT}_{\text{res}}}{2\sigma_{\text{res}}} = \frac{0.24}{0.08} = 3$$

In this case, the proportion of rejected parts would be 0.0027 or about 3 per thousand. This result is obtained without sorting parts in production. If sorting were executed, e.g. for $\mathrm{T}_k = 6\sigma_k$, the proportion of rejected parts would be even smaller.

The tolerance of each component can be taken as

$$\mathrm{IT}_k = 6 \times 0.02 = 0.12$$

instead of 0.06 by the arithmetic solution.

Example 2
When considering a continuous production, which is checked by taking samples of parts from time to time, the question arises whether the average value calculated on the sample – which, of course, is never exactly equal to the average value of the total production – has to be readjusted? Taking as an example a production with a mean value of $m = 40$, adjusted at the beginning of manufacturing, it is observed after a certain time that, for a sample of $n = 17$ parts, the mean and the standard deviation are:

$$m_1 = 39.99 \quad \text{and} \quad s_1 = 0.02$$

Statistical laws show that the real mean M_1 of the complete series will be bound by the following limits:

$$\frac{m_1 - t_\alpha s_1}{\sqrt{(n-1)}} < M_1 < \frac{m_1 + t_\alpha s_1}{\sqrt{(n-1)}} \tag{4.38}$$

where t_α is a confidence parameter which is given in Student's statistical tables. Taking a probability value of 0.95, the mean value M_1 should be included in the following limits of confidence:

$$39.9794 < M_1 < 40.0006 \tag{4.39}$$

As shown by (4.39), the initial value of the average $m = 40$ is included in the limits of confidence, so no change of setting is necessary in production.

If, for example, the result of sampling would have been given limits of confidence as follows:

$$39.9594 < M_1 < 39.9806 \tag{4.40}$$

the conclusion would have been that there is a chance of 0.95 that the dimensions of the parts should be readjusted.

These two simple examples of the application of statistical laws to production settings show the efficiency which can be obtained. They should therefore be used as much as possible in manufacturing planning and control, as can be seen in more detail in Chapter 12.

4.10 CONCLUSION

The importance of correctly determining tolerancing and setting in production cannot be over-emphasized. A process plan which cannot guarantee the manufactured dimensions required by the design department would be meaningless for a manufacturing industry.

The use of correct methods of tolerancing, therefore, as presented in this chapter, should be of great help to process planners. The computerization of tolerancing, as explained here, should act as a powerful aid for defining optimal tolerances, as well as being a user friendly tool for process planners who are not experts in tolerancing technology.

REFERENCES

Fainguelernt, D. (1986) Development of a Computer Program for the Optimisation of Tolerance Transfer from Design to Manufacturing, Master thesis, March, Technion, Israel Institute of Technology, Haifa, 74–85.

Fainguelernt, D., Weill, R. and Bourdet, P. (1986) Computer aided tolerancing and dimensioning in process planning, *CIRP Annals*, **35**(1), 381–6.

Lehtihet, E.A. and Gunasena, N.U. (1991) On the composite position tolerance for patterns of holes, *CIRP Annals*, **40**(1), 495–8.

Weill, R. (1991) *Computer Aided Tolerancing and Tool Setting for Machining Operations*, International Conference on Innovative Metal Cutting Processes and Materials, October 2–4, COREP, Politecnico di Torino, Italy.

CHAPTER 5

General selection of primary production processes

5.1 FROM DESIGN TO PROCESS PLANNING

The process planner defines in detail the process that will transform raw material into the desired shape. The shape is defined by the product designer and is expressed in engineering drawings and GDT – geometric dimensioning and tolerances. The process planner is bound by the defined drawing.

The designer is a problem solver who applies such fields as physics, mathematics, hydraulics, pneumatics, electronics, metallurgy, strength of materials, dynamics, magnetics and acoustics in order to find a solution, namely, the new product. His/her main responsibility is to design a product that meets customer specifications. A parallel target is to design a high-quality, low-cost product.

There is no single solution to a design problem, but rather a variety of possible solutions which surround a broad optimum. The solution can come from different fields of engineering and apply different concepts. The designer is bound by constraints that arise from physical laws, the limits of available resources, the time factor, company procedures and government regulations.

Among all these possible solutions, the designer selects the most suitable.

To insure against failure, the designer provides a margin of safety. Strength failures, for instance, are protected by a factor of safety. For mechanical components, it is customary to use a factor from 4 to 40. To insure against potential errors in manufacturing, the designer specifies the permissible deviations, that is, the acceptable range of tolerances. All too often, designers specify excessively tight tolerances in order to be conservative and avoid risk.

Product designers are not process planners. However, what they have in mind during the design stage significantly affects the manufacturing process and the process planning. They do not go into details of the manufacturing process, but usually work by intuition. However, parts that were designed with a specific manufacturing process in mind might turn out to be very difficult to manufacture if the process has to be changed. In such cases, it should be remembered that parts are designed subject to functional, strength or manufacturing constraints. Part drawing should always by seen as a constraint;

it might be an artificial constraint if the manufacturing process is the controlling factor in part design.

Studies have indicated that the cost of the engineering stages, i.e. product design, detail design, testing and process planning, is about 15% of the product cost, while the production stage accounts for 85%. However, since the committed cost of the product is about 90% established in the engineering stages, it is worthwhile not to rush but to increase thinking time in design, before making decisions.

The product designer should bear in mind the manufacturing process that will produce the designed part. Each manufacturing process has its advantages, capabilities and limitations. The cost of a part can be kept to a minimum if its features, dimensions and tolerances match the capabilities of one of the available processes. Otherwise, the cost might be excessively high or the production might even be impossible. Designers do not define the process plan, but rather steer toward utilization of existing processes, preferably to one available in their own plant.

5.2 CLASSIFICATION OF MANUFACTURING PROCESSES

Manufacturing processes can be broadly divided into the following categories.

5.2.1 Forming from liquid (casting, molding)

To form a part by liquid casting or molding, the raw material is heated to its liquid state and then poured, or pushed, into a mold of the desired form (Fig. 5.1).

This is an economical method of producing complicated shapes. However, it is susceptible to internal porosity resulting from shrinkage and the presence of gas. The flow of material in the mold in thin channels is also a serious problem

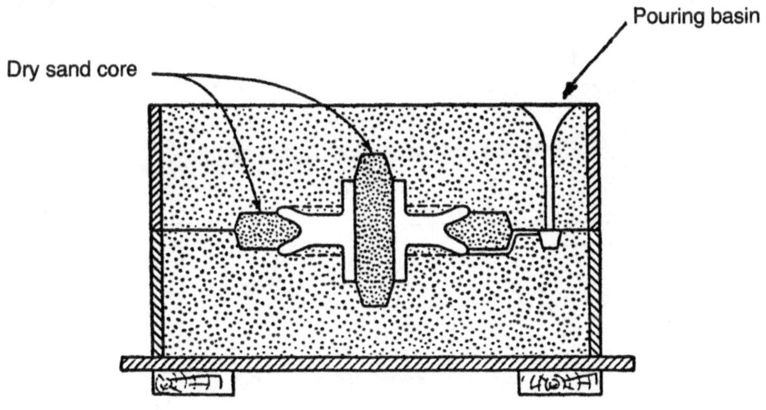

Fig. 5.1 Cross-section view of a mold for sand casting.

that should be considered in the design. The cost of the mold is high, and the dimensions and surface finish often cannot be kept to tight tolerances. On the whole, casting is a good process for mass production.

5.2.2 Forming from solid by deformation

This type of forming can be divided into three subgroups:

1. *hot working*, including hot rolling, forging and extrusion;
2. *cold working*, stamping, bending, spinning, stretch forming, shearing, cold rolling, extrusion, deep drawing etc.; and
3. *forming from powder*, such as powder metallurgy and plastic molding.

Rolling (Fig. 5.2) is the cheapest method of shaping materials. Rolling into bars, plates or sheets is executed by passing the ingot between rolls that grip

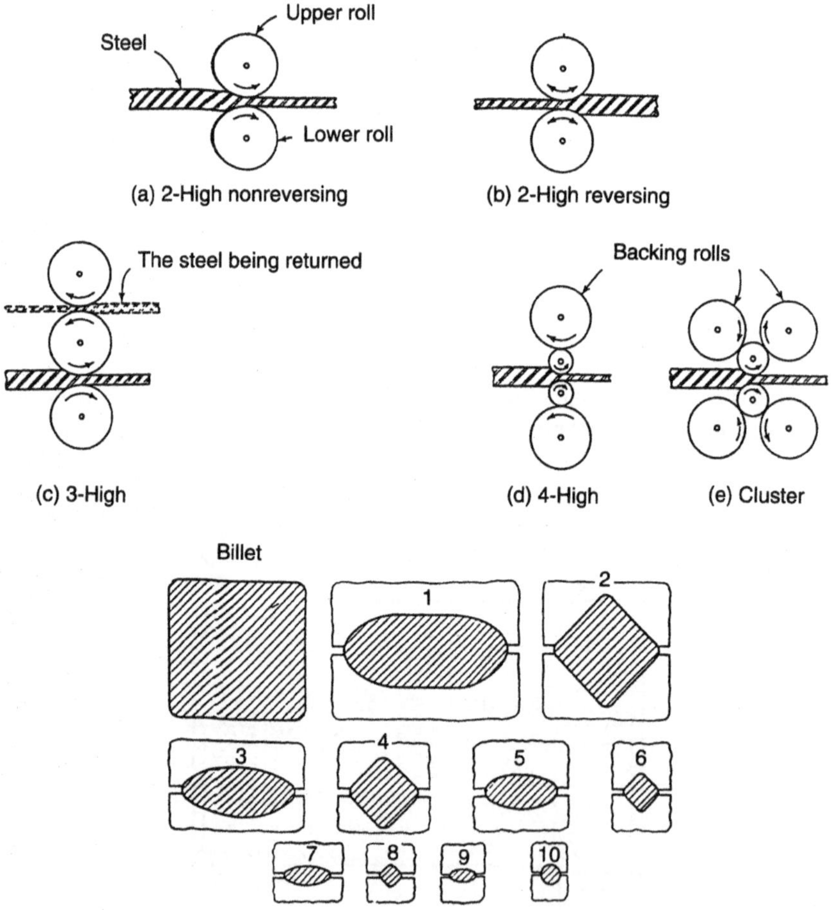

Fig. 5.2 The arrangement and shape of the rolls on a rolling mill.

the material and draw it through, compressing it and reducing its cross-section, which remains constant while its length increases.

Not all materials are formable. The process is limited to simple shapes and is susceptible to changes in such material properties as tensile strength, hardness and ductility. Cracks developing during the process can be a problem. Dimensions can be kept to tight tolerances.

Forging (Fig. 5.3) is defined as the working of a piece of material into a desired shape by hammering or pressing, usually after heating, to improve its plasticity. The material may be shaped by drawing it out, which decreases the cross-sectional area and increases the length; by upsetting, which increases the cross-sectional area and decreases its length; or by squeezing it in close impression dies.

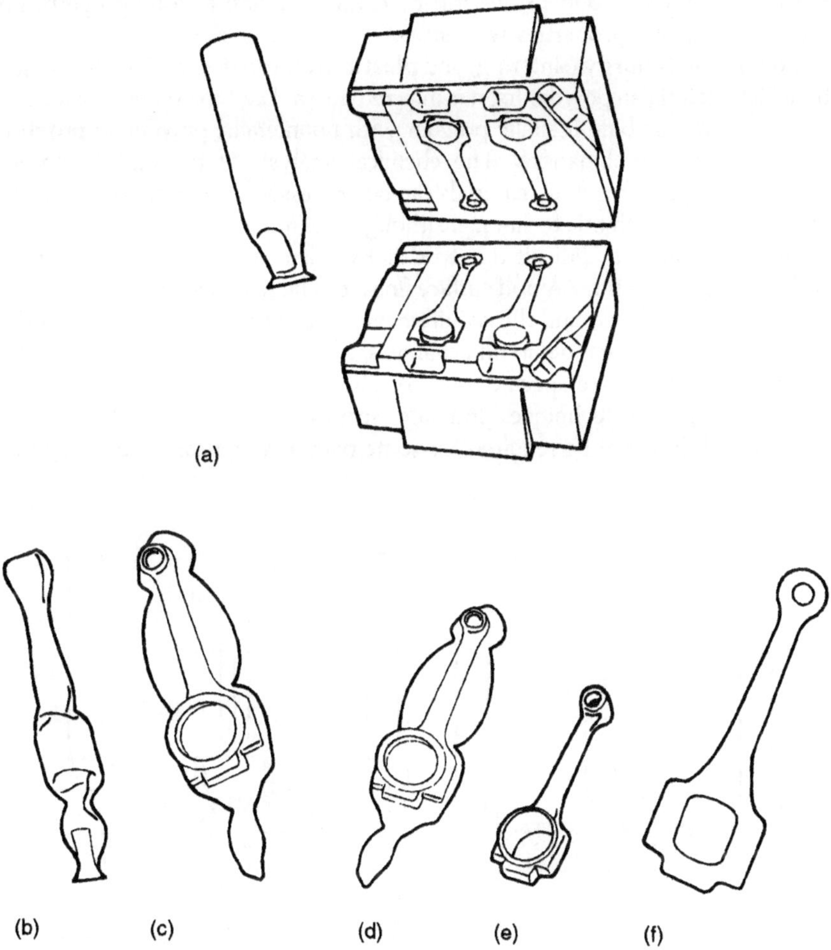

Fig. 5.3 Dies for drop-forging.

Forging has better mechanical properties than casting, so parts that have to withstand severe stresses should be made by forging. Forging has its disadvantages, however. Die costs are generally higher than casting and molds; many intricate and cored shapes possible in casting cannot be forged; closed impression die forging is limited with respect to size, and dimension control is difficult due to shrinkage and die wear, or the die may strike out of alignment.

Press work of cold material may be used for shearing, **stamping** or drawing (Fig. 5.4). Close dimensional control may be achieved, although special dies are required for each part. Bending along straight lines (Fig. 5.5) uses standard brake press tooling; this may keep close dimensional control, although it has limitations as regards the shape produced by this method.

Spinning (Fig. 5.6) and **stretch forming** (Fig. 5.7) are better suited for small quantity production; in fact some of the work cannot be done by other processes. Thin wall dimensions can be controlled, and if done properly no change in material properties will occur.

Powder metallurgy (sintering) and **plastic molding** (Fig. 5.8) is a technique by which material in powder form is injected into a die of the required shape. It is possible to combine metallic powder with nonmetallic powder to produce useful material combinations. The chemical analysis of the part is closely controlled, although not all materials can be processed. A high production rate can be obtained with this technique, although the initial cost of the dies is high. Dimension control is difficult due to shrinkage and die wear, or the die may strike out of alignment. A good surface finish can be maintained. There is a size limitation due to press and die requirements, and there is a problem with the flow of material in the mold in thin channels, which should be considered at the design stage. There are specific rules to follow when designing a part to be produced by this technique. Intricate shapes cannot be produced, and complicated dies may be required for some parts that appear suitable for this

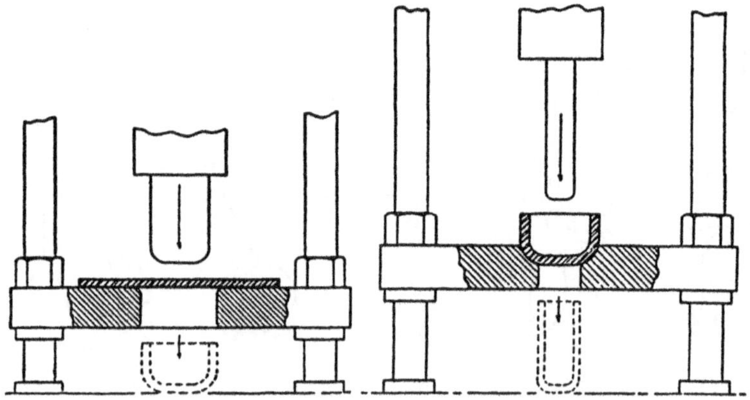

Fig. 5.4 Cold drawing.

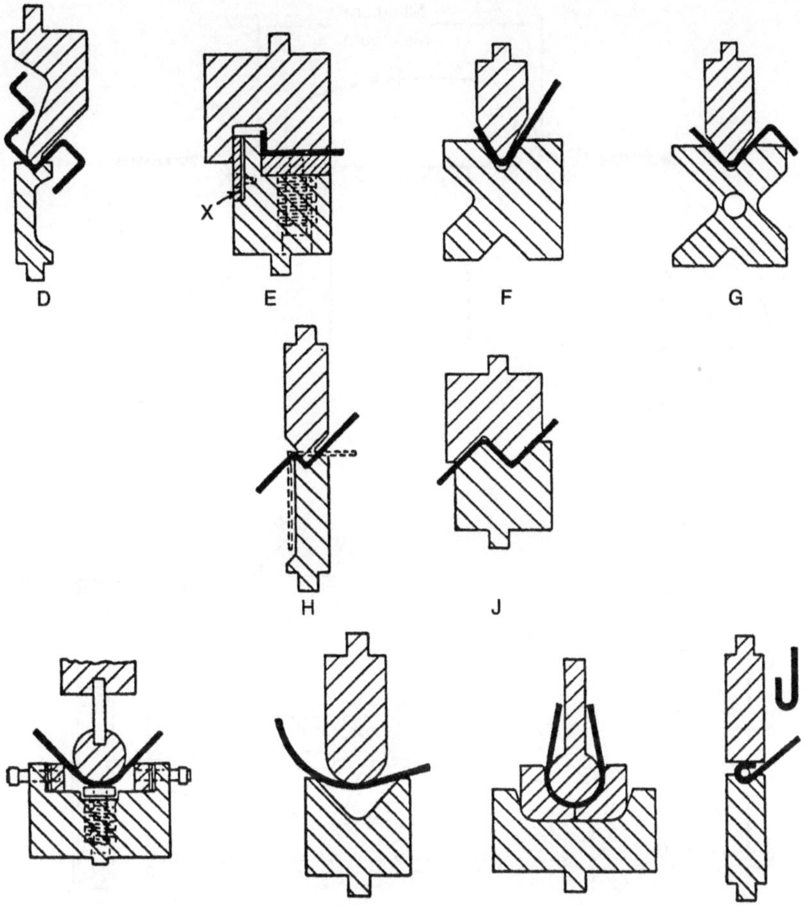

Fig. 5.5 Standard bending dies used on brake press.

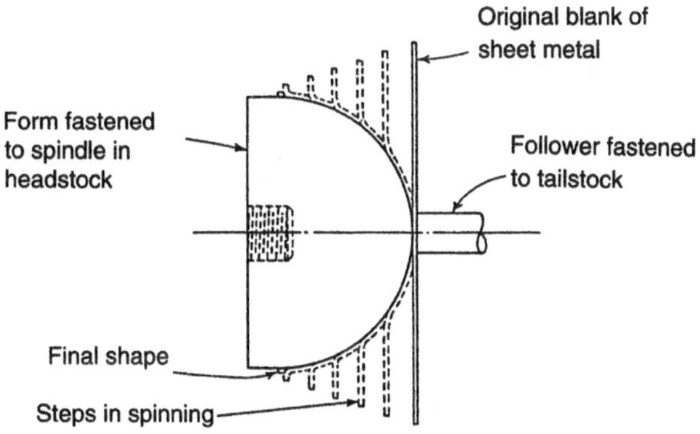

Fig. 5.6 Spinning operation.

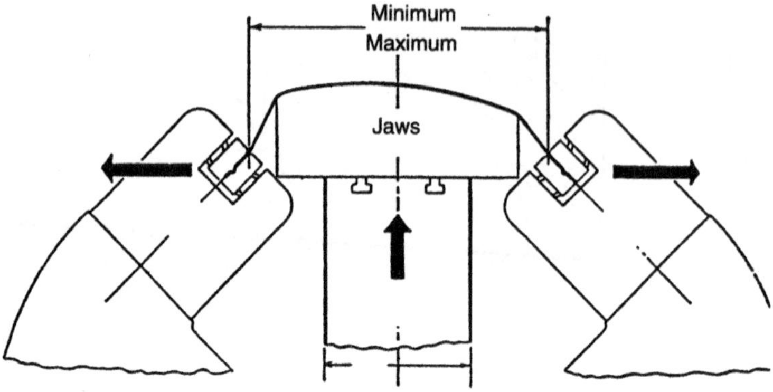

Fig. 5.7 Stretch forming.

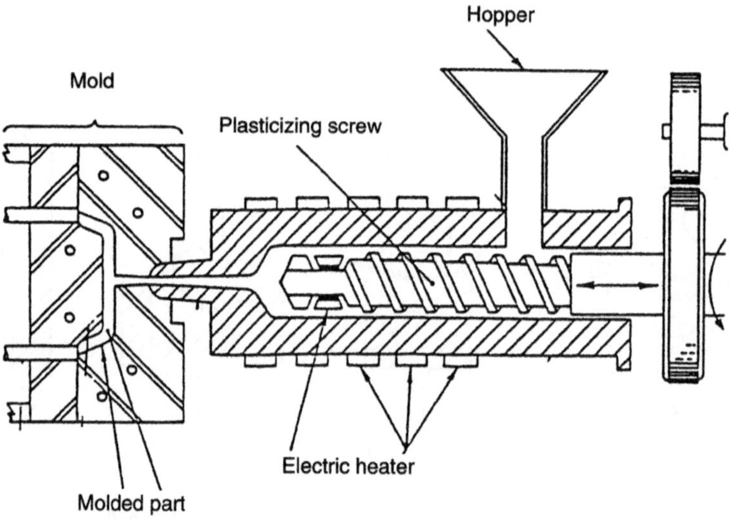

Fig. 5.8 Plastic molding machine.

technique. Some materials, such as hard metals, can only be produced by sintering because of their refractory properties.

5.2.3 Forming from solid by material removal

This technique encompasses all types of cutting operations that transform the workpiece into its final geometry by removing pieces of material from it. Some of the most common machining processes are illustrated in Fig. 5.9, and described below.

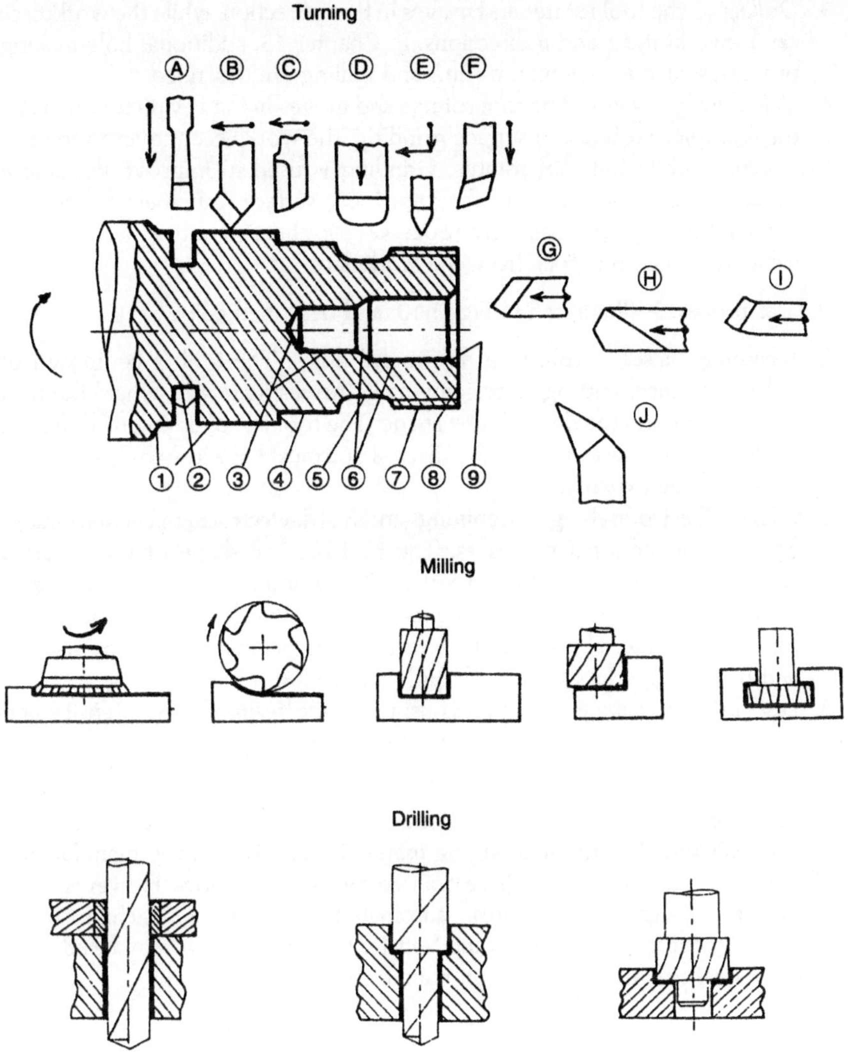

Fig. 5.9 Common metal removing processes.

1. *Milling* The tool, which has several cutting edges, rotates and the workpiece moves in one of several given directions, according to the shape required. For complex shapes (sculpture surfaces) three additional degrees of freedom are added. In Chapter 14, additional milling operations are described.
2. *Turning* The tool, usually with one cutting edge, moves in one x–z direction, while the part rotates around its line of symmetry. Therefore, only round symmetrical parts can be produced.

3. *Drilling* The tool rotates and moves in the z direction, while the workpiece can move in the x and y directions. In Chapter 13, additional hole-making processes such as reaming, boring and milling are described.
4. *Cylindrical grinding* The tool rotates and moves in the x–y direction, while the workpiece rotates. In surface grinding, the workpiece moves in the x–y direction while the tool rotates. Grinding is used to improve workpiece surface roughness. Processes for improving surface roughness further are honing and lapping. These two processes use abrasive rather than chipping tools for removing stock from the workpiece.

A laser process is illustrated in Fig. 5.10, and described below.

1. *Broaching* Used to produce non-round holes. The tool is composed of successive linear cutting edges, gradually increasing in size, where the final cutting edge has the desired hole shape. The tool is moved in one direction while the workpiece is stationary. Each shape must have its own special tool, which can be expensive.
2. *EDM* Electrodischarge machining (and ECM, electrochemical machining) are non-conventional processes. The tool has the shape of the required workpiece and moves in the z direction. The workpiece is stationary. Sparks vaporize a small spot on the workpiece material which is then washed away by an electrolyte. This method is effective for machining very hard material and producing odd shapes with high surface finish.
3. *Laser beam machining* A process using a light beam of high intensity and single frequency to burn and vaporize workpiece material.

The machines employed are usually universal machines that use universal and commercial tools. Tight dimensional tolerances and good surface finishes can be maintained, without changing material properties during manufacture. This material removal process is best suited for small quantities. However, with special tools, jigs and fixtures, automated versions of machine tools, numerically controlled machines and transfer lines, the process may be adapted to large quantity, mass production manufacturing.

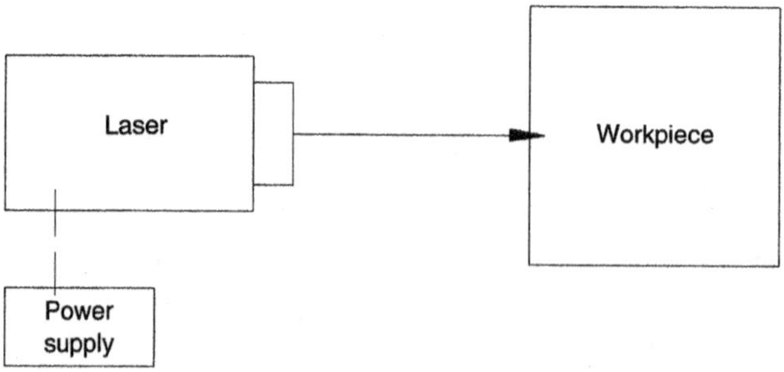

Fig. 5.10 Laser use in metal removal processes.

5.2.4 Forming by joining parts

This technique involves the use of welding, brazing, soldering and adhesive to join several components into one complex part. It is often used in conjunction with the material removal process to eliminate excessive waste of raw material.

Welding results in a strong bond between components; however, since it is a high temperature process, with the material being heated to melting point, distortions, shrinkage and chemical changes may occur, therefore not all materials are suitable for welding. If tight dimensional tolerances and a good surface finish are required, additional processes should be applied.

Good weld joints will have at least the same strength as the components themselves. Design for welding may use relatively thin walled materials, with ribs to strengthen the part. Industrial robots may be used to perform this operation.

Soldering and brazing produce a weaker joint than welding. They use filler materials whose melting points are lower than those of the component materials, although both can produce a wide range of material combinations.

Adhesives can be used to join different types of materials; for example, metal to wood, plastic to metal, plastic to plastic. A strong joint with no distortion or shrinkage can be obtained. Rubber based adhesives produce elastic and vibration-resistant joints. Special surface treatment is required, and the designer must realize that the strength of the joint is not uniform in all directions (usually, the joint is strong in shear and weak in peeling).

5.2.5 Forming by assembly

This technique involves the mechanical joining of parts by threads, bolts, rivets, assembly by press, etc.

5.2.6 Forming by material increase (incress)

This is a new technique applying a kind of selective solidification or binding of liquid or solid particles by glueing, welding, polymerization or chemical reaction. Material is progressively added until the whole part is created.

Most techniques build up parts in successive two-dimensional layers created one on top of the other. This is done by slicing a CAD model of the part to reveal horizontal layers, usually a few tenths of a millimeter deep. A single layer is created by scanning and solidifying it in a point-to-point fashion by means of a laser beam.

There are several methods of solidifying the layer. Figure 5.11 shows the stereolithography method. The main advantage of this technique is that parts can be produced directly from a CAD system with no extra tooling or dies. As it is a slow and expensive process, it is used mostly to build prototypes and one-of-a-kind parts. This process is also called **rapid prototyping**.

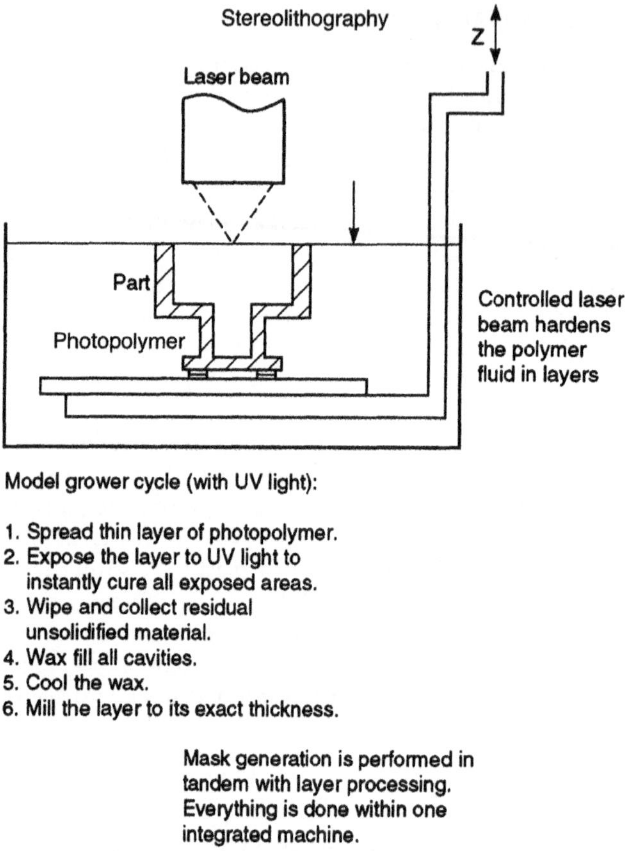

Fig. 5.11 Rapid prototyping.

5.3 DESIGN FOR MANUFACTURING

A part can often be manufactured by any of the available processes. However, the part shape and dimensions will be different according to the process selected.

For example, a machine frame can be cast or constructed from profile bars that are welded, riveted or bolted. The stress and strain computations are different in each case. The strength constraint is not always the controlling one; it is possible that the selected process capabilities will control the design. Figure 5.12 illustrates examples of design for manufacturing. Some general considerations are applicable as a function of the process:

(a) *Casting or molding* The thickness of the fin and the adjoining fillets have to be large in order to allow material flow in the die. The thickness will probably be determined by process rather than strength considerations.

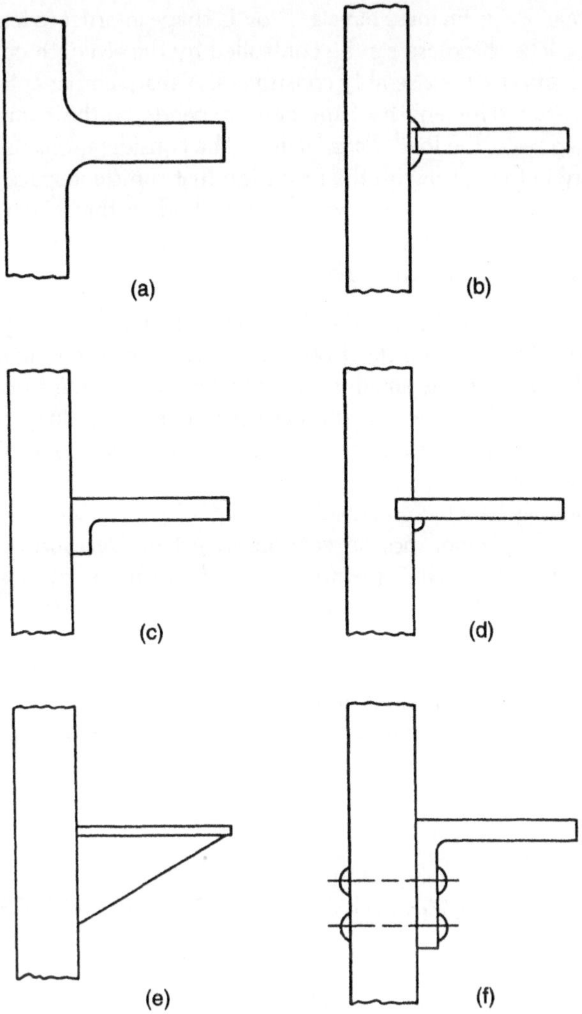

Fig. 5.12 Examples of design for manufacturing.

(b) *Welded joints* The thickness of the part may be controlled by strength constraints and will probably be significantly smaller than that in casting. There is a minimum thickness required for welding, but it is much less than in casting or molding. The design problem lies in the joint itself. Butt welding is weak in this case, and in order to increase welding strength and ease of part positioning, a bend of the fin (c) or slot (d) in the body is recommended.

(e) *Welded joint with support plate* For added strength and reduced fin thickness, support plates can be used.

(f) *Riveted joint* The fin must have a 'T' or 'L' shape in order to leave room for the rivets. The thickness can be controlled by the strength constraint and the shape and size by assembly constraints. A sharp corner at the top of the fin may result. The length of the bend depends on the number of rivets required to resist the load. *Bolted joint* The considerations used for rivets also apply in this case, with the provision that sufficient space be available for a thread in the body and a wrench. If the body is thin, either a screw and nut or self-tapping screws can be used, depending on body thickness and accessibility for a wrench or screwdriver.

In machine–metal removal, the thickness of the fin will probably be determined by strength consideration rather than process considerations. The adjoining fillets have to be small in order to save machining time.

Any one of the above design alternatives will meet product specifications. Costs, manufacturing time and facilities required for any one of the above alternatives will, however, vary.

The decision as to which method to select is the product designer's. He is neither a process planner, nor an economist, yet his decision constrains the process planner. It should be remembered that if the process the designer had in mind is not the one selected by process planning, a review of part design should be made.

CE (concurrent engineering) proposes that the decision should be mutual and should be taken after discussions between all disciplines involved in or affected by the decision. A good design for manufacturing should strike a balance between material cost, manufacturing cost and assembly cost.

5.4 SELECTING PRIMARY MANUFACTURING PROCESSES – ROUGH RULES

The following design factors have a bearing on the choice of a manufacturing process:

1. Quantity.
2. Complexity of form.
3. Nature of material.
4. Size of part.
5. Section thickness.
6. Dimensional accuracy.
7. Cost of raw material, possibility of defects and scrap rate.
8. Subsequent processes.

The choice of process should be made initially with economic factors in mind. The differences in direct manufacturing time can be quite significant. For example, the direct time taken to mold a part of moderate complexity with a metal die is about 25 seconds; to produce the same part using a material removal process might take about an hour. However, the cost of the metal die is high, probably in the neighborhood of $25,000. Assuming that the direct labor cost of the material removal process is about $15 per hour (ignoring indirect hourly rate and set-up costs, which will probably be higher for the molding process) the economic quantity should be at least $25{,}000/15 = 1666$ pieces in order to reach break-even point.

The quantity required will be a major determining factor of process selection. As was shown in Chapter 1, thinking time should be restricted to an economic level. Therefore, rules of thumb are needed.

As a general rule, the manufacturing processes may be ranked in order of economic consideration as follows:

High quantity (2000 or more)

1. Forming from solid by deformation
2. Forming from liquid – casting, molding
3. Forming by joining parts
4. Forming from solid by material removal
5. Forming by assembly

NB Forming by material increase is not suitable.

Low quantity (up to 50)

1. Forming from solid by material removal
2. Forming by joining parts
3. Forming from solid by deformation
4. Forming by assembly
5. Forming by material increase

NB Forming from liquid – casting, molding – is not suitable.

Medium quantities should be analyzed separately in each case. However, in order to save thinking and computation time, additional rules may be added relating to quantity and part complexity:

For a simple part, the lower quantity should be increased to 150 pieces and the higher quantity should be reduced to 1000 pieces. For complex parts, the higher quantity may be reduced to 1500 pieces.

A more specific ranking of primary process selection is given in the next section.

5.5 SELECTING PRIMARY MANUFACTURING PROCESSES – REFINED RULES IN RELATION TO THEIR CAPABILITY AND QUANTITY

Selecting the primary manufacturing forming process should be the first decision. Then, to check which process is most suitable for manufacturing the required part, refined rules are considered.

These decisions should be based on the parameters of quantity and shape complexity. Table 5.1 may assist the process planner in making this decision. Shape complexity is divided into four categories:

1. *Mono* This shape consists of a constant cross-section of the part along the main axis with no lateral features. (Fig. 5.13).
2. *Open* This shape consists of parts that can be divided into a maximum of two separate parts by a plane in such a way that no cross-section from the

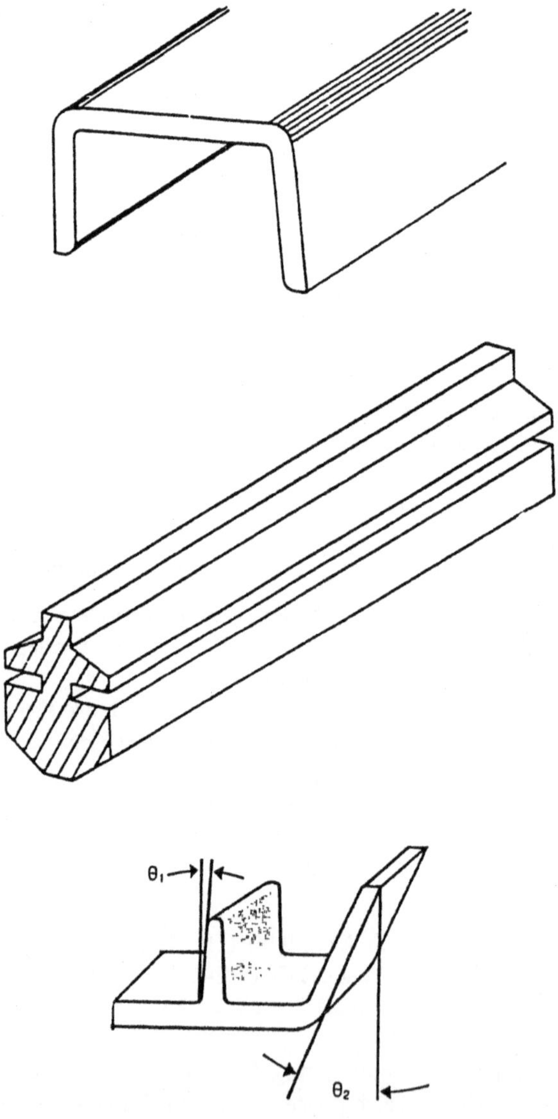

Fig. 5.13 Example of mono shape type.

parting plane up or down will be larger than the parting plane in any direction. Many steps are allowed, but their peak must be in a decreasing order. No features on the side walls of the part are allowed. (Fig. 5.14.)

3. *Complex* This shape consists of open parts but lateral features are allowed. (Fig. 5.15.)
4. *Very complex* This shape consists of parts with any shape and having hidden hollow spaces.

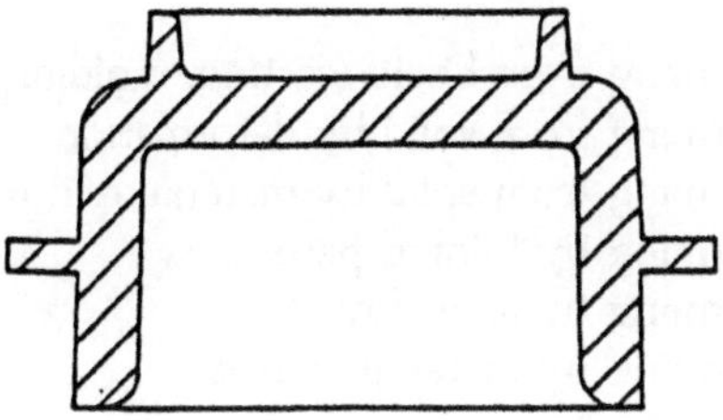

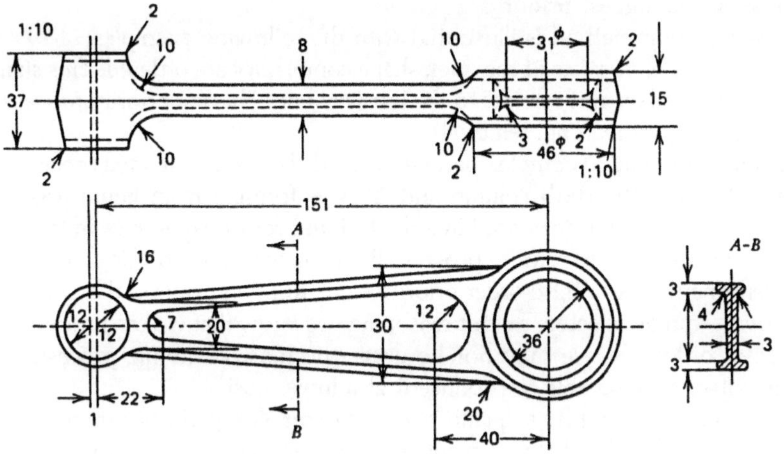

Fig. 5.14 Example of open shape type.

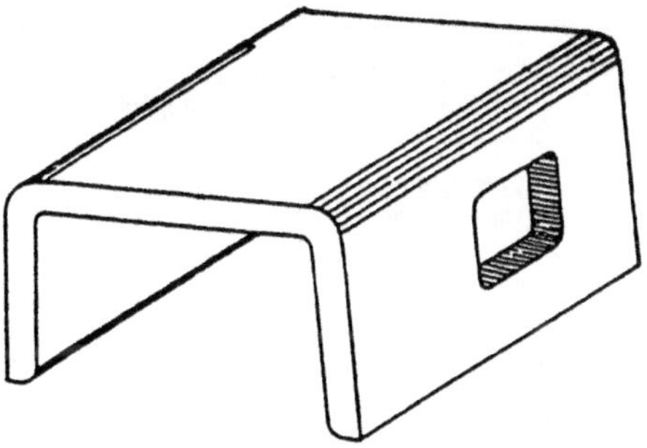

Fig. 5.15 Example of complex shape type.

As each preceding shape is included in the following one, the selection should be the least complex shape, i.e. the selection of shape complexity should start with the mono and proceed toward the very complex shape.

The quantity of parts is divided into low and high numbers. The quantity values are a function of the shape complexity. The values for the low and high quantities are given in Table 5.1.

In the appropriate row of the table, the code of the basic forming technique is given as follows:

Code A = forming from liquid (casting, molding)
Code B = forming from solid by deformation
Code C = forming from solid by material removal
Code D = forming by joining parts
Code E = forming by assembly
Code F = forming by material increase

The entries in the column are arranged in sequence of economic priorities, the top being the highest priority.

The selection method is illustrated with the following examples.

A quantity of 80 parts of the open shape complexity are ordered. This shape complexity and quantity points to the third column. The first priority process is 'C' – forming by material removal.

A quantity of 3000 complex parts is ordered. The recommended process for this order is in the sixth column and is 'A' – forming from liquid (casting, molding). If the plant does not have the facilities for or experience in forming from liquid, the second priority process, 'B' can be selected – forming from solid by deformation. As a third order priority, 'C' – forming from solid by material removal – can be selected. By this reasoning, we know that we can produce the required order, but we may not be using the most economic process, and competitors may be able to produce it at a lower cost.

In a further example, a quantity of 800 complex parts is ordered. This quantity is higher than the low quantity and lower than the high quantity in the

Table 5.1 Selecting basic forming techniques as a function of shape complexity and quantity

Shape complexity							
Mono		*Open*		*Complex*		*Very complex*	
Quantity		*Quantity*		*Quantity*		*Quantity*	
< 180	*> 1000*	*< 150*	*> 2000*	*< 50*	*> 1500*	*< 100*	*> 1000*
D	B	C	B	C	A	E	B
E	E	D	A	D	B	D	D
B	D	B	D	B	C	C	E
C	C	E	C	E	D	A	C
A	A	F	E	F	E	B	A
—	—	A	—	A	—	F	F

table. The first alternatives are 'C' forming by material removal, or 'A' forming from liquid (casting, molding). If our plant does not have the facilities for forming from liquid, then forming by material removal has to be used. If both alternatives are available, economic computations should be done in order to select the best process.

5.5.1 Selecting among forming from liquid processes

There are several methods of forming from liquid. The commonly used processes are:

- sand casting
- permanent casting
- die casting
- investment casting.

Each molding method has certain inherent advantages and limitations. Selecting the molding method should be based on the following parameters:

- part material
- size of part
- section thickness
- dimensional accuracy
- cost of raw material, possibility of defects and scrap rate.

Table 5.2 Selecting the molding method

	Sand	*Permanent*	*Die*	*Investment*
Material	All	Almost all	Light materials	[1]
Size or weight	Any size	Limited up to 25 kg	Limited[2] up to 10 kg	Limited[3] up to 40 kg
Section thickness	3–5 mm minimum	2.5 mm minimum	0.6 mm minimum	0.75 mm minimum
Accuracy[4]	± 1.5 mm	± 0.4 mm	± 0.05 mm	± 0.12 mm
Surface R_a finish μm	6.25–25	2.5–6.25	1–2.5	0.25–2
Tool costs	Low	Medium	High	High

Notes:

1 Materials that may be produced by investment casting belong to the following families: aluminium alloys, copper alloys, zinc alloys, magnesium alloys, tin alloys, lead alloys.

2 Usually under 8kg.

3 Best results are for parts around 1kg.

4 The accuracy tolerance depends on the scatter of the skrinkage factor and on the dimension of the mold. It can be computed by the equation:

$$\text{Tolerance} > \text{dimension}\left(\frac{S_{max} - S_{min}}{2}\right)$$

where S is the shrinkage factor, a function of the material.

Table 5.2 may assist the process planner in making a decision. The dimension refers to the mold, not the part. The transformation of the part dimension to mold dimension is computed by the following equation:

$$\text{Mold}_{\text{dimension}} = \frac{\text{Part}_{\text{dimension}}}{1 - S_{\text{mean}}}$$

where

$$S_{\text{mean}} = \tfrac{1}{2}(S_{\text{max}} + S_{\text{min}})$$

Figure 5.16 shows the relationship between the dimension and tolerance of the mold and the part. The value of $S_{\text{max}} - S_{\text{min}}$ is not a controllable value and will change randomly.

Selecting a molding process should start by attempting to match part specifications to the least cost process capability. Textbooks or manufacturing handbooks can supply the detailed data needed for evaluation and specifying process parameters such as parting lines, gate location and size, runners, venting, mold temperature and pressure, etc. These topics relate to process parameters in the way that we relate to process planning.

Example

A quantity of 2500 complex parts made of bronze, weight 4 kg, (8.8 lb) minimum section thickness 1.2 mm (approx. 0.040 in), minimum dimension tolerance ± 0.10 mm, and minimum surface finish of 2 μm R_a is ordered. Recommend a process to manufacture this part and prepare a quotation.

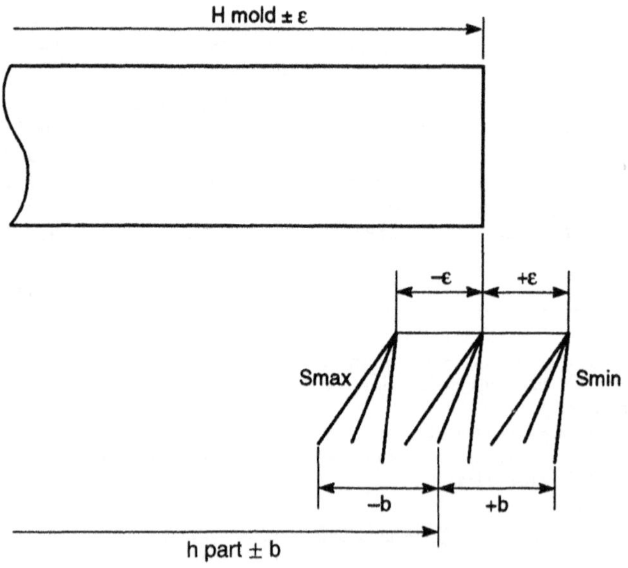

Fig. 5.16 The relationship between the dimension and tolerance of the mold and the part.

The first step is to use Table 5.1 and decide on the recommended process. Refer to the sixth column of the table for complex shapes and quantities higher than 1500 pieces; it recommends forming type 'A' – forming from liquid (casting, molding). If the plant is familiar with this type of forming, the recommendation is accepted.

Next, refer to Table 5.2 to determine the specific forming by liquid process. A sand casting process is not suitable because it is not accurate enough. Permanent casting can provide the accuracy, but not the surface finish and section thickness. Die casting, on the other hand, can accommodate all part specifications, so this is the process to choose. Available die casting machines have a capacity of 10 kg (22 lb) and can accommodate two parts in a die.

As a rough estimate, the die will cost about $25 000 and can operate at a rate of 100 shots per hour. The hourly rate at the plant, including overheads, is $60 per hour, and the raw material costs $0.8 per kg.

The estimated cost to manufacture the order is therefore:

Cost of the die. $25 000
Labor cost 60 × 2500/(100 × 2) $750
Raw material cost 2500 × 4 × 0.8 × 1.2 . . . .$9600 (20% waste)
Total. $35 350

or $14.14 per part.

Forming this part by a metal removal process will take 25 minutes. With the same hourly rate and raw material cost, the order will cost:

Labor cost 60 × 2500 × (25/60) $62 500
Raw material cost $9600
Total. $72 100

or $28.84 per part.

Subsequent processes

The parameters relating to material, size of part and section thickness are mandatory. If the candidate process does not match any of them it should be disregarded and the next alternative considered. However, if the candidate process does not meet the parameters of dimensional accuracy, i.e. surface finish or tolerance, a cost analysis by using a subsequent process should be done.

Usually, the surface finish or tolerances are selected because of the requirements for specific segments of the part. Segments that do not match any other part or are not external ones will not require a good surface finish or tolerance.

A check should be made to determine how many segments or dimensions are outside the capability of the selected process. Usually, these segments may be processed to a wider dimensional accuracy, and then the part is finished to the desired specifications by adding a metal removing process. This increases manufacturing costs, but overall manufacturing costs are likely to be lower than if alternative processes are used. If in doubt, check the viability of using an alternative process or a subsequent manufacturing one.

The well-known strategy called group technology presents the idea of creating families of parts, thereby increasing total production capability. This principle may be used to build a mold or die which is not specifically suited to any of the required parts. However, the larger quantity justifies the cost of the die and additional machining will produce parts to the required shape.

Example

A quantity of 2500 complex parts made of bronze is ordered, weight 4 kg, minimum section thickness 1.2 mm, minimum dimension tolerance ± 0.10 mm, with minimum surface finish of 0.8 μm R_a on one dimension, and 1.5 μm R_a on the others. Recommend a manufacturing process for this part and prepare a quotation.

The example is similar to the one in section 5.5.1 except that a close surface finish is needed on one dimension.

Table 5.2 indicates that this part cannot be formed by die casting, or any of the forming by liquid processes. By its shape, it cannot be formed from solid (second choice 'B'), which leaves metal removal. The cost of metal removal, as seen in the previous example, is $28.84 per part and $72 100 per order, compared to $14.14 per part by die casting. Therefore the process will be die casting with subsequent processing by metal removal for the dimension that requires a fine surface finish of 0.8 μm R_a.

It is estimated that this metal cutting operation will take three minutes. The extra costs are:

Total die casting cost	$35 350
Labor cost 60 × 2500 × (3/60)	$7500
Total	$42 850

or $17.14 per part instead of $28.84 where all the parts are formed by a metal removal process.

5.5.2 Selecting forming from solid by deformation process

There are several methods of forming from solid (hot or cold), and each one has its advantages and limitations.

Each method is usually restricted to a limited number of shapes. For example, extrusion can produce a profile according to its die cross-section and can be a very complex, closed shape. Along the length of the part, it will maintain a uniform shape. Forming from sheet metal may create a variety of shapes. However the wall thickness of the part will maintain the thickness of the sheet metal used for deformation.

There are many other forming methods. The most common ones are:

- G11 – rolling

- G12 – extrusion
- G13 – swaging
- G14 – forging
- G21 – spinning
- G22 – deep drawing
- G23 – bending (brake press)
- G24 – press work (power press, with progressive die).

Each of these methods has many variations, although these are not detailed here.

Tables 5.3, 5.4 and 5.5 may assist the process planner in making process selection decisions.

The first rows in each table indicate the process capability for dimensional accuracy and surface finish. They also list data for cold and hot work.

The following rows are organized by shape complexity: mono, open, complex and very complex, as defined at the beginning of this chapter. The numbers in the table represent the priority of a process in terms of economy, where (1) is the first priority. An 'X' instead of a number indicates that this process is unable to perform the job.

Although quantity was the main parameter in selecting a forming mode (Table 5.1), there might exist several alternative processes in the same group, such as rolling, extrusion and swaging, all of which may be used to produce a rod. Rolling and extrusion can both produce a 'T' or 'U' or similar profiles. Rolling is a faster operation than extrusion, but its tooling is more expensive. The selection of a process is thus dependent on the required quantity. The basic quantity was defined in Table 5.1 as the lowest limit. A refinement of the quantity influences is given as a factor of relative quantity.

Relative low quantity – the quantity defined in Table 5.1
Relative medium quantity – 50% higher then the low quantity
Relative high quantity – twice the low quantity

To select a recommended process, determine the relative quantity and move to other tables:

Table 5.3 – for relative high quantity
Table 5.4 – for relative medium quantity
Table 5.5 – for relative low quantity

To select a recommended process, follow these steps:

Step 1: Determine the relative quantity coefficient. Go to the appropriate decision table (Table 5.3, 5.4 or 5.5).
Step 2: Determine shape complexity type: mono, open, complex or very complex. Enter the row of the part shape complexity.
Step 3: Under the shape complexity, there are several alternatives of shape geometry. Select the appropriate row.

Table 5.3 Selecting forming from solid processes for relative high quantity

	G11	*G12*	*G13*	*G14*	*G21*	*G22*	*G23*	*G24*
Cold work								
Accuracy[3] (mm)	0.20	0.25	0.10	X	1.1[4]	0.20	0.35	0.2
Surface finish (μm) R_a	1.5	1.0	2.5	X	1.5[5]	1.0	0.8[5]	1.0
Hot work								
Accuracy[3] (mm)	0.30	0.38	0.15	0.20	2.0	X	X	X
Surface finish (μm) R_a	12.5	1.5	12.5	3.2	4.0	X	X	X
Mono								
Long parts with:								
uniform cross								
thick section	1	2	3[1]	X	X	X	X	X
thin section	1	2	4[1]	X	X	X	3	X
variable cross								
thick section	1	2	3[1]	X	X	X	4	X
thin section	2	1	4[1]	X	X	X	3	X
Short parts with:								
uniform cross								
thick section	1*	2*	X	X	X	X	3	4
thin section	2*	3*	X	X	X	X	1	4
variable cross								
thick section	3*	2*	X	X	X	X	1	4
thin section	4*	3*	X	X	X	X	1	2
Open								
Thick section	X	X	X	1	3[2]	2	5*	4
Thin section	X	X	X	2	3[2]	1	5*	4
Complex								
Thick section	X	X	X	1	X	2*	4*	3
Thin section	X	X	X	2	X	1*	4*	3
Very complex								
Thin section	X	X	X	X	X	X	2*	1

Notes:
* – a subsequent process is required.
[1] – only for symmetrical cross-sections (bars, rods, tubes).
[2] – usually open only to one side of the parting line.
[3] – this is an average value. Actual values depend on size.
[4] – special applications can go down to 0.3 mm.
[5] – depends on the raw material surface finish.

Table 5.4 Selecting forming from solid processes for relative medium quantity

	G11	*G12*	*G13*	*G14*	*G21*	*G22*	*G23*	*G24*
Cold work								
Accuracy[3] (mm)	0.20	0.25	0.10	X	1.1[4]	0.20	0.35	0.2
Surface finish (μm) R_a	1.5	1.0	2.5	X	1.5[5]	1.0	0.8[5]	1.0
Hot work								
Accuracy[3] (mm)	0.30	0.38	0.15	0.20	2.0	X	X	X
Surface finish (μm) R_a	12.5	1.5	12.5	3.2	4.0	X	X	X
Mono								
Long parts with:								
uniform cross								
thick section	2	1	3[1]	X	X	X	X	X
thin section	2	1	4[1]	X	X	X	3	X
variable cross								
thick section	2	1	3[1]	X	X	X	4	X
thin section	2	1	4[1]	X	X	X	3	X
Short parts with:								
uniform cross								
thick section	2*	1*	X	X	X	X	3	4
thin section	3*	2*	X	X	X	X	1	4
variable cross								
thick section	3*	2*	X	X	X	X	1	4
thin section	4*	3*	X	X	X	X	1	2
Open								
Thick section	X	X	X	2	3[2]	1	5*	4
Thin section	X	X	X	3	2[2]	1	5*	4
Complex								
Thick section	X	X	X	2	X	1*	4*	3
Thin section	X	X	X	2	X	1*	4*	3
Very complex								
Thin section	X	X	X	X	X	X	2*	1

Notes:
* – a subsequent process is required.
[1] – only for symmetrical cross-sections (bars, rod, tubes).
[2] – usually open only to one side of the parting line.
[3] – this is an average value. Actual value depends on size.
[4] – special applications can go down to 0.3 mm.
[5] – depends on the raw material surface finish.

Table 5.5 Selecting forming from solid processes for relative low quality

	G11	*G12*	*G13*	*G14*	*G21*	*G22*	*G23*	*G24*
Cold work								
Accuracy[3] (mm)	0.20	0.25	0.10	X	1.1[4]	0.20	0.35	0.2
Surface finish (μm) R_a	1.5	1.0	2.5	X	1.5[5]	1.0	0.8[5]	1.0
Hot work								
Accuracy[3] (mm)	0.30	0.38	0.15	0.20	2.0	X	X	X
Surface finish (μm) R_a	12.5	1.5	12.5	3.2	4.0	X	X	X
Mono								
Long parts with:								
uniform cross								
thick section	3	2	1[1]	X	X	X	X	X
thin section	4	3	1[1]	X	X	X	2	X
variable cross								
thick section	3	2	1[1]	X	X	X	4	X
thin section	4	3	2[1]	X	X	X	1	X
Short parts with:								
uniform cross								
thick section	3*	2*	X	X	X	X	1	4
thin section	3*	2*	X	X	X	X	1	4
variable cross								
thick section	3*	2*	X	X	X	X	1	4
thin section	4*	3*	X	X	X	X	1	2
Open								
Thick section	X	X	X	3	2[2]	1	5*	4
Thin section	X	X	X	4	3[2]	1	2*	5
Complex								
Thick section	X	X	X	2	X	1*	4*	3
Thin section	X	X	X	2	X	1*	4*	3
Very complex								
Thin section	X	X	X	X	X	X	2*	1

Notes:
* – a subsequent process is required.
[1] – only for symmetrical cross-sections (bars, rod, tubes).
[2] – usually open only to one side of the parting line.
[3] – this is an average value. Actual value depends on size.
[4] – special applications can go down to 0.3 mm.
[5] – depends on the raw material surface finish.

Step 4: Move in the selected row across the columns up to the column with the lowest priority value. Start with priority value of one (1).
Step 5: Check if the candidate process meets dimensional accuracy and surface finish specifications.

If a candidate process meets the specifications, it is the selected process. If not, return to step 4.

Example 1

A quantity of 1600 'U' shape profile bars with a wall thickness of 3.5 mm and 10 m long, dimension accuracy 0.5 mm and surface finish 13 μm R_a is ordered. Recommend a process plan for this order.

The first step is to use Table 5.1. A 'U' shape is a mono shape complexity, and for a quantity greater than 1000 pieces forming type 'B' – forming from solid – is recommended.

The second step is to follow the five steps of selecting a specific forming process from solid as described above.

Step 1: Relative quantity is (1600/1000=)1.6, therefore it is a relative medium quantity. Go to Table 5.4.
Step 2: Shape complexity is mono.
Step 3: This part is a long part with uniform thin section.
Step 4: Priority 1 is in column G12 – extrusion.
Step 5: Dimension accuracy and surface finish are below the specification. Therefore this is the recommended process.

Example 2

For this example, we use the same specifications as in the previous example but the order is 4000 parts and the surface finish is 6 μm R_a

Step 1: Relative quantity is (4000/1000=)4.0, therefore it is a relative high quantity. Go to Table 5.3.
Step 2: Shape complexity is mono.
Step 3: This part is a long part with uniform thin section.
Step 4: Priority 1 is in column G11 – rolling.
Step 5: Dimension accuracy and surface finish are above the specification, so this process is not recommended. Next priority is 2.
Step 4: Priority 2 is in column G12 – extrusion.
Step 5: Dimension accuracy and surface finish are below the specification, so this is the recommended process.

Subsequent processes

Rolling and extrusion processes produce a long bar (about five meters) of constant cross-section. If a shorter part is required, a cutting operation must be added. For high quantity, the cutting operation might be by a parting die in a press, or by a sawing operation. For low quantity, a sawing operation is more appropriate.

Forging might need a subsequent operation in case the dimensional accuracy, i.e. surface finish or tolerance, cannot be met in a few part segments. A check should be made to determine how many segments or dimensions are outside the capability of forging. Such segments may be forged to a wider dimensional accuracy, then, by adding a metal removal process operation the part is brought to the desired specifications. It is going to increase manufacturing cost, but it is highly probable that it will incur a lower cost than any alternative process suitable for manufacturing the part. If in doubt, check the viability of an alternative when selecting subsequent processes.

Drawing, bending and spinning produce a part with no side features. However, side features can be added, before or after, with a subsequent operation. This subsequent operation might involve die cutting operations such as perforating, slotting, notching, trimming or metal removal.

5.5.3 Forming from solid by material removal

This type of forming is the most common of processes, about 80% in machine and production volume. Moreover, metal removal has an inherent flexibility which enables its use for a wide range of applications, with a much higher number of possible solutions.

The following chapters deal with this forming group in depth.

5.5.4 Forming by joining parts

Parts to be joined can be produced by one of the forming techniques previously described. The recommended joining process is by adhesives. However, not all materials can be joined by this method.

5.5.5 Forming by assembly

The product to be manufactured is divided into several parts. Each part will be produced by one of the forming techniques previously described. Next, the parts are assembled to create the product. The assembly may be manual, mechanized or by industrial assembly robot. If robots are to be used, this should be taken into consideration at the design stage, as design for assembly with robots calls for special attention.

5.5.6 Forming by material increase

This is a new forming technique. It is very slow, and thus is mostly suitable for preparing prototypes.

FURTHER READING

Dallas, D.B. (ed.), (1976) *Tool and Manufacturing Engineering Handbook*, McGraw-Hill, SME.

Glusti, F., Santochi, M. and Dini, G. (1991) Robotized Assembly of Modular Fixture, *Annals of the CIRP*, Vol **40**(1), 17–20.

Goetsch, D.L. (1991) *Modern Manufacturing Processes*, Delmar.

Kochan, D. (1992) Solid Freeform Manufacturing – Possibility and Restrictions. *Computers in Industry* **20**, 133–40.

Nnaji, B.O. (1992) *Theory of Automatic Robot Assembly and Programming*. Chapman & Hall.

CHAPTER 6

Selecting detailed methods of production

6.1 FORMING BY MATERIAL REMOVAL

Forming material removal is a most comprehensive process. There is almost an infinite number of combinations of machines and tools that will produce the part as specified by the drawing. However, cost and machining time will vary substantially according to the selected process. Therefore, it requires a skillful handling of the operating conditions in order to arrive at the economic optimum. In this respect, the sensitivity of the machining conditions in relation to the time and cost of machining, can be seen in the following examples.

Example 1

The difference in machining times resulting from the recommendations of 37 process planners when asked to produce a hole – 30 mm (1.181 in) in diameter and 30 mm (1.181 in) long, with a tolerance on the diameter ± 0.15 and 7.5 R_a (μm) surface roughness (R_a = arithmetic average height; see section 6.4) – is shown in Table 6.1. The machining time range is 10:1, although all the recommendations are technically feasible.

Table 6.1 Comparison of 37 expert recommendations

Operation	*Number of experts*	*Time of machining (minutes)*
Drill 30	9	0.13–0.58
Drill 28 + bore 30	9	0.22–0.65
Drill 20 + drill 30	7	0.49–0.84
Drill 15 + drill 30	1	0.81
Drill 10 + drill 30	2	0.78
Drill 5 + drill 30	1	0.81
Drill 8 + drill 28 + bore 30	1	0.86
Drill 8 + drill 18 + bore 30	1	0.77
Drill 10 + drill 20 + drill 30	2	1.04
Drill 10 + drill 28.7 + ream 30	1	1.07
Drill 10 + drill 20 + drill 28 + bore 30	2	1.13
Drill 5 + drill 13 + drill 22 + drill 30	1	1.29

Example 2

This is described in Table 6.2. It shows the number of cuts recommended by several process planners to remove a 2.5 mm (0.100 in) depth of cut using a face milling operation with different values of tolerance and surface roughness.

In this case, the machining time differences when one or two passes are used can be as high as 25%.

Many more similar examples can be presented, stressing the point that a part can be produced by many alternative processes. The recommended process usually reveals the past experience of the process planner in question. It is of interest to note that as the part complexity increases, the number of proposed process alternatives is reduced.

The recommended process is not only a result of the process planner's experience, but also an outcome of the sequence of decisions made. Once a decision is made, it imposes constraints on the decisions following it. In many cases, such constraints are artificial ones, existing only because of the sequence of decisions chosen. Such decisions reduce both flexibility and economy of processing and should be avoided.

A method to avoid artificial constraints will be given in the chapters following this one.

Table 6.2 Comparison of process planners' recommendations

Surface roughness R_a *(μm)*	*Tolerance (mm)*	*Recommended number of cuts*	*Number of planners*
8.75	0.35	1	57
		2	4
3.2	0.35	1	33
		2	20
		3	2
0.8	0.35	1	5
		2	42
		3	7
3.2	0.15	1	24
		2	32
3.2	0.08	1	2
		2	63
		3	12
0.8	0.08	1	0
		2	36
		3	18

6.2 DECISIONS AND CONSTRAINTS

Process planning is a decision-making process. The objective is to devise an economic process plan.

The parameters to consider are:

- part geometry
- part raw material
- part dimensional accuracy
- part surface finish
- part geometric tolerances
- part heat treatment
- quantity required.

The constraints are:

- the part specification and strength
- the available machines
- the available tools
- the available fixtures, chucks, clamping devices, etc.
- the available technology.

The criteria of optimization are economic ones and might be:

- maximum production of parts
- minimum cost of parts
- maximum profit rate during a given period of time.

The decisions to make are:

- select type of metal removal process
- select machine for the job
- select chucking type and location
- select detail operations
- select tooling for each operation
- select path for each operation
- select cutting conditions for each operation.

A wrong sequence of decisions may result in artificial constraints, because if the sequence of decisions were different, the constraints might not have existed. For example, if the first decision is to select a machine, then its power, spindle torque moment, force, stability, available speed range and feed rates act as constraints in selecting the cutting parameters. If another machine is selected, another set of constraints would arise. A first decision of selecting chucking location and type imposes constraints on the allowed cutting forces, and thus on depth of cut, feed rate and machine size. Similarly, a selected tool imposes constraints on the maximum cutting speed, depth of cut, feed rate and tool life.

The real constraints should be technological constraints and should be independent of the sequence of decisions. For example:

- A boring operation cannot be the first operation in making a hole.
- Twist drills have constraints on dimensional tolerance and surface finish of the produced hole.
- There is a relationship between the exerted force on a part and its deflection.
- The allowed deflection of the part during metal cutting is a function of the dimension tolerance.
- The allowed cutting forces in a metal cutting operation is a function of the allowed part deflection.
- Cutting forces are a function of cutting conditions.
- There is a minimum depth of cut, below which there will not be a chip removal process.
- There is a spring back in elastic bodies.
- There is a relationship between feed rate and surface finish.
- Tool wear is a function of cutting speed and cutting feed.

To demonstrate the complexity and the importance of making decisions in a sequence that will not reduce economical productivity in machining, let us analyze the simple equation for computing the direct machining time.

$$T = \frac{L}{n \times f} \tag{6.1}$$

where: T = machining time
L = length of cut (mm)
n = revolutions per minute, rotational speed
f = feed rate (mm per revolution).

At first sight, it looks easy to handle. However, if the factors that affect each one of these variables are carefully examined, it is found that:

L (length of cut) depends upon:

- part geometry
- part specifications
- depth of cut
- chucking locations.

n (cutting speed) depends upon:

- chosen tool
- machine power
- available speeds of the machine
- cutting forces
- part material
- feed rate
- depth of cut.

f (feed rate) depends upon:

- surface finish
- tool constraints
- cutting forces, which depend upon:
  - (a) torsion stress in the part
  - (b) gripping forces
  - (c) part deflection, which depends upon:
    - – tolerance
    - – chuck type
    - – gripping location
    - – depth of cut
  - (d) machine spindle moment (maximum torque)
  - (e) depth of cut, which depends upon:
    - – chatter
    - – part length
    - – gripping location
    - – chucking type
    - – part specifications.

These factors are thus all interrelated. The depth of cut for example affects each of the parameters – each differently. The conventional mathematical optimization techniques are difficult to employ in this case because:

- there are many variables;
- there are few auxiliary equations;
- there are many limiting factors;
- the relationships are neither linear nor continuous functions; and
- not all relationships are reliable.

Another way to handle this type of problem is to use an algorithm based on metal-cutting technology. The main problem in constructing the algorithm is to decide which parameter to begin with; once a parameter is set, it limits the others. This is an artificial constraint: if a different value is assigned to the parameter, it will result in different constraints. This type of limitation prevents the possibility of reaching optimum operating conditions easily.

For example, a true absolute constraint is the part's mechanical strength. It is clear that if the forces acting in the metal removal process break the part, the part cannot be produced. If the forces deflect the part to such a degree that it becomes impossible to keep the required tolerances, it is a real constraint. However, if the forces exceed the part gripping forces, this might be an artificial constraint, meaning that the previous decision on gripping selection was not good and must be altered.

Any decision – except those concerning part geometry, specifications, and material – can be changed in order to overcome a constraint that arises due to a previous decision. This may result in an endless loop of computations, decisions and changes of decisions, since one decision affects many parameters, each in a different manner. A careful study should be made in order to establish the best sequence of decisions; that is, the one that results in a minimum number of changes and a short decision loop. The following chapters propose a sequence of decisions which guarantees such a result.

6.3 BASIC TYPES OF MATERIAL REMOVAL PROCESSES

The first decision should be to select the types of material removal processes from among the many basic processes.

To assist in making this decision, the basic material removal processes are classified according to their capability to machine a group of parts to a required shape. This classification is given in Table 6.3.

The planner has to review the part drawing and classify the shape according to the columns of Table 6.3, i.e. round symmtrical parts, prismatic or free form (sculptured) surface parts.

On top of these shapes, special features such as

- holes
- threads
- slots
- flats

can be superimposed.

A part might have features that belong to more than one group. The process selection will handle each one separately, as each one requires a different machine.

The final selection of the basic processes depends on the accuracy of the part. Concerning selection of machining technology the most important parameter is the surface roughness required, followed by the geometrical and dimensional

Table 6.3 Classification of basic processes by shape of part groups

Main shape groups		*Superimposed shapes*
Round symmetrical	*Prismatic*	*Holes and threads*
Turning	Milling	Drilling
Grinding	Grinding	Reaming
Honing	Honing	Boring
Lapping	Lapping	Peripheral milling
Polishing	Polishing	Grinding
		Burnishing
		Broaching
		Tapping
		Milling

tolerances. The capabilities and limits for surface roughness for each process are given in Table 6.4.

The ability of a process to produce a specific surface roughness depends on many factors. For example, in turning, the final surface depends on parameters such as feed rate, cutting speed, tool condition, coolant and machine rigidity. In grinding, the final surface roughness depends on the peripheral speed of the wheel, the speed of the part, the feed rate, the grit size and the bonding material. Therefore, the data in Table 6.4 covers a range rather than citing precise values.

The sequence of the proposed basic processes in each section of a shape group is arranged by priority and technical constraints. In other words, if the basic process first proposed does not meet the surface roughness requirements, an additional basic process might be added, though it should not replace the first basic process.

The dimensional tolerance has an effect similar to the surface roughness. To consider the dimensional tolerance, there is an empirical relation between the dimensional tolerance and surface roughness (finish), so that a minimum dimensional tolerance requires a minimal surface roughness. The smaller of both surface roughness measurements (given and translated in Table 6.5) will be used to enter into Table 6.4 for selecting the basic process.

Table 6.4 Surfaces roughness range of basic processes

	Surface roughness R_a (μm)		
Process	*(min)*	*(max)*	*Machine type*
Round symmetrical shapes			
Turning	0.8	25.0	Lathe
Grinding	0.1	1.6	Grinding
Honing	0.1	0.8	Honing
Polishing	0.1	0.5	Polishing
Lapping	0.05	0.5	Lapping
Prismatic shapes			
Milling	0.8	25.0	Milling
Grinding	0.1	1.6	Grinding
Honing	0.1	0.8	Honing
Polishing	0.1	0.5	Polishing
Lapping	0.05	0.5	Lapping
Holes, Threads, Misc.			
Drilling	1.6	25.0	Lathe, milling Drill press
Reaming	0.8	6.3	Lathe, milling Drill press
Boring	0.8	10.0	Lathe
Peripheral milling	0.8	15.0	Milling
Grinding	0.1	1.6	Grinding
Burnishing	0.2	0.4	Burnishing
Broaching	0.8	6.3	Broaching
Milling	0.8	25.0	Milling

Table 6.5 should assist the planner in the translation of tolerances into surface roughness and surface roughness into dimensional tolerance.

The geometric tolerances should be considered as the final criteria to be checked with regard to the capability of the process. Meeting the specified geometric tolerances involves, in turn, many other criteria, such as chucking, machine selection, etc. Decisions on this are discussed in Chapter 7. However, there are limits to process capability, which are listed in Table 6.6.

The use of these tables is as follows. Select the basic processes as outlined earlier by using Tables 6.4 and 6.5. Check if the last basic selected process meets the required geometric tolerances. If it does, no additional process is needed. If it does not, another basic process from the list in Table 6.6 should be added, while

Table 6.5 Conversion of dimension tolerance to surface roughness

Tolerance $\pm$ *(mm)*	*Surface roughness* R_a *(μm)*
<0.005	>0.20
0.010	0.32
0.015	0.45
0.020	0.80
0.030	1.0
0.040	1.32
0.050	1.60
0.060	1.80
0.080	2.12
0.100	2.50
0.150	3.75
0.200	5.00
0.250	6.25
0.350	9.12
0.600	12.50
1.000	25.00

Table 6.6 Geometric tolerances capability of basic processes

	Geometric tolerance type (mm)			
Basic process	*Parallelism*	*Perpendicularity*	*Concentricity*	*Angularity*
Turning	0.01–0.02	0.02	0.005–0.01	0.01
Milling	0.01–0.02	0.02	—	0.01
Drilling	0.2	0.1	0.1	0.1
Boring	0.005	0.01	0.01	0.01
Grinding	0.001	0.001	0.002	0.002
Honing	0.0005	0.001	0.002	0.002
Superfinish	0.0005	0.001	0.005	0.002

retaining the previous basic process. Note: in certain cases you can substitute the last chosen basic process with one that meets the geometric tolerances specifications. In this instance, the lower limit (higher roughness) of the initial chosen process is higher than that of the one recommended for substitution. However, their upper limits (lower values) are the same.

When more than one basic process is required for machining a part, production drawings should be prepared for each one. These working documents are needed because of technological constraints and economic considerations. Dimensions have to be changed so as to leave material to be removed by the following process, as each process must have a minimum depth of cut. (If there is no material to remove, the process will compress the material instead.) Dimensional tolerances and surface roughness can be increased as they are performing a rough cut operation which does not directly affect the final part dimensions.

Accurate computations to establish the economic division of dimensions and shapes between the final part drawing and the production drawing can be made. However, for manual and fast decisions the following rules of thumb may be employed:

- In the production drawing, increase the surface roughness of the dimensions which do not meet process capability to the maximum capability value of the surface roughness.
- Increase, in the case of external dimension, or decrease, in the case of internal dimension, the basic dimension by a value of 10 times the tolerance equivalent to the new surface roughness. (Use Table 6.5 to determine its value.)
- Increase the dimensional tolerance and multiply the equivalent tolerance by a factor of 2. If the result is bigger than the original tolerance, use it as the tolerance for the modified dimension. If it is smaller, then retain the original tolerance.

Example 1

One of the external sizes of a round symmetrical part has to be:

$$60 \pm 0.15;\ 0.5\ R_a\ \mu m$$

What is the recommended basic process used to manufacture this dimension?

1. Refer to the round symmetrical section in Table 6.4.
2. Note that the first basic process is turning.
3. The equivalent surface roughness of the tolerance (from Table 6.5) is 3.75. As it is bigger than 0.5 R_a, the controlling surface roughness will be 0.5 R_a.
4. Turning minimum surface roughness is 0.8 R_a. Therefore an additional process is needed.
5. The next basic process is grinding.

6. Grinding minimum surface roughness is 0.1 and maximum 1.6, and it is done on a different machine.
7. Grinding can be used to produce the required part.
8. A preceding turning operation is also required.
9. The surface roughness for turning will be 1.6 R_a (maximum grinding).
10. From Table 6.5 the equivalent tolerance for 1.6 R_a is 0.05 mm.

$$0.05 \times 10 = 0.5 \text{ mm}$$

11. The new basic dimension for turning should be (60 + 0.5 =)60.5 mm.
12. The equivalent tolerance doubled is (0.05 × 2 =)0.10 mm.
13. The original tolerance was 0.15 which is greater than 0.10, therefore it remains.

The decisions reached with regard to the basic process are therefore:

(a) turn the part to dimension of 60.5 ± 0.15; 1.6 R_a

(b) grind the part to 60.0 ± 0.15; 0.5 R_a.

Example 2

One of the dimensions of a flat prismatic part has to be

$$85 \pm 0.01;\ 0.5\ R_a\ \mu m$$

and has to be parallel to another flat surface within 0.0008 mm.

What recommended basic process is used to manufacture this surface?

1. Look up the prismatic shapes section in Table 6.4.
2. The first basic process is milling.
3. The equivalent surface roughness of the tolerance (from Table 6.5) is 0.32 R_a. As it is smaller than 0.5 R_a, the controlling surface roughness will be 0.32 R_a.
4. Milling minimum surface roughness is 0.8 R_a. Therefore, an additional process is needed.
5. The next basic process is grinding.
6. Grinding minimum surface roughness is 0.1 and maximum 1.6. It is done on a different machine.
7. Grinding can be used to produce the required part.
8. Check for geometric tolerance. Table 6.6 indicates that grinding can maintain a parallelism of 0.001 mm. This value is higher than the required 0.0008 mm. Therefore an additional process is needed.
9. The next basic process (Table 6.4) is honing.
10. Honing can maintain parallelism of 0.0005 mm (Table 6.6) and can therefore machine the part to the required specifications.
11. The recommended basic processes are: milling, grinding, honing.

Note: from Table 6.4 it would appear that the grinding operation might be replaced by honing, instead of honing being added. However, the minimum of milling is 0.8 R_a and the maximum of honing is 0.8 R_a. Since there is not enough overlap between these two operations, it is not recommended. The grinding operation fits neatly between milling and honing, with a good overlap.

We now need to prepare a production drawing for each basic process.

12. The surface roughness for grinding will be 0.8 R_a, the maximum of honing, (Table 6.4).
13. From Table 6.5 the equivalent tolerance for 0.8 R_a is 0.02 mm

$$0.02 \times 10 = 0.2 \text{ mm}$$

14. The new basic dimension for grinding should be (85 + 0.2 =)85.2 mm.
15. The equivalent tolerance doubled is (0.02 × 2 =)0.04 mm.
16. The surface roughness for milling will be 1.6 R_a, the maximum of grinding (Table 6.4).
17. From Table 6.5 the equivalent tolerance for 1.6 R_a is 0.05 mm.

$$0.05 \times 10 = 0.5 \text{ mm}$$

18. The new basic dimension for milling should be (85.2 + 0.5 =)85.7 mm (approx. 3.350 in).
19. The equivalent tolerance doubled is (0.05 × 2 =)0.10 mm.

The decisions with respect to the basic process are therefore:

(a) mill the part to dimension of 85.7 ± 0.10; 1.6 R_a
(b) grind the part to 85.2 ± 0.04; 0.8 R_a
(c) hone the part to 85.0 ± 0.01; 0.32 R_a.

6.4 MATERIAL REMOVAL AS A SUBSEQUENT PROCESS

As was discussed in the section on general process selection, forming from liquid or from solid are far more economically sound than metal removal processes. However, there are two main restrictions to their general use. The first is that the quantity has to be large in order to cover the high tooling cost and realize the economic benefits. The second concerns the accuracy of parts that can be produced by these processes.

To overcome the first restriction, group technology proposes to increase the quantity artificially by forming families of parts. A family of parts is defined, for this purpose, as a collection of similar related parts. They are related by geometric shape or to a number of shape elements which are contained within the basic (master) part family shape. In other words, you are recommended to examine the individual dies or castings and determine how to modify them to make a single casting or die that will serve the needs of several parts. Thus, the total quantity is the sum of the required quantity of all individual members of the family, although the produced parts have the family basic shape. Additional, subsequent operations have to be made to transform the family shape into the shape of the individual member of the family.

To overcome the second restriction, add a subsequent metal removal process which is capable of meeting the accuracy requirements of the parts. By this method, an economic compromise is obtained: an approximate shape of the required part is produced using an economic process. The weight and thus the cost of raw material is reduced, compared to that required for a pure material removal process. The amount of material to be removed is smaller, which saves machining time and cost.

Whenever a subsequent basic process is needed, a production drawing should be prepared for any intermediate process. The method of preparing such a drawing is similar to that described earlier, although there are some additional factors to consider.

Parts produced by forming from liquid or from solid usually have wider dimensional and geometrical tolerances and high values of surface roughness and surface integrity (surface hardening). Furthermore, there are certain DFM (design for manufacturing) rules that must be observed in order to obtain good conforming parts. These rules (corner radius, steps, relative wall thickness) are used to ensure a good flow of material in the die or casting.

Accurate computations to establish the economic division of dimensions and shapes between the final part drawing and the working part drawings can be made. These computations will involve the translation of part dimensions to die dimensions, shrinkage of part in the dies, machine accuracy, die location in the machine, surface hardening, geometric tolerances and process parameters variations such as temperature, pressure, etc. However, for fast decisions, the following rules of thumb may be employed:

- Observe all DFM rules relating to the selected process.
- Change only those dimensions that do not meet part accuracy, surface finish and/or dimension tolerance.
- Increase the surface finish to the maximum value of the surface finish of the process as given in Table 6.7.

Table 6.7 Accuracy of primary basic processes

Process	*Surface finish* R_a *(μm)*	*Accuracy (mm)*
Sand casting	6.25–25	±1.5
Permanent casting	2.5–6.25	±0.4
Die casting	1.0–2.5	±0.05
Investment casting	0.25–2.0	±0.12
Rolling – cold work	1.5	±0.2
Rolling – hot work	12.5	±0.3
Extrusion – cold work	1.0	±0.25
Extrusion – hot work	1.5	±0.38
Swaging – cold work	2.5	±0.1
Swaging – hot work	12.5	±0.15
Forging – hot work	3.2	±0.2

- Increase, in case of external dimension, or decrease, in case of internal dimension, the basic dimension by a value of 2.5 times the accuracy tolerance or to 0.8 mm, whichever is the larger.
- Increase the dimensional tolerance to the one given in Table 6.7.

6.5 AUXILIARY TABLES

In the previous section a reference to surface roughness and dimensional tolerances was made. Tolerances were defined by millimeters (mm) and surface roughness by R_a in micrometers (μm). However, process planners might come across other scales of definitions. To assist the planner, there follow some conversion tables.

Table 6.8 gives the values in millimeters of standard tolerances.

The most common method of specifying surface roughness is by R_a (arithmetic average roughness) which is also referred to as arithmetic average heights (AA) or center line average heights (CLA).

However, in the past, several other methods have been used, such as root mean square of average heights (RMS), and max peak-to-valley roughness height (R_t). Table 6.9 gives conversions of the different scales.

For practical purposes, it may be assumed that RMS = AA = CLA = R_a.

Table 6.8 Standard tolerances (ISO grades)

Nominal size up to (mm)	*Grade*									
	4	5	6	7	8	9	10	11	12	13
	Tolerance (mm)									
3	0.003	0.004	0.006	0.010	0.014	0.025	0.040	0.060	0.100	0.140
6	0.004	0.005	0.008	0.012	0.018	0.030	0.048	0.075	0.120	0.180
10	0.004	0.006	0.009	0.015	0.022	0.036	0.058	0.090	0.150	0.220
18	0.005	0.008	0.011	0.018	0.027	0.043	0.070	0.110	0.180	0.270
30	0.006	0.009	0.013	0.021	0.033	0.052	0.084	0.130	0.210	0.330
50	0.007	0.011	0.016	0.025	0.039	0.062	0.100	0.160	0.250	0.390
80	0.008	0.013	0.019	0.030	0.046	0.074	0.120	0.190	0.300	0.460
120	0.010	0.015	0.022	0.035	0.054	0.087	0.140	0.220	0.350	0.540
180	0.012	0.018	0.025	0.040	0.063	0.100	0.160	0.250	0.400	0.630
250	0.014	0.020	0.029	0.046	0.072	0.115	0.185	0.290	0.460	0.720
315	0.016	0.023	0.032	0.052	0.081	0.130	0.210	0.320	0.520	0.810
400	0.018	0.025	0.036	0.057	0.089	0.140	0.230	0.360	0.570	0.890

Table 6.9 Conversion of different surface roughness scales

$CLA = AA = R_a$ R_a (μm)	R_a (μin)	RMS (μm)	R_t μm	Surface quality	French surface quality	Triangles
0.4	15.8	0.44	1.58	N4	5	▽▽▽
0.6	23.7	0.66	2.37	N4	6	▽▽
0.8	31.5	0.88	3.16	N5	6	▽▽
1.0	39.4	1.11	4.00	N5	7	▽▽
1.2	47.2	1.33	4.75	N4	7	▽▽
1.4	55.1	1.55	5.53	N5	7	▽▽
1.6	63.0	1.77	6.32	N6	8	▽▽
1.8	70.8	2.00	7.11	N6	8	▽▽
2.0	78.7	2.22	7.95	N6	8	▽▽
2.3	90.5	2.55	9.10	N6	8	▽▽
2.5	98.4	2.77	9.90	N6	8	▽▽
2.8	110.2	3.11	11.10	N6	8	▽▽
3.0	118	3.33	11.85	N6	8	▽▽
3.5	138	3.88	13.85	N7	9	▽
4.0	157	4.44	15.85	N7	9	▽
4.5	177	5.00	17.85	N7	9	▽
5.0	197	5.55	19.85	N7	10	▽
6.0	236	7.66	23.75	N7	10	▽
7.0	276	7.77	27.75	N8	10	▽
8.0	314	8.88	31.60	N8	10	▽
9.0	354	10.00	35.60	N8	10	▽
10.0	394	11.10	39.75	N8	11	▽
11.0	433	12.21	43.55	N9	11	▽
12.0	472	13.32	47.40	N9	12	≈
13.0	511	14.43	51.50	N10	12	
14.0	551	15.54	55.50	N10	12	
15.0	590	17.65	59.50	N11	12	
16.0	630	17.76	63.50	N11	13	

CHAPTER 7

Elements of positioning and workholding

7.1 THE PROBLEM OF POSITIONING AND CLAMPING

In order to produce a mechanical part correctly, it must be on a suitable set-up (or jig) which guarantees a well-defined location (position and orientation) in space. The surfaces of the jig represent data references for the coordinate system in which the different features of the part are defined. As an example, Fig. 7.1 shows a prismatic part located on three orthogonal planes simulating datum surfaces existing in the physical jig. These planes are used to set the manufacturing dimensions of the features in the part. As mentioned already in Chapter 2, the exact definition of datum features is found in a recent standard

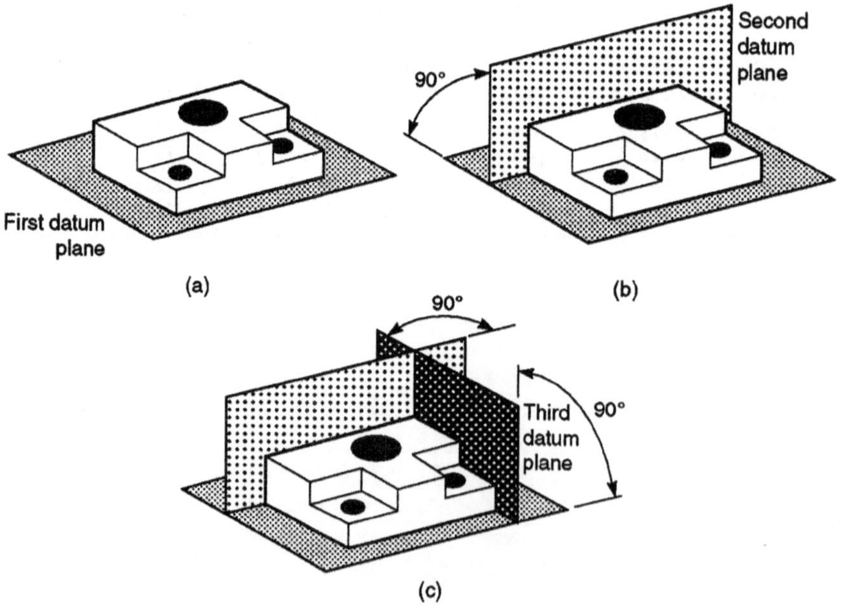

Fig. 7.1 Three orthogonal planes simulating datum rules.

(ISO Standard 5459, 1981) distinguishing between simulated and real datum surfaces.

In addition to defining the positioning of datum surfaces in a jig, it is necessary to design a way of holding a part firmly in position under the combined effects of external forces such as gravity, cutting forces, vibrations, centrifugal forces and so on. This is the role of clamping devices known as 'fixtures' or 'workholders' and which should not influence the positioning function defined earlier, but rather help to ensure that the part is stable in its set-up. In addition, clamping devices should not damage the part by excessive pressure at the contact points. Section 7.6 gives a more detailed analysis of the clamping problem.

The positioning problem is closely related to the concept of isostatism, already mentioned in Chapter 2. The principle of isostatism means essentially that positioning a part on an isostatic jig confers upon it a unique location in space which remains unchanged by removing and replacing the part on its support. In the case of a batch of parts, assuming that the parts are of ideal form, it means that each part will be in an identical position in space. Of course, parts in production will be affected by diverse errors and their theoretical position will never be repeated. The calculation of the errors in the dimensions of the part due to this slightly changed location is the subject of section 7.5.

Parts made by material removal machines are only as good as the workholding and positioning equipment permits. Therefore, workholdings fixtures should comply with the following:

- *Rigidity and stiffness* The acting forces on the part by the cutting tool may be of significant values. The part should not:
  - break under the applied loads;
  - deflect beyond the allowable amount (tolerance);
  - change position (slide) under the applied load;
  - be damaged by clamping forces on its machined surfaces – the forces should have limiting values.
- *Accuracy* The quality of the fixture, from the point of view of positioning function, has a direct impact on the scatter of dimensions observed in manufacturing. The selection of the contact points between part and fixture which determine the datum for manufacturing dimensions is of utmost importance and deserves special attention (section 7.4). They must fulfill the following conditions:
  - They should be cost effective.
  - The fixture setup and teardown time should be as short as possible.
  - The fixture and its clamps should not act as obstacles in the way of machining a part segment. Surfaces and segments of the part that are covered by the fixture or its clamps cannot be machined with that setup. An additional setup might be required. However, each additional setup increases the cost of manufacturing.
  - The fixture and its clamps should not obstruct tool travel, which would increase tool path length and motion time.

The decision on workholding fixtures thus affects:

- the surfaces and segments that can be machined in a setup;
- the accumulation of tolerances and thus the accuracy of each cutting operation;
- the allowable cutting forces and thus the cutting conditions; and
- the tool path, and thus the tool shape and size.

Therefore, after selecting the basic process, as discussed in Chapter 6, the first decision to be taken is how to define the workholding reference points. This is discussed in section 7.4.

The emphasis in this chapter will be on aspects of accuracy. Other aspects of fixture design can be handled by normal design procedures and through strength of material computations. Some of these topics are referred to in the appropriate sections.

7.2 THEORETICAL POSITIONING OF A SOLID IN SPACE (ISOSTATISM)

It is well known that a perfect solid can be positioned in space by six parameters, normally three positioning parameters (T_x, T_y, T_z, dimensional coordinates) and three orientation parameters (R_x, R_y, R_z, angular coordinates) (Fig. 7.2). It is equivalent to say that a solid has six degrees of freedom which characterize its location. A simple way to prove this assertion is to consider a solid in space and to fix one of its points (three degrees of freedom cancelled), then to fix a second point and a last point to cancel all degrees of freedom. It would appear that the position is defined by nine parameters (three points × three degrees of freedom). However, because of three relations between the nine parameters (the three distances between the points), the final number of parameters is only six.

In the field of mechanical engineering, this method of cancelling the degrees of freedom to locate a part, called the **principle of isostatism**, has been defined

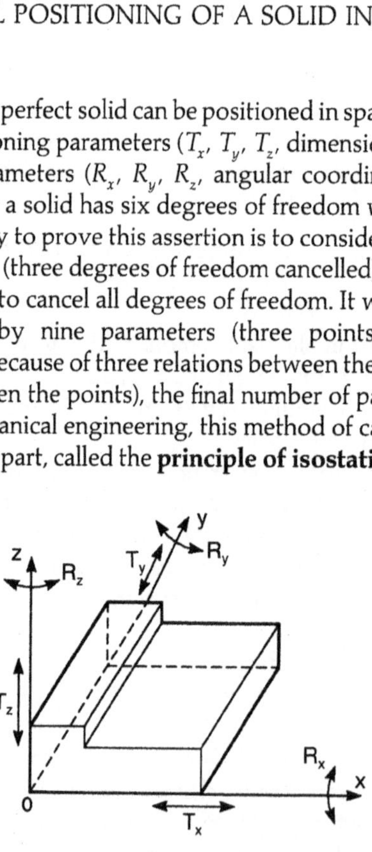

Fig. 7.2 Dimensional and angular coordinates. Redrawn from Karr. J., *Methodes et Analyses de Fabrication Mecanique*, published by Dunod, Bordas, Paris, 1979.

as the six point principle, i.e. positioning a part by six points of contact on its faces (without friction). This principle was first proposed by physicists of the nineteenth century such as Lord Kelvin, Maxwell, Curie, Michelson and others, who needed to position objects exactly for their measuring instruments. An equivalent way of defining isostatism is to say that the reaction forces on the six points of contact can be determined by the six equations of classical statics.

Examples of the application of the six point principle are shown in the following Figs 7.3a,b,c for a prismatic part, 7.4 for a cylindrical part and Fig. 7.5 for a part of any form.

In Fig. 7.3a, a prismatic part is represented with its dimensions and tolerances. In Fig. 7.3b, its positioning on six points is represented in accordance with standardized symbols, explained in section 7.3. Its real positioning and clamping is represented in Fig. 7.3c, which also shows how to process its surfaces using a set of milling cutters.

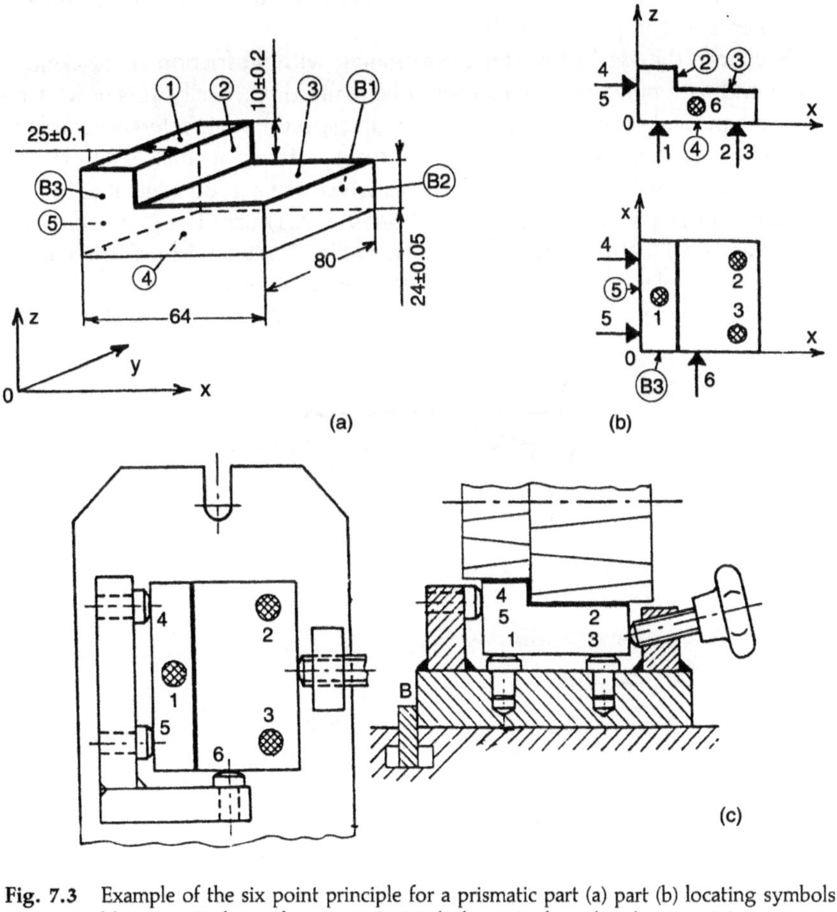

Fig. 7.3 Example of the six point principle for a prismatic part (a) part (b) locating symbols (c) practical locating. Redrawn from Karr. J., *Methodes et Analyses de Fabrication Mecanique*, published by Dunod, Bordas, Paris, 1979.

In Fig. 7.4, a cylindrical part is located on its surface using four points (1, 2, 3 and 4) and four degrees of freedom have been cancelled. Two other degrees of freedom are cancelled by points 5 and 6. Whereas point 5 is cancelling the translation in the direction of the axis of the cylinder, point 6 cancels the rotation around this axis. However, this latter positioning is not strictly isostatic because point 6 acts by friction, which could be accepted for clamping but not for isostatic positioning. A sixth point far from the axis of the cylinder and cancelling its rotation would be acceptable.

For a part of any shape, Fig. 7.5 proposes two isostatic solutions. In Fig. 7.5a, the part has three degrees of freedom in translation cancelled by a sphere captive in a pyramidal hole. Two degrees of freedom of rotation are cancelled by a prismatic slot aligned with the hole and the last degree of freedom is cancelled by a little planar face. A variant of such an isostatic device is shown in Fig. 7.5b where the six degrees of freedom are cancelled by concurrent slots of prismatic section. Both concepts are widely used in precise measuring instruments and in precise machining.

Of course, the ideal of isostatic positioning without friction at the contact points and with punctual contact cannot be realized practically. It is however a useful means of designing fixtures which will respect the data references and the production dimensions effectively. In reality, the punctual contacts are approximated by very small contact faces which are a good simulation of the isostatism principle. The ISO Standard 5459 (1981) uses the same six point principle, called 'datum targets' as shown in Fig. 7.6, with a prismatic part.

Other examples of the application of the six point principle are illustrated in section 7.7 where designs for real fixtures for manufacturing typical parts are

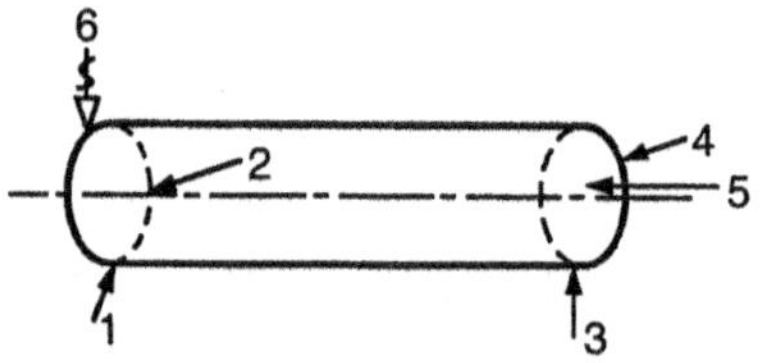

Fig. 7.4 Example of the six point principle for a cylindrical part.

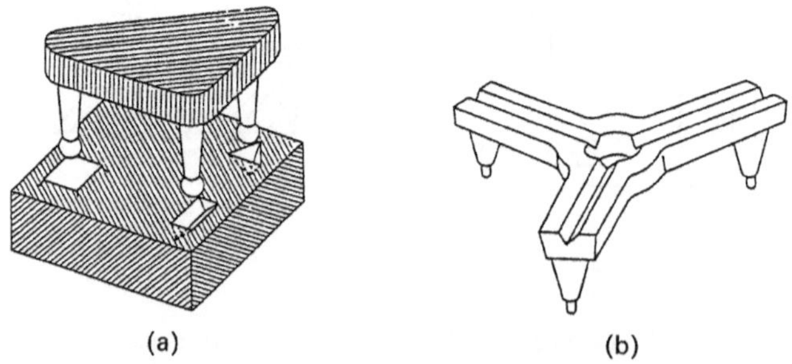

Fig. 7.5 Example of the six point principle for a part of any form.

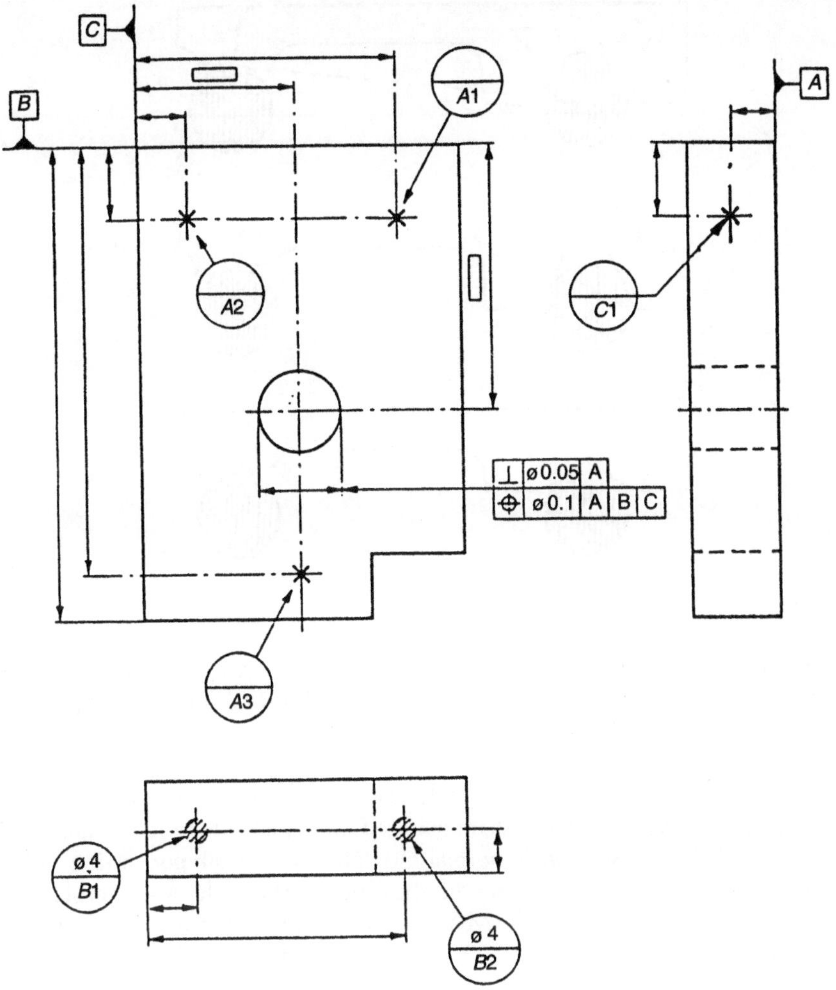

Interpretation:

Datum targets *A*1, *A*2 and *A*3 establish datum *A*
Datum targets *B*1 and *B*2 establish datum *B*
Datum target *C*1 establishes datum *C*

Fig. 7.6 The principle of 'datum targets'.

shown. Its use is also very important for the calculation of position errors which are discussed in section 7.5.

Finally, it should be added that the independence conditions for the locations of the six points on the part have to be respected in order to cancel the six degrees of freedom effectively. For example, if four points were placed on a plane, a redundance would result and the isostatism would not be respected.

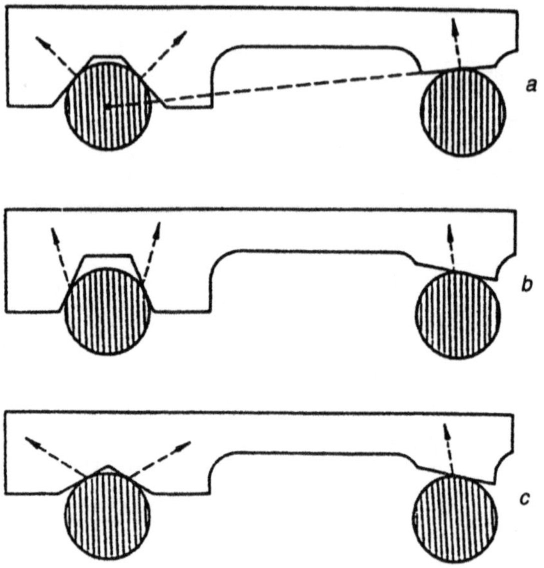

Fig. 7.7 Maxwell's principle.

Another consideration relates to the optimization of the locations of the six contact points so that the fixture can guarantee the best location in spite of errors in the contact points. One solution to this problem is proposed by Maxwell's Principle which says that each contact should be such that when removing the contact, the motion of the part will be normal in relation to the contact. Maxwell's principle is demonstrated in Fig. 7.7 where rotation of the part around the left support is normal in relation to the support at the right when the contact facet is collinear with the center of the circle. Also, translation of the left side contacts fulfills the Maxwell condition in the case of a vee contact with an angle of 90 degrees, and only in this case. Needless to say, the application of this principle in the general case raises difficult questions regarding optimization of the six locating points.

7.3 STANDARDIZATION OF LOCATORS AND FIXTURING ELEMENTS

In order to create a common language for locating symbols, a tentative standardization has been developed shown in Figs 7.8, 7.9, 7.10 and 7.11 (French Standard NF 04-013, 8.85). It is useful for the preparation of a process plan and, therefore, has been included in this section.

The symbolization of locators has been divided into two categories:

1. symbolization for the cancellation of the degrees of freedom of the part in order to design a first draft of a process plan; and

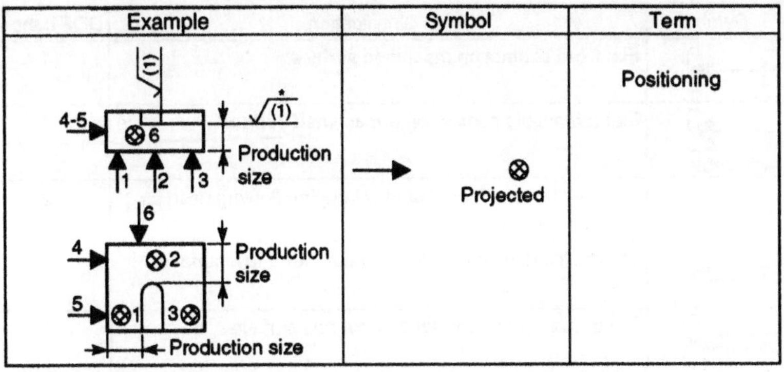

Fig. 7.8 Example of positioning symbols: to cancel the necessary degrees of freedom.

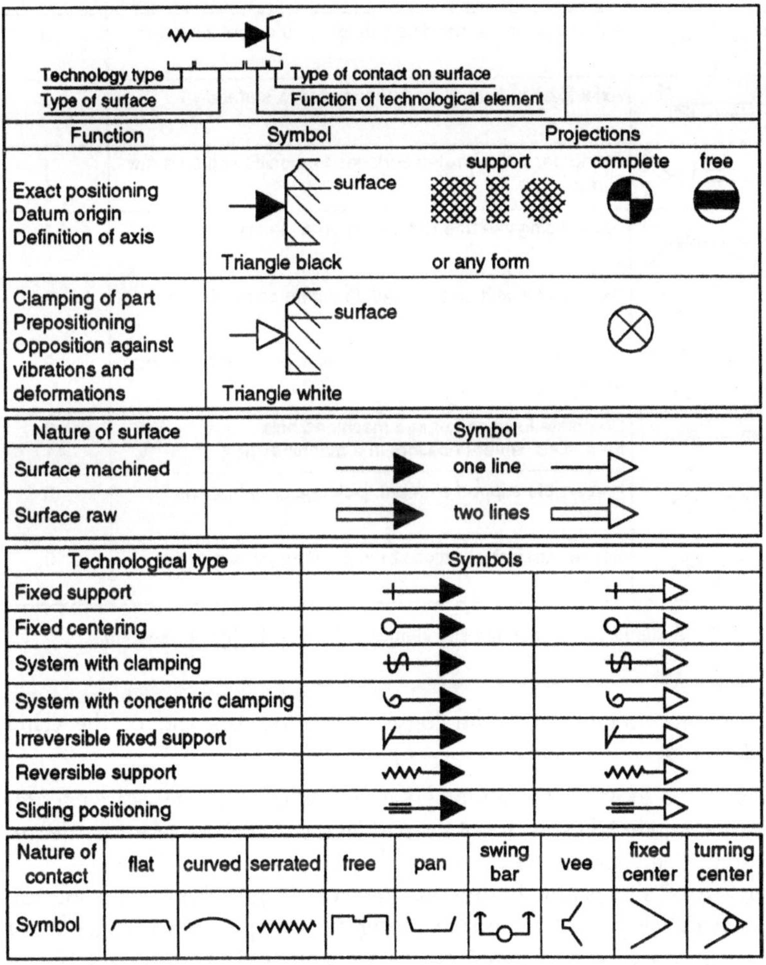

Fig. 7.9 Symbolization of technological elements.

Symbol	Signification	DOF cancel
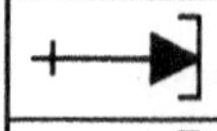	Flat fixed contact on machined surface	1
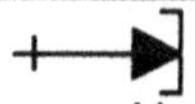 removable	Flat removable contact on a machined surface	1
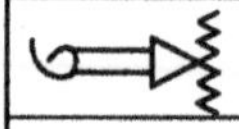	Serrated jaws with concentric clamping floating used as driver on raw surface	2
	Curved fixed contact used as datum for a raw surface	1
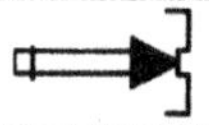	Fixed free contact for datum on a raw surface	1
	Axial pan used as contact point for remachining a machined surface	2
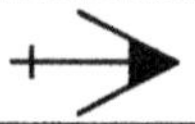	Axial fixed center used as support on a machined center	3
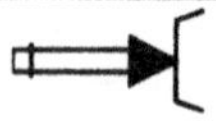	Axial fixed pan used as datum on a raw surface	2
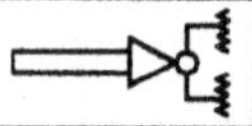	Swing bar with serrated surfaces for positioning on a raw surface	1
	Axial sliding vee used as datum on a machined surface	1
	Axial center with spring used as datum on machined center	2
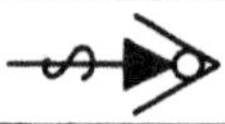	Axial rotating center of tail stock used as short centering and locking	2
	Complete fixed center in a machined hole Free fixed center (locating) in a machined hole	2 1
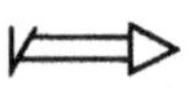	Irreversible support element, jack against vibrations	0
	Spacer against vibrations or support for parts with thin walls	0

Fig. 7.10 Symbolization of composite elements (DOF = degree of freedom).

2. symbolization of the technology of the locators and of the clamping elements.

Figure 7.8 shows the positioning symbols which must be located in order to meet the required dimensions of the part and to cancel the necessary degrees of freedom. The symbols are represented in both front and plan view.

Figure 7.9 shows how to describe the technological type of the locator and includes details on its function (centering pin, clamping function, nature of the surface of contact, etc.). Figure 7.10 gives examples of composite symbols, e.g. a fixed locator with a flat contact intended for positioning a machined surface, and finally, Fig. 7.11 shows examples of real parts with their positioning symbols, as

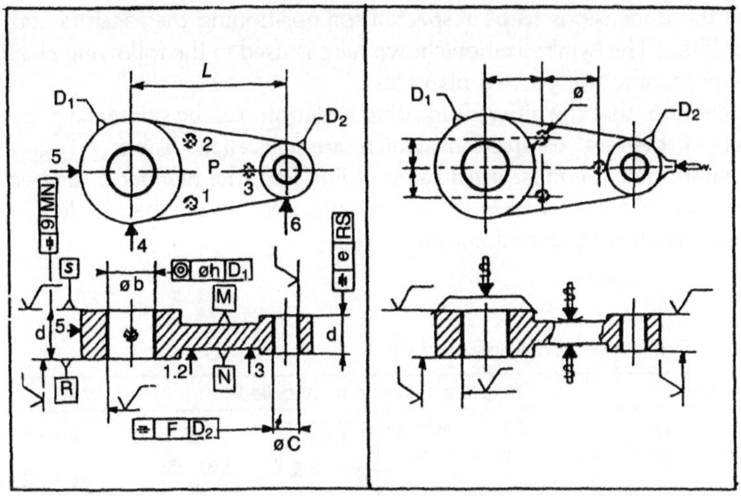

Three degrees of freedom eliminated on plane N
Two degrees of freedom eliminated on cylinder D1
One degree of freedom eliminated on cylinder D2
Five production sizes and four tolerances

Clamping on planes M and N by symmetrical system with curved contacts for raw surfaces
Orientation by removeable pan on raw cylinder (to permit machining)
Orientation by reversible sliding vee on raw cylinder D2

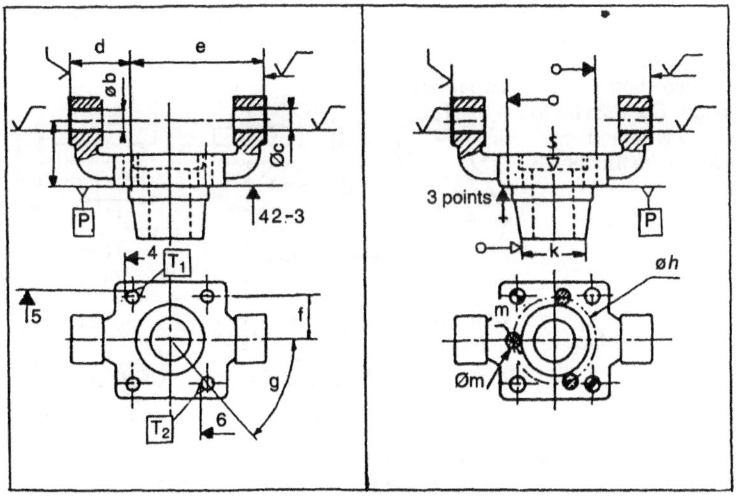

Three degrees of freedom eliminated on plane P by three points
Two degrees of freedom eliminated by hole T1
One degree of freedom eliminated by hole T2
Seven production sizes

Fixed contacts on surface P in three points
Øm (machined surface)
Complete centering on machined hole T1
Free centering on machined hole T2
Precentering on cylinder k
Clamping by split washer axially.

Fig. 7.11 Applications of the symbolization.

well as the dimensions to be respected for positioning the locators and the workholding. The symbolization shown here is used in the following chapters for the preparation of process plan files.

Finally, note that the different locating situations can be summarized by the number of degrees of freedom which are cancelled. Table 7.1 gives a representation of the most usual cases of jigs with the number of degrees of freedom cancelled. This table is used to evaluate the number of degrees of freedom cancelled by partial liaisons in jigs.

Table 7.1 Definition of partial liaisons in jigs

Liaison type	Degrees of freedom cancelled
Plane support Three non aligned on a plan	One translation Two rotations 
Orientation Two points on a line in a plane	One translation One rotation 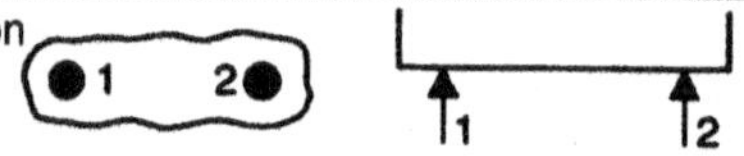
Short centering Two points on a circular line	Two translations 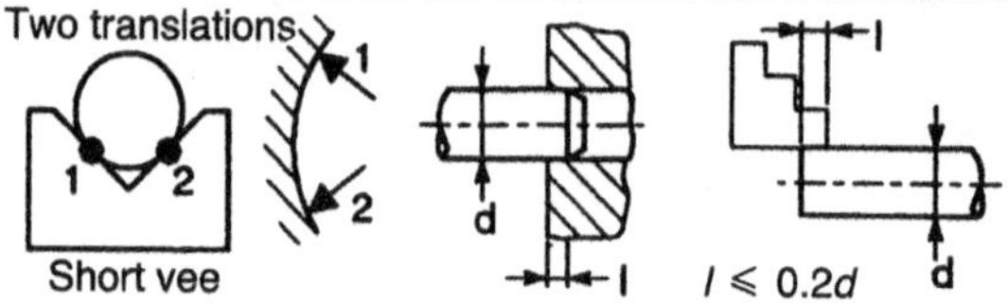
Long centering Equal to two short centerings. The ratio l/d to define a long centering depends on the roughness of the surface.	Two translations Two rotations 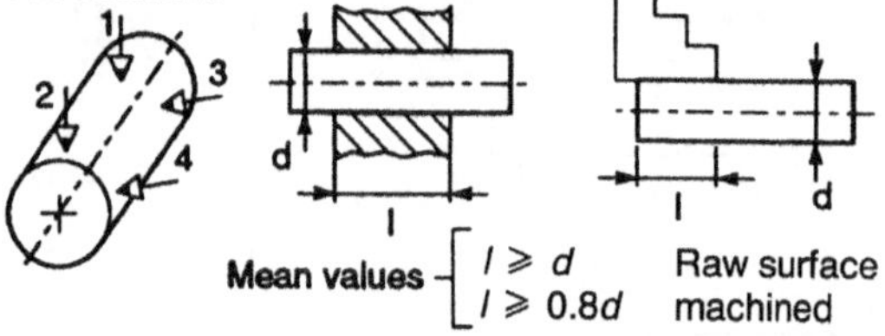
Stop One point on a surface	One translation or one rotation 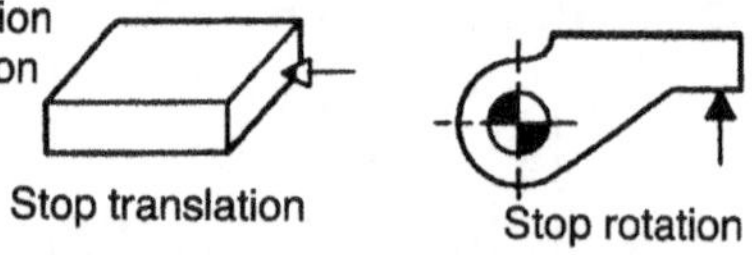
Locating or free centering Locating cancels one degree of freedom, but in two directions. No stop in normal direction	One translation or one rotation 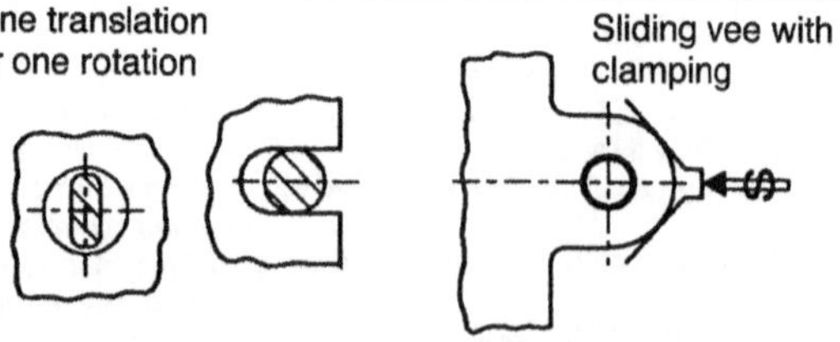

Source: Karr, J., *Methodes et Analyses de Fabrication Mecanique*, published by Dunod, Bordas, Paris, 1979.

7.4 SELECTION OF LOCATING SURFACES AS DIMENSIONAL DATA REFERENCES

The function of locating surfaces is to serve as data or dimensional references for the manufacturing of features in the part. Their selection is of primary importance for the execution of correct processing operations and the fulfillment of dimensional requirements. The order of precedence in choosing datum surfaces has already been considered in Chapter 2 where geometrical anteriorities have imposed locating constraints. However, special considerations may be necessary when selecting datum surfaces in relation to location requirements. The general rules applied in these cases are now briefly reviewed.

The most commonly used criteria are as follows:

1. The first datum surfaces, also called 'starting surfaces', make use of the surfaces existing on the raw blank. They should be of sufficient size to ensure stability and be free of holes and slots. In many cases, the drawings give indications called 'liaison with raw material' (Fig. 7.12), which are a great help when positioning the first surface to be processed in relation to the raw material. For example, surface (D) in relation to raw surface (B) has to be at a minimum distance of 15. As a consequence, different liaisons with the raw blank can result in dimensional conditions on the raw part (e.g. dimension X must be limited to 35.8).
2. The datum surfaces defined by accurate dimensions should be given priority to avoid the stacking up of tolerances in following phases. The precedence of surfaces has to be ordered according to the anteriorities already defined in Chapter 2.
3. The choice of contact points is carried out as shown in Chapter 2 for the part in Fig. 2.1 (Fig. 2.11 and Table 2.3), and is a consequence of precision requirements between surfaces in the part.

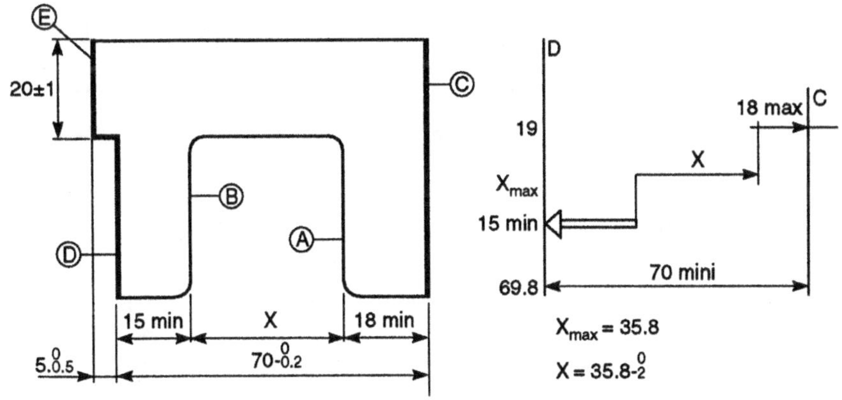

Fig. 7.12 Calculation of raw material dimension X to respect liaison to blank.

4. The choice of a hole as a datum is less accurate than the choice of a plane, as shown in Chapter 2.
5. The stability of fixturing has to be guaranteed against the actions of cutting forces and other external disturbances. Fastening on surfaces with defects or welding joints should be avoided.
6. Good accessibility to the machine tool is necessary so that the part can be positioned in its fixture properly.
7. Finally, the choice of fixture should be made on economic grounds. Where possible, modular elements should be used which can be combined in suitable clamping devices.

As stated earlier, starting and reference surfaces should be large so that errors of form or position will not influence the correct locating of the part. Figure 7.13 shows typical examples of the influence of such errors and their reduction by making sure the datum references are large. In the case of a cylindrical datum, the length of the cylinder should be sufficient to guarantee the accuracy of positioning, and, at the same time, provide good stability.

To give an idea of the errors commonly encountered in jigs and fixtures, Table 7.2 gives the order of magnitude of inaccuracies for different types of workholding. It is important to note that the highest figures relate to the part fixtures, whereas the other errors (tool holder, machine kinematics, etc.) are an order of magnitude smaller.

Concerning starting surfaces on raw material, it is important to emphasize the role of punctual target data references which define data surfaces accurately, whereas the complete surfaces in the raw blank will generate large positioning errors. Of course, these points should be separated by the maximum distances possible.

The preceding considerations are combined as precision criteria and illustrated in Table 7.3, based on the part in Figs 7.3a and 7.3b. Assuming that

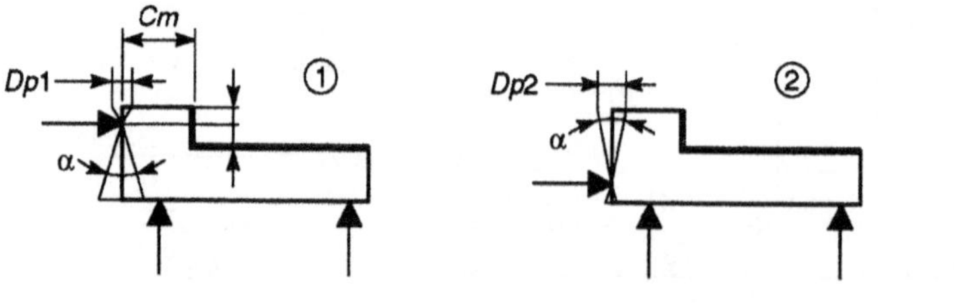

Fig. 7.13 Example of error in selecting position location. Redrawn from Karr. J., *Methodes et Analyses de Fabrication Mecanique*, published by Dunod, Bordas, Paris, 1979.

Table 7.2 Estimated values of factors contributing to inaccuracy

Type of error	*Estimated error value (mm)*	
Kind of fixture		
Chuck three hard jaws	0.1	
Chuck three soft jaws	0.05	
Mounting with pressure against the support	0–0.05	
Precision chuck collet	0.00127 TIR*	
Error of form in the fixture surface		
– fixture surface rough	0.5	
– fixture surface machined	0.05	
Tolerance due to tool wear	0.02–0.05	
Tool position tolerance		machining
– limit top fixed	0.02–0.05	finishing
	0.1–0.2	rough
– limit stop with clutch	0.05–0.1	finishing
	0.1–0.2	rough
Error of form in the tool	0.05	
Factors contributing to inaccuracy		
– machine slide – tracking	0.005	
– machine lead screw – absolute and repeatable	0.005	
– turrent index – repeatability	0.00125	
– spindle runout	0.005	
Machine control system		
– resolution	0.005	
– response	0.0025	
Tooling		
– tool holder repeatability	0.0025–0.025	
– insert accuracy of index	0.01–0.05	
– tool setting	0.0025–0.025	

*TIR – total indicator reading.

surfaces (4) and (5) have been processed in previous phases, they are used as data references for the execution of surfaces (3) and (2). The positioning of the part is therefore defined in the following way, cancelling the six degrees of freedom:

- three points (1, 2, 3) cancel the three degrees of freedom of plane (4);
- two points (4, 5) cancel the degrees of freedom of plane (5), i.e. one less than normally required; and
- one point (6) cancels the degree of freedom in the direction of axis y.

This distribution of the contact points is justified in Table 7.3 where the dimensional and angular accuracies of the different surfaces are stated. The maximum number of contacts and the degrees of freedom cancelled are also indicated in column 5. However, because the dimensional accuracy of the data

Table 7.3 Considerations in precision criteria

Surface machined	*Manufactured dimensions*	*Surface datum*	*Dimension tolerances*	*No. of contacts chosen and d.o.f. cancelled* *Maximum*	*Normal*	*Type of datum*
(3)	24 ± 0.05	(4)	Dim.: 0.1 Ang. for R_x: $0.1/80 = 0.00125$ Ang. for R_y $0.1/39 = 0.0026$	$3\begin{cases} T_z \\ R_y \\ R_x \end{cases}$	$3\begin{cases} T_z \\ R_y \\ R_x \end{cases}$	Plane on (4)
(2)	25 ± 0.10	(5)	Dim.: 0.2 Ang. for R_z: $0.2/80 = 0.0025$ Ang. for R_y: $0.2/10 = 0.02$	$3\begin{cases} T_x \\ R_z \\ R_y \end{cases}$	$2\begin{cases} T_x \\ \\ R_z \end{cases}$	Direction on (5)

Source: Karr, J., *Methodes et Analyses de Fabrication Mecanique*, published by Dunod, Bordas, Paris, 1979

references and the angular accuracy are different, a priority choice has been made: surface (4) has a dimensional accuracy of 0.1 compared to 0.2 for surfaces (5) and an angular accuracy of 0.0026 compared to 0.02 for R_x. Therefore surface (4) is chosen as the primary datum with three points of contact and surface (5) as the secondary datum with two points of contact. An equivalent way to justify this choice would consist in giving priority to surface (4) because of its larger size.

Finally, the distribution of the six contact points is based on precision requirements relating to simple tolerance considerations taken from the information on the drawing (e.g. orientation tolerances derived from linear tolerances). A more precise evaluation of the influence of positioning errors on the accuracy of the part is developed in the next section by using the theory of small screw displacements.

7.5 CALCULATION OF POSITIONING ERRORS DUE TO GEOMETRIC ERRORS IN JIGS AND/OR IN PARTS

7.5.1 General analysis of positioning errors

When a mechanical part is positioned on a fixture, ready to be machined or measured, its real position and orientation in space is never precisely known because of errors in its locators in the fixture and geometric errors in the surface of the part. These errors are generally the most important because they have been accumulating since the design stage. Errors in locators are normally smaller, but they can still influence the accuracy of the part because the tool settings in the machine tool are related to data references which are now only theoretical; the real ones have been changed because of errors in the locators.

The two eventualities can be formalized in a similar way as shown in Figs 7.14a and 7.14b. Figure 7.14a shows, in model form, the effect of errors of form in a part on its localization in a fixture. The error ξ_i is defined in relation to the ideal form of the theoretical part without errors of form as the deviation along the normal to the surface n_i (Bourdet and Clement, 1974).

Similarly, the error in a setup (Fig. 7.14b) is shown as the shift of the contact along the normal n_i of the locator, i.e. ξ_i. The workpiece is represented in two positions (S) and (S') corresponding to this shift. As a consequence of the six errors ξ_i on a complete isostatic fixture (Fig. 7.15), the part is slightly shifted according to its six degrees of freedom. A reference surface belonging to the part reference system, as indicated in Fig. 7.15, will be moved to a new location. Therefore, the geometric relationship between the features of the part to this datum will be changed because the features processed in this phase will be referred to the theoretical machine datum. In other words, the part reference

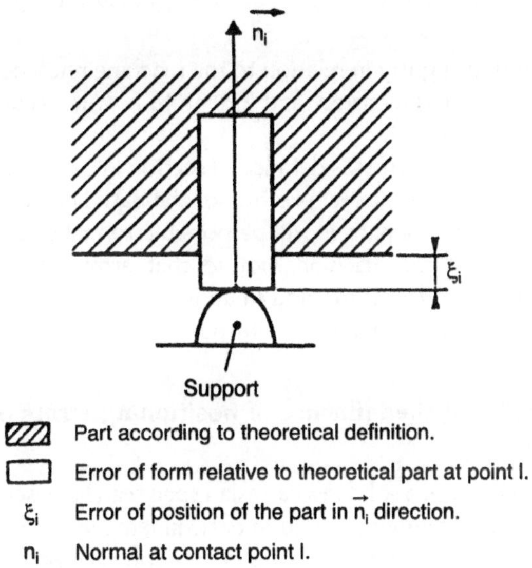

Fig. 7.14a Model of positioning error due to form errors in the part.

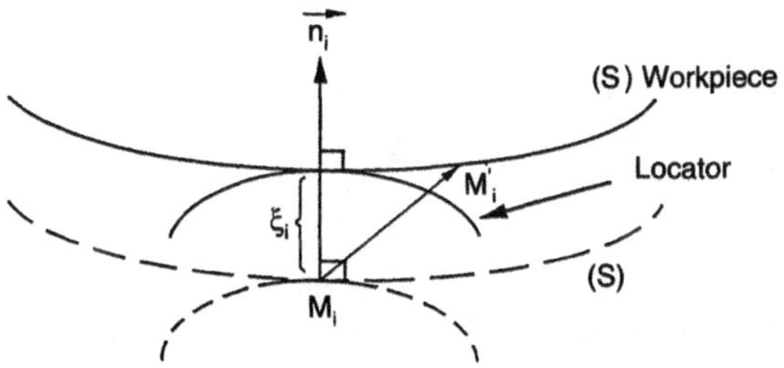

Fig. 7.14b Model of position error due to locator deviation.

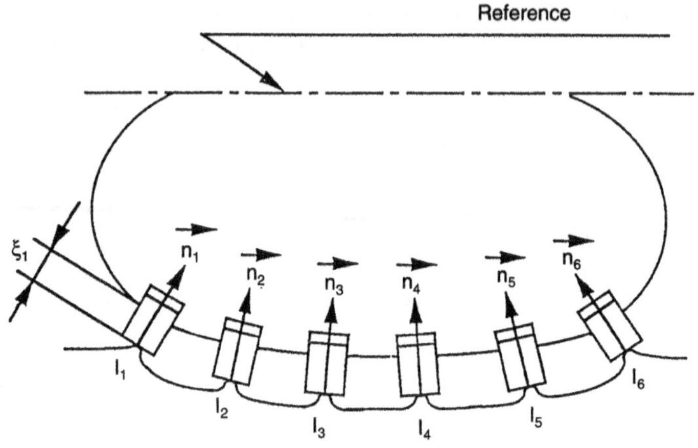

Fig. 7.15 Model of errors of form of a part positioned on an isostatic jig.

system will be shifted slightly in relation to the machine reference system. As a consequence, errors of dimension and orientation will occur between the different features of the part.

The calculation of such errors is detailed in section 7.5.2. It is also interesting to note that by identifying the real position of the part, i.e. the real position of the reference system of the part, it will be possible to make corrections in the machining program of the machine tool so that errors in the part can be corrected using the calculation method in an inverse manner. This strategy is discussed in more detail in Chapter 4, on the setting of tools.

7.5.2 Calculation of the influence of positioning errors on the real location of a part

The small displacement of a solid in a cartesian space can be described by a small screw displacement as commonly used in basic kinematics. The total motion is composed of a translation (vector $\bar{V}$) and a rotation (vector $\bar{\Omega}$) which form a vector space also called 'twist'. Therefore, the displacement of a point M_i of a solid in movement (Fig. 7.14b) is given by:

$$\overline{D_i} = \overline{V} + \overline{\Omega} \times \overline{OM_i} \tag{7.1}$$

where: $\bar{\Omega}$ (α, β, γ) is the rotational vector
$\bar{V}(u, v, w)$ is the translation vector
$\bar{D}$ is the displacement vector of a point M_i on the part
O is an arbitrary origin, conveniently chosen for calculation

Every displacement is therefore a function of six parameters (u, v, w) and (α, β, γ) which completely describe the motion of a solid.

Bourdet and Clement (1974) used this model to determine the parameters (u, v, w) and (α, β, γ) which define the small motion of a part due to errors in its localization. Any feature in the part, in particular a datum surface, will undergo a displacement which is a function of the parameters (u, v, w) and (α, β, γ).

Derivation of the six equations defining the small screw is based on the deviation of each contact point M_i (Fig. 7.14b) to a new position M'_i. The projection of the displacement $\overline{M_iM'_i}$ can be approximated by its projection on the normal n_i which is equal to ξ_i, the error of positioning:

$$\bar{\xi}_i = \bar{V}.\bar{n}_i + \bar{\Omega} \times \overline{OM_i}.\bar{n}_i = \bar{V}.\bar{n}_i + \overline{OM_i} \times \bar{n}_i.\bar{\Omega} \qquad (7.2)$$

Using the Plucker coordinates (u_i, v_i, w_i) and (l_i, m_i, n_i, moments at O of vectors $\bar{n}_i$), this relationship becomes for the six points a system of six linear equations which deliver the solution of the small screw motion in the following form:

$$\begin{bmatrix} u_1 & v_1 & w_1 & l_1 & m_1 & n_1 \\ '' & & & & & \\ '' & & & & & \\ '' & & & & & \\ '' & & & & & \\ u_6 & v_6 & w_6 & l_6 & m_6 & n_6 \end{bmatrix} \begin{bmatrix} u \\ v \\ w \\ \alpha \\ \beta \\ \gamma \end{bmatrix} = \begin{bmatrix} \xi_1 \\ \xi_2 \\ \xi_3 \\ \xi_4 \\ \xi_5 \\ \xi_6 \end{bmatrix} \qquad (7.3)$$

If $\bar{\delta} = (\bar{V}, \bar{\Omega})$ is the small screw, the equation can be written as follows:

$$\bar{\bar{M}} \,.\, \bar{\delta} = \bar{\xi}_i \qquad (7.4)$$

with $\bar{\bar{M}}$ being the matrix of Plucker for the vectors $\bar{n}_i$.

As a consequence of the displacement, every reference in the part coordinate system will be shifted and generates a geometric error in the workpiece. To hold this error to a minimum value compatible with specified tolerances, it becomes necessary to search for optimum locations of the supporting contact points. This problem is defined as an optimization problem of finding the best position of six points on the faces of a part, i.e. the optimization of a function of 12 variables bounded by the edges of the faces of the part. The resolution of this problem is beyond the purpose of this book, but see Weill, Darel and Laloum (1991).

7.5.3 Application of the small screw calculations to practical examples

To show how the theory developed above can be aplied to real problems in production, let us take first the example of a round part positioned on a V-shaped jig (Fig. 7.16a). Three solutions are presented for the positioning of the contact points. The surface to be machined is situated at a distance *BE*2 +/− from the bottom of the cylinder of diameter ∅*BE*1 ±. Because of tolerances in the diameter ∅*BE*1, the dimension *BE*2 will change by a certain amount

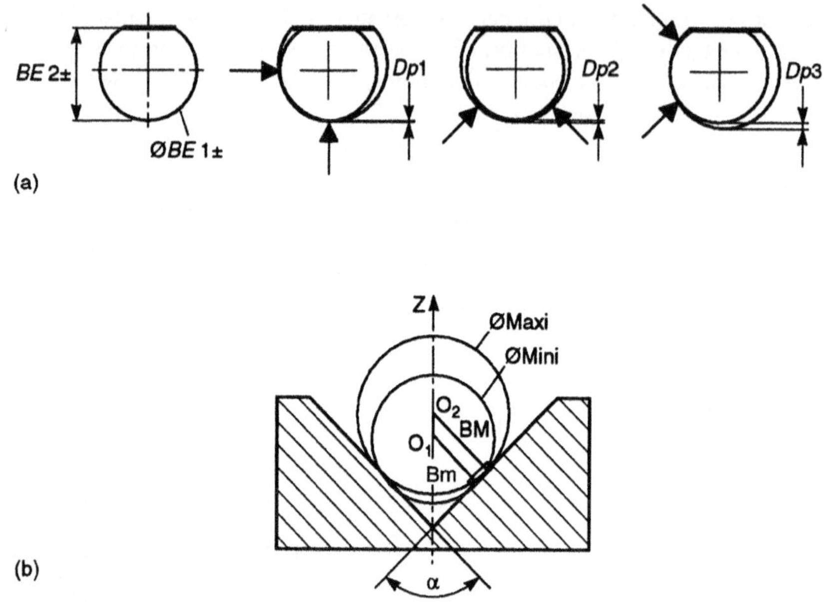

Fig. 7.16 Example of a round part positioned on a vee-shaped jig (a) different positioning situations (b) cylinder on plain vee. Redrawn from Karr. J., *Methodes et Analyses de Fabrication Mecanique*, published by Dunod, Bordas, Paris, 1979.

depending on the type of positioning. The calculation of this error is shown in Fig. 7.16b in the case of a plane vee as follows:

$$\Delta BE2 = \frac{\Delta \varnothing BE1}{2}\,\frac{1}{\sin\frac{\alpha}{2}} - \frac{\Delta \varnothing BE1}{2}$$

$$= \frac{\Delta \varnothing BE1}{2}\left(\frac{1}{\sin\frac{\alpha}{2}} - 1\right) \tag{7.5}$$

The error on *BE*2 depends on the value of the angle α of the vee, a larger angle causing a smaller error in *BE*2 for a plane vee.

The same result can be obtained by applying the general theory of positioning errors. Because of the symmetry of the figure, the only displacement is a translation $\bar{v}$ in the vertical direction. The value of the translation $\bar{v}$ is given by (7.2):

$$\frac{\Delta \varnothing BE1}{2} = \bar{v}\,.\,\sin\frac{\alpha}{2}$$

and the final error is:

$$\Delta BE2 = \frac{\Delta \varnothing BE1}{2}\left(\frac{1}{\sin\dfrac{\alpha}{2}} - 1\right)$$

because of the change of diameter $\varnothing BE1$. The result is identical to (7.5).

A more comprehensive application of the general theory is shown in Fig. 7.17 where a part is positioned on three points (A, C and E) in a plane. Because of variations in the dimensions of the part (diameters of circles U and V, and of the distance EV), the high point on circle V is shifted by an amount equal to ΔR when projected on the normal $\bar{n}_R$ in the vertical direction. The value of ΔR is calculated according to (7.3) which in this case contains only the variables u, v and γ. After transformations, the result is:

$$\Delta R = -0.31A - 0.48C + 1.25E$$

with A, C and E being the values of the ξ_i in the general equation (7.2) and the coefficients being functions of the dimensions of the part. Numerically, if $A = 0.33$; $C = 2.43$ and $E = 4.28$ because of variations of $\Delta\varnothing u = 3$, $\Delta\varnothing v = 2$ and $\Delta EV = 0$, the value of ΔR is 4.1 mm (0.158 in).

The method can be applied to any set of errors in positioning provided the setup used is really isostatic.

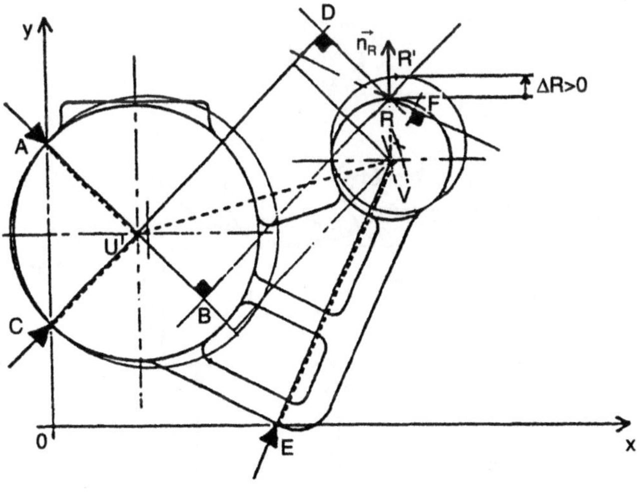

Fig. 7.17 Part positioned on three points (A, C and E).

7.6 CALCULATION OF CLAMPING POSITIONS AND CLAMPING FORCES

This section is devoted to finding optimal clamping positions and to calculate clamping forces in order to ensure a good stability of the part in its fixture.

As shown in Fig. 7.18 relating to the processing of the part from Fig. 7.3, the clamping of the part on its setup has to resist the cutting forces (shown in two components F (resultant) and F_{yz}).

In this case, the cutting forces apply the part on the fixture and, on condition that they are located between the supporting points, the stability of the part is guaranteed. However, the situation can change during the trajectory of the tool, and to ensure stability, it is necessary to analyze the new configuration. For example, in Fig. 7.18, it could happen that the direction of force F_{yz} becomes upwards and, therefore, the clamping force should be able to oppose this force by friction.

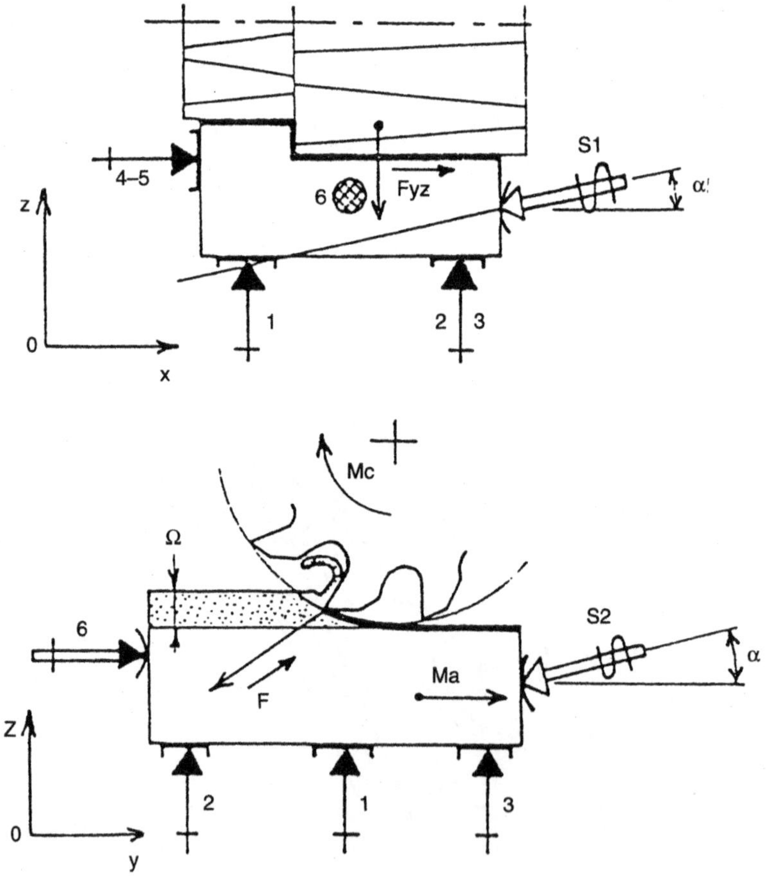

Fig. 7.18 Acting forces on the clamping system.

Calculations of clamping forces in different typical cases are explained below. Turning and milling/drilling have been used for this example because these two methods are basically different.

7.6.1 Chucking type on a lathe

There are several methods of chucking round symmetrical parts in a lathe. Below are some rules for selecting the most economical chucking type.

Three jaw chuck

- *Segment shape* A part can be chucked only on cylindrical segments.
- *Segment length* It is recommended that the length of contact between the jaws and the chucked segment be 1.2 times the segment diameter. This limit is to make sure that the part lines of symmetry coincide with the spindle line of symmetry. Short contact length might result with a certain angle between these two lines of symmetry. In any case, the length of the contact between the jaw and the part should not be less than 5 mm. When the length is lowered, for reasons of practicality, a check for concentricity of the part should be made at the far free end of the part.
- *Part diameter* The chuck jaws can be adjusted to accommodate any part diameter (within chuck specifications). However, the chuck through hole and the spindle bore controls the part diameter in cases where the part has to be inserted into them. These cases occur when machining long bars (uncut pieces) and when the chucking location is not at one end of the part, (some can be in the middle of the part). In order to have maximum part stability, it is recommended that the extended portion of the part be as short as possible. Therefore, it is advisable that the part be inserted into the chuck and, if the length is not sufficient, for it to be inserted into the spindle bore as well.
- *Type of jaws* The jaw exerts compression stresses upon the part. These forces should counterbalance the cutting forces in the axial and tangential direction. The chucking forces are transformed into tangential forces by the coefficient of friction between the part and the jaws. The acting moment might cause the part to rotate in the chuck if it is bigger than the chucking moment. The equilibrium point is when

$$F_{\text{cutting}} \times R_{\text{part}} \leqslant F_{\text{chuck}} \times R_{\text{chuck}}$$

therefore

$$F_{\text{cutting}} \leqslant F_{\text{chucking}} \times \left(\frac{R_{\text{chucking}}}{R_{\text{cutting}}} \right) \tag{7.6}$$

where R is the radius, F_{cutting} is the cutting force and F_{chuck} is the friction force in the jaws.

This equation points to the recommendation that the chucking should be done on the biggest diameter possible. F_{chuck} is the tangential chucking force and is equal to the chucking force $F_{ch} \times \mu$ (the coefficient of friction). Estimated coefficients of friction are given in Table 7.4.

An estimated equation to compute the compression chucking force F_{ch} (Fig. 7.19) is:

$$F_{ch} = N \times b \times r \times \left[2 \sin^{-1}\left(\frac{a}{2r}\right) \right] \times \sigma \qquad (7.7)$$

where: N = number of jaws
b = length of chuck (mm)
r = radius of part under the chuck (mm)
a = width of jaws (mm)
σ = allowable compression stress (N/mm^2).

Table 7.4 Coefficient of friction in chuck jaws

Surface of work piece	*Gripping surface at jaws* *Smooth*		*Diamond style*		*Serrated*	
	Axial	*Radial*	*Axial*	*Radial*	*Axial*	*Radial*
Smooth machined, finished ground	0.07	0.13	0.13	0.24	0.24	0.40
Rough to medium machined finish	0.11	0.22	0.22	0.40	0.40	0.66
Unmachined	0.15	0.31	0.31	0.55	0.55	0.92

Material factor: Steel 1.0
Aluminium alloy 0.97
Brass 0.92
Cast iron 0.8

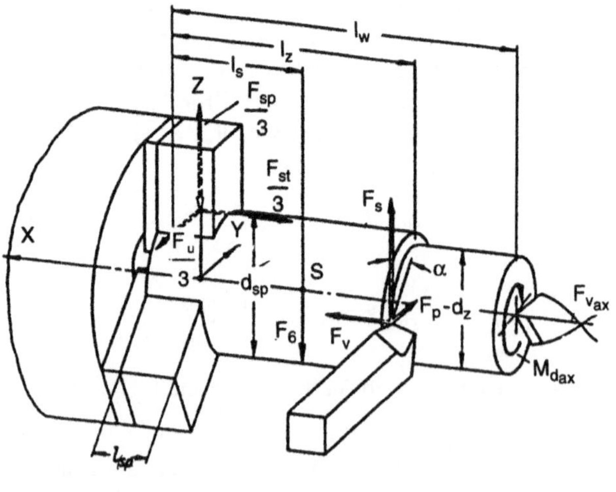

Fig. 7.19 Forces on a three-jaw chuck.

This equation points to the recommendation that the jaw's length and width should be as large as possible.

For economical considerations, it is advisable to use maximum depth of cut and feed rates. However, such cutting conditions will result in a large cutting force, and thus large chucking forces will be needed. Such forces might distort the part. It is therefore recommended to reduce the chucking forces by increasing the coefficient of friction. This can be accomplished by using rough jaws (diamond or serrated) for rough cuts, and smooth (machined jaws) for the finish cuts.

Three-jaw chuck without support

This type of chucking is the most economical. The part is inserted in the chuck, up to a stop (datum) and the jaws are closed. Three jaws guarantee the concentricity of the part in relation to the machine spindle.

However, this type of chucking has some limitations. The extended section of the part can be considered as a beam fixed at one end and free at the other, with a local load at the point of the tool. Such a type of support is weak at the load end and suffers from a large amount of deflection. As a rule of thumb, it is recommended not to use this type of chucking in cases where the ratio of length to diameter is higher than 3.5. However, this figure should be treated with reservation. The equation for maximum deflection (e) at the end of such a supported beam is:

$$e = \frac{F \times L^3}{3 \times E \times I} \tag{7.8}$$

where: F = load (N)
L = free (extended) length (mm)
E = modulus of elasticity of part material (N/mm^2)
I = moment of inertia of part cross-section (mm^4).

In turning, the cutting forces are the load and can be expressed by:

$$F = C_y \times a \times f^{uy}$$

where: C_y = coefficient
a = depth of cut (mm)
f = feed rate (mm per revolution)
d = diameter before cutting (mm).

The moment of inertia of a round bar is $I = 0.049 \times d^4$. The allowable value for deflection should be controlled by the diametrical tolerance and some of the geometrical tolerances.

Equation (7.8) can therefore be rearranged as follows:

$$e = K \times \left(\frac{L}{d}\right)^3 \times a \times \frac{f^{uy}}{d} \tag{7.9}$$

where K represents all the constants.

Rearranging (7.9), it comes:

$$\left(\frac{L}{d}\right)^3 = \frac{e \times d}{K \times a \times f^{uy}} \tag{7.10}$$

Thus, the ratio of length to diameter (L/d) is a function of the tolerance, the diameter of the machined segment (d) before cutting and the cutting conditions. For mild steel the following values are computed for L/d:

d	e	a	f	L/d
20	0.1	3	0.3	2.86
40	0.1	5	0.5	2.68
20	0.05	1	0.1	4.31
40	0.1	2	0.2	4.57

The figure of 3.5 is recommended in order to save computation time, but it should be remembered that it is only an approximation.

Three-jaw chuck with center support

This type of chucking helps to overcome the length-to-diameter ratio limits. By adding the center support (assuming the equation for deflection to be approximated by the case of a beam fixed at one and supported at the other end with load at any point), the L/d ratio can be increased by three to six times.

The disadvantage of this type of chucking is that the end face surface has to be prepared either by a countersink or by machining which adds machining operations and thus increases machining time and cost. This extra time might be compensated by the increase in allowable cutting forces (and thus increase in depth of cut and feed rate) used with this type of chucking.

In cases where the drawing calls for countersinks, or if there is a bore at the end of the part, it can be used as the center support holder. It is therefore recommended to machine the inner features of the part first, and for any part with $L/d > 2.5$ to use a center support chucking type. Furthermore, center supports are recommended whenever the chucking length is less than 5 mm, or the ratio of length of chuck to chucking diameter is less than 0.8.

Chuck with collet

The reasons for using collet are similar to those for the three-jaw chuck, except that the collet is more accurate. The part to be chucked must have a smooth surface.

Chuck with four-jaw chuck (independent)

Four contact points do not necessarily result in a circle. Indeed, it is quite difficult to adjust the movement of the jaws (each one separately) to chuck a part concentrically with the spindle bore. The four-jaw chuck is therefore used for non-symmetrical parts that must be adjusted for a certain centerline on the part.

A high degree of accuracy can be achieved with this type of chucking, but more time will be needed to center the part.

Face plate

This is actually a fixture plate (as in milling) suitable for any type of part and fixture aid. The chucking accuracy and speed is left to the fixture designer.

In cases where the extreme segments have geometric tolerances that must be machined in one fixturing, the following chucking types can be employed:

- *Chucking with a three-jaw chuck with a through hole* This is a good method in cases where there is a hole and the part is short (see above, three-jaw chuck without support).
- *Chuck with mandrel* For a long part with a through hole, it is recommended to use a one- or two-piece mandrel. This type of chucking is excellent for thin walled parts and accurate profile machining.
- *Chuck between centers* This type of chucking is useful for parts without through holes or holes that cannot be used for chucking and it produces accurate parts. However, the allowable cutting forces are limited, therefore increasing machining time. To overcome these limits, a steady rest is added to support the back side of the cutting tool edge.

7.6.2 Chucking type on milling and drilling fixtures

Designing the jigs and fixtures to chuck the workpiece in the machine is no different from any other design. Chucking location, discussed earlier, has to comply with accuracy requirements and respect the relationship between part segments. It must also ensure that the part will not move while machining, that the clamping parts will not interfere with the tool movement or increase tool travel and that there is easy access for chip removal.

Some additional points to consider are that:

- clamps should be placed and firmly set directly over the supporting surface of the fixture;
- clamps should always contact the part at its most rigid point;
- the cutting forces should be directed toward the locators and not the clamps;
- the direction of the clamping forces should be directed toward the solid locators and in such a way as to keep the part in the fixture;
- the mounting surface of the part must be flat and supported uniformly, without distortion by the fixture clamping forces;
- fixture locators should be as far apart as practically possible;
- locators should be positioned to avoid chips and foreign matters whenever possible;

- the part should be capable of being loaded into the fixture in only one position; and
- drilling bushings should be used whenever possible.

7.7 PRACTICAL EXAMPLES OF JIG AND FIXTURE DESIGNS

Before describing typical examples of jigs and fixtures, it is useful to identify for each support its contribution to the isostatic positioning of the part. Table 7.1 reviews the principal cases of partial liaisons encountered in mechanical fixtures. Schematically, the practical realization of these supports is also indicated. For example, in the case of chucking in the spindle of a lathe, the equivalent positioning is identical to a short cylindrical jig cancelling two degrees of freedom in translation, yet not opposing motions in the three directions of orientation as well as in one direction of translation. In other words, this chucking is equivalent to a short vee.

Different examples of practical fixtures are presented in relation to the principles of isostatism and with solutions for the clamping problem:

- Figure 7.20a relates to drilling operations in a rotating fixture, with the ability to process several holes. Drilling bushings are used for greater accuracy.
- Figure 7.20b shows an example of a milling fixture having a multidirectional clamping device.
- Figure 7.20c shows an example of clamping for the turning of a complex part.

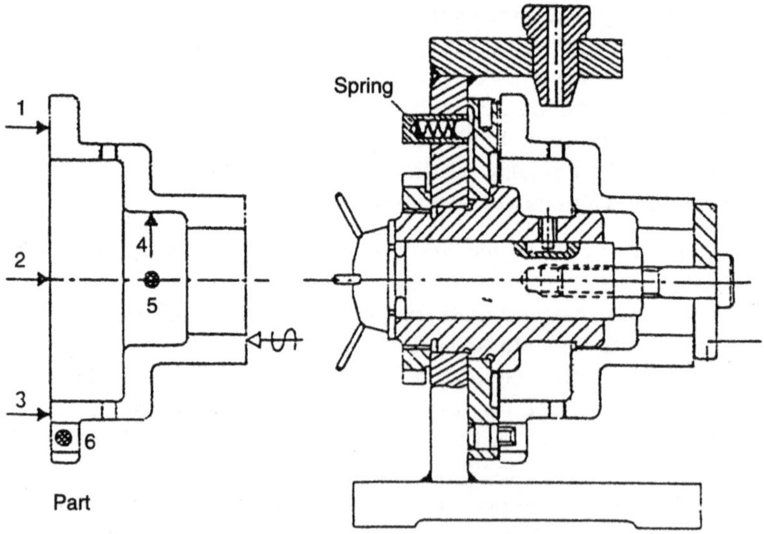

Fig. 7.20a Fixture with revolving support and positions fixed by fingers.

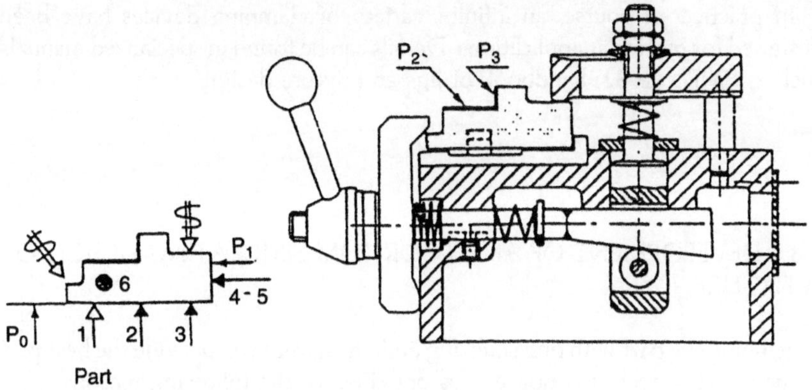

Fig. 7.20b Fixture with multidirectional clamping.

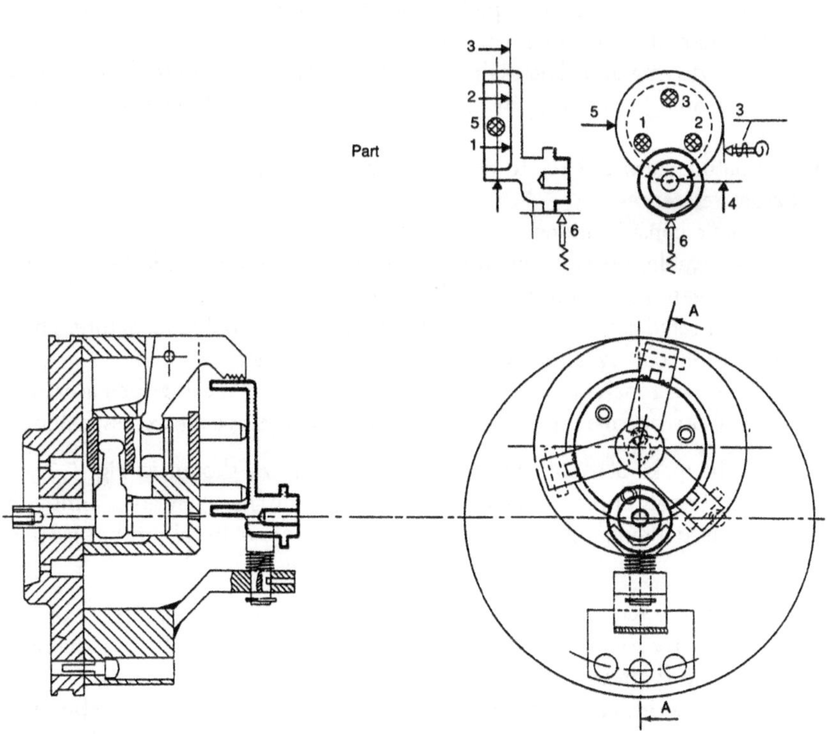

Fig. 7.20c Fixture for a complex part in turning.

In practice, of course, an infinite variety of clamping devices have been designed for particular applications. Details can be found in specialized manuals such as SME's (1982) handbook of jigs and fixture design.

7.8 DEVELOPMENT OF AN ALGORITHM SIMULATING DESIGN OF A FIXTURE

Machining the part with one chucking and one fixture will provide the best part accuracy. However, it is not always possible, for the following reasons:

- There are blind holes on both sides of the part.
- Tool entry considerations dictate that the part should be turned over.
- There is a through hole, but it is stepped on both sides in such a way that it does not allow tool entry from only one side.
- It is a round symmetrical part with slots or special features on both sides.
- The raw material shape is not conducive.
- The material removal follows a casting or a forging process. The datum surface might be between part edges so that the fixture conceals some segments that have to be machined.
- In a round symmetrical part where raw material has been cut to length before the machining operation, there is no clearance for tool entry beyond the chuck.
- The datum surfaces have to be machined initially in order to provide good datum references.
- Economic considerations.
- The allowable forces when chucking in only one location can be low, and thus increase machining time and cost.
- A complicated fixture design is needed in order to use only one fixture. Thus, the fixture cost and the fixturing time will be higher.
- Adding extra material in order to allow only one chucking increases raw material cost and may require extra machining to remove the excess material.

In order to make a sound decision, the algorithm as shown in Fig. 7.21 is proposed. The algorithm was initially devised for turning operations, although it may be used for milling as well. The main advantages of the algorithm are that it:

- considers the required tolerances and accuracies as an integral part of the decision process, i.e. it does not check the validity of the accuracy specifications after designing the process; and
- eliminates artificial constraints, i.e. forces can be controlled by selecting the appropriate cutting conditions (discussed in Chapters 8 and 9). Costs can be computed and decisions based upon figures rather than intuition alone.

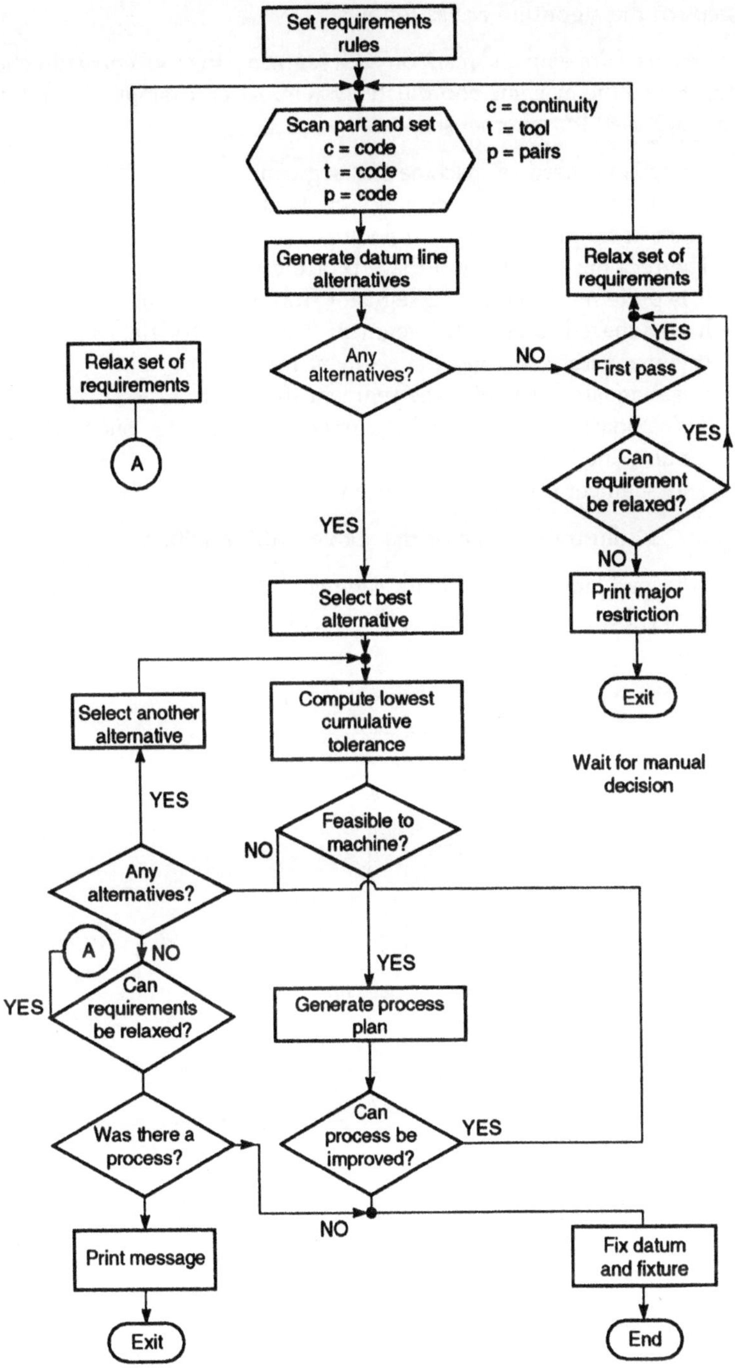

Fig. 7.21 Chucking type and chucking location algorithm.

The steps of the algorithm are as follows:

1. Scan the part drawing and mark on each segments the tool entry direction in reference to the machine coordinate system. As an example the following codes are used for external shapes:

   a – There is no need to machine this segment.
   b – Chucking on this segment is not allowed.
   c – There is no preference to tool entry.
   d – The tool must enter from the top of the segment.
   e – It is preferred to enter the segment from left to right.
   f – It is preferred to enter the segment from right to left.
   g – It is mandatory to enter the segment from left to right.
   h – It is mandatory to enter the segment from right to left.
   i – It is mandatory to machine the segment in two phases, one from right to left and one from left to right.
   j – This segment cannot be produced from either side.

   Figure 7.22 illustrates some of the above code conditions.

2. Scan the part drawing and mark segments that must be processed with continuity, i.e. without changing the tool. Let us assume that the drawing

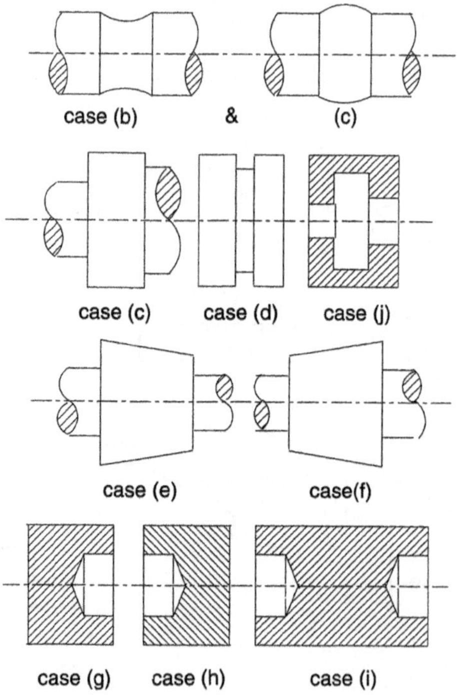

Fig. 7.22 Example of condition codes.

specifies two or more segments on one surface (or line) with only one dimension. It looks as if only one surface would be machined, then slots, pockets or other features would be superimposed on that surface.

3. Scan the part drawing and mark pairs of segments that must be machined from the same datum (because of geometric tolerances).
4. Generate alternative data references, i.e. those segments where any pair or continuity marks are located on one side of the alternate datum and where the tool entry marks change direction at the borders of this segment (Fig. 7.23).
5. If such alternatives exist, go to step 8.
6. If no such segments exist, relax the requirements of the following. Go to step 4 after each attempt.

   - *Tool entry direction* Remember that pairs and continuity might affect part accuracy, while tool entry affects economics. Choosing an alternate tool entry direction might increase machining time and cost, but will produce the part as specified.
   - *Pairs* In step 3 it was assumed that if a geometric tolerance is specified, the reference surface and the specified surface must be machined in one fixturing. The value specified was not considered. Relaxing means that the value should be examined, and a more precise decision be made.
   - *Continuity* In step 2, it was assumed that if a continuity is specified in the part drawing, the relevant segments must be machined in one fixturing.
     The tolerance of the dimension was not considered. By relaxing, it means that the tolerance of straightness, flatness etc. values should be examined, and a more precise decision made.

Tables 7.5, 7.6 and 7.7 give more detailed data concerning inaccuracies caused by refixturing a part. If the values in the tables are lower than those in the part specification, then the need for pairs or continuity might be waived.

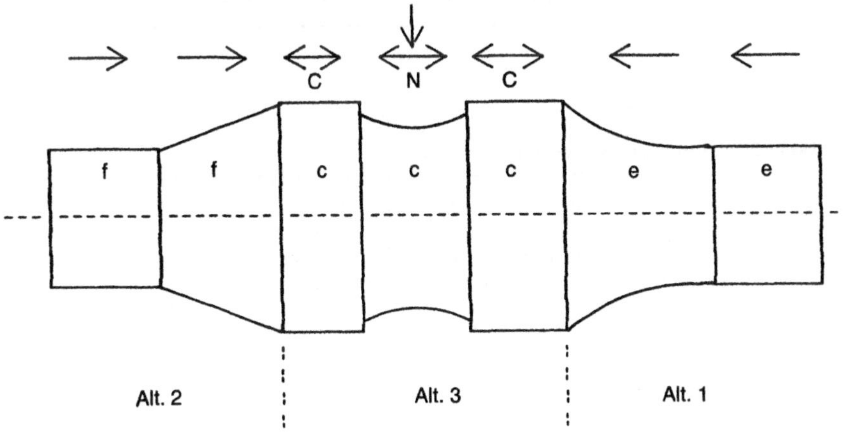

Fig. 7.23 Generating alternative segments.

Table 7.5 Radial inaccuracies in chucking

Type of material	*Diameter (mm)*					
	−10	*10–18*	*18–30*	*30–50*	*50–80*	*80–100*
Three-jaw hard chuck						
Forged	0.12	0.17	0.20	0.25	0.30	0.40
Precision casting	0.06	0.07	0.085	0.10	0.12	0.14
Rolled bar	0.04	0.045	0.055	0.06	0.075	0.085
Three-jaw soft chuck						
Machined surface	0.025	0.035	0.04	0.05	0.06	0.07
Collet						
Bar	0.04	0.05	0.06	0.07		
Machined	0.02	0.025	0.03	0.035		

Table 7.6 Axial inaccuracies in chucking

Type of material	*Diameter (mm)*					
	−10	*10–18*	*18–30*	*30–50*	*50–80*	*80–100*
Three-jaw hard chuck						
Forged	0.07	0.08	0.09	0.10	0.11	0.11
Precision casting	0.05	0.06	0.07	0.08	0.09	0.10
Rolled bar	0.04	0.05	0.06	0.07	0.08	0.09
Three-jaw soft chuck						
Machined surface	0.025	0.035	0.04	0.05	0.06	0.07
Collet						
Bar	0.02	0.03	0.04	0.05		
Machined	0.02	0.03	0.04	0.05		

Table 7.7 Inaccuracy values of form errors

Form inaccuracy in length cross-section machined with chuck.................	0.012 mm
Form inaccuracy in length cross-section machined between centers	0.020 mm
Form inaccuracy in width cross-section machined between centres............	0.010 mm
Form inaccuracy in width cross-section machined between centers............	0.015 mm

	Allowable toolwear in mm		
Diameter (mm)	*Rough cut*	*Semi-finish*	*Finish cut*
<30	0.03	0.015	0.002
30–80	0.045	0.02	0.003
80–160	0.06	0.03	0.007
160–250	0.075	0.04	0.01

7. If no relaxation is possible, or does not resolve the problem, then pinpoint the restricting requirements and suggest drawing changes. Exit this module. Note: The recommended algorithm considers only standard technology. It is possible that a skilled planner will come up with a better specific fixturing solution.
8. From among the alternatives, select a fixture datum with:
   - a criterion of the lowest stack-up of tolerances; and
   - the criterion of part maximum stability, i.e. maximum cutting conditions may be applied during machining of the part. If needed, add strengthening features to the fixture, as long as they do not interfere with tool access.
9. The purpose of this algorithm is to select the datum line or datum surface. It is the basis of the process plan and the design of the fixture. However, it does not specify all the details required for fixture design. In order to finalize fixture design decisions:
   - compute a process plan with the assumption of a given fixture datum and mechanical resistance; and
   - scan the proposed process play to determine whether cutting conditions are lowered from maximum values due to fixturing constraints. If so, add strengthening features to the fixture.

The aim of this algorithm is essentially to help process planners to make reasonable decisions. The final process plan must be completed according to the personal expertise of the process planner.

7.9 CONCLUSION – ECONOMIC CONSIDERATIONS IN FIXTURE DESIGN

Jigs and fixtures are built as an accessory to part processing, and their task is to hold the part firmly in the machine. In many cases, especially when working with small lots, their cost might be higher than the machining cost itself. To make manufacturing more economical, some methods reduce both the direct costs of jigs and fixtures and the cost of setting them up:

- *Group technology method* (GT) (Ham, Hitomi and Yishida, 1985) There are many definitions of GT. Here it means that parts with similar shapes, dimensions, producing technologies and functions are grouped together so that small and medium batches can be produced using mass production technologies. This is achieved by using common tooling and fixturing and processing the parts sequentially. In other words, group technology consists in forming groups of machines to process component families on a flow line basis or in a 'manufacturing cell'.

  To use toolings to the full, the operations must be arranged so that the maximum number of parts in the family can be processed in one setup, which means that jigs accepting all the members of the family have to be designed. For example, the design of a master jig with additional adapters is one way of dealing with changes in size, numbers and location of features.

- *Modular fixturing* (Fig. 7.24) is also a method of dealing with groups of parts. In this case, the fixture is composed of building blocks assembled in a suitable manner. A base plate has T-slots or locating holes machined at right angles. The slots are parallel to each other to ensure the accurate alignment of fixturing elements. Large structural elements such as angle plates, clamping plates and blocks are used and have the same T-slots to align, locate and attach other elements of the set. Precision position holes along with tapped holes to align, locate and secure fixturing elements are also part of the modular fixturing system. The individual blocks are very accurate: figures of 0.01 mm over a distance of 1016 mm in angularity and parallelism have been reported. However, the cost of the building blocks is quite high. Users report that the economic break-even point should be 8.2 fixtures used in production when an individual fixture costs $3000 and the initial investment is $20 000. The blocks are torn apart after use and reassembled for another fixture. Some users use a CAD sytem with a block library to design the fixtures on the screen and keep the design for future use. Others report using a polaroid camera to store the design of modular fixtures. It is reported that rebuilding a fixture after a long period of time allows a precision of positioning of 0.08 mm.

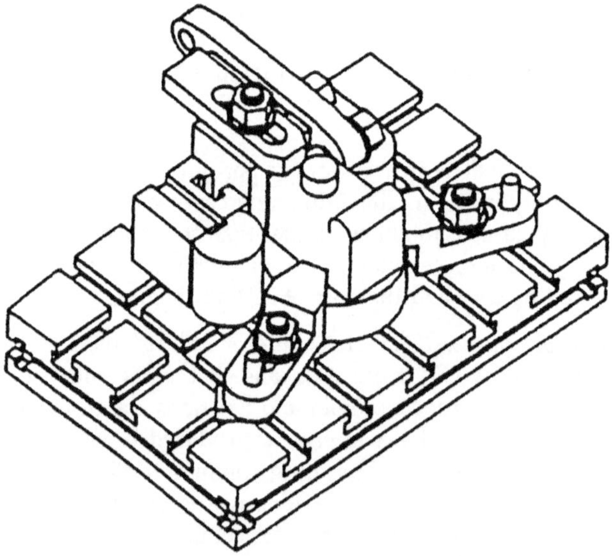

Fig. 7.24 Modular fixture.

- *Setup time reduction* The GT method and modular fixtures can also reduce the setup times. Using rotary tables allows machining of the four sides of a cube in one setup. Part accuracy improves because the part does not have to be reset. Quick clamping devices, special eccentric cams, slotted bolts and hydraulic or pneumatic clamping can also reduce handling and setup times and are therefore efficient means of improving overall productivity.

REFERENCES

Bourdet, P. and Clement, A. (1974) Optimalisation des montages d'usinage. *L'Ingenieur et le Technicien de l'Enseignement Technique*, **7**(8), 35–7.

French Standard NF 04-013, 8.85, *Symbols of workpiece holding on operation drawings.*

Ham, I., Hitomi, K. and Yishida, T. (1985) *Group technology. Application to production management*, Kluwer, Nijhoff Publishing, Boston.

ISO Standard 5459 (1981) Geometrical tolerancing, Datum and datum systems for geometrical tolerances.

SME's Handbook of jigs and fixture design. (1982) One SME Drive, P.O. Box 930, Dearborn, Michigan 48121-0930, USA.

Weill, R., Darel, E. and Laloum, M. (1991) *Influence of Positioning Errors in a Fixure on the Accuracy of a Mechanical Part*, CIRP Conference on Precision Engineering and Manufacturing Engineering, Tianjin, China, Sept. 12–14, pp. 215–55.

CHAPTER 8

How to determine the type of operation

Much energy has been devoted over the years in trying to establish theories and algorithms for complete metal cutting optimization. However, most of them have been concerned with the cutting speed parameter. It is an important parameter and it is directly proportional to the machining time. However, if the operation being performed is superfluous, is not an optimum one, or is not carried out with the correct tool, then not only is an unessential operation being performed, but it is uneconomic.

The required operations and their sequence should therefore be determined before the cutting conditions and machine have been selected. Our proposed method for doing this holds true for all metal cutting processes, although there are some variants which are unique to a specific process. In this chapter, these basic methods and their specifics are discussed.

8.1 BOUNDARY LIMIT STRATEGY

Operation selection should be based upon the capabilities of the basic process used and its boundary limits. The depth of cut is one of the most influential parameters in making this decision.

The boundary limit strategy is based on the concept that there are technical constraints and economic considerations in selecting the optimal cutting operations. The method proposed is to set the technical constraints as boundary limits, and then, bearing in mind economic considerations, to select the working point within these limits.

The following constraints are considered:

- technological constraints, based on metal cutting theories
- part constraints
- material constraints
- machine constraints
- tool constraints
- user constraints.

These constraints are used to establish boundaries of depth of cut, feed rate and cutting speed. The use of these boundary limits is discussed in section 8.4.

8.1.1 Definition of technological constraints

Based on metal cutting theories, it is known that there are minimum and maximum values for:

- depth of cut
- feed rate
- cutting speed.

Below a certain depth of cut, the metal compresses instead of forming chips and will spring back when the tool has passed. With a very low feed rate, an abrasive forming will result instead of a chip forming process. There are also maximum values for feed rate and depth of cut. Above a certain value of feed rate, the tool wear process changes and the crater wear becomes the dominant factor instead of the flank wear (Fig. 9.3). As there is no in-depth knowledge concerning crater wear and it is advisable to avoid working where there is uncertainty, one should not use a feed rate in excess of the limiting value.

Similar effects are observed for cutting speed. Above a certain cutting speed, the temperature becomes so high that the wear mechanism of the tool changes from abrasive to diffusion wear. This imposes an upper limit to cutting speed. At low cutting speeds, due to low temperature chip formation, small workpiece fragments escape from the chip and lodge in the contact area between tool and workpiece. Because of extremely high forces in this area the particles become pressure welded to the cutting edge. This effect is called BUE (built up edge) and is shown in Fig. 8.1. Occasionally, the workpiece particles break off the cutting edge, pulling carbide particles along with them and causing rapid tool wear, leaving an irregularly shaped part which may introduce vibrations and scratch the surface. Therefore, there is also a low limit to cutting speed.

The above limits are functions of the workpiece material and the cutting tool. In principle, the raw material is given, whereas the choice of cutting tool material is a decision which has to be made (discussed in Chapter 11).

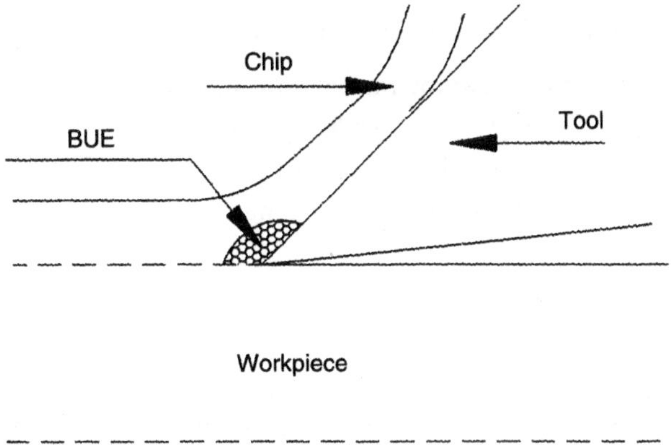

Fig. 8.1 BUE – built up edge effect.

The proposed optimization technique is to use extreme values and, in a later stage, to adjust values to the specific and most practical machine and tool.

The following boundary values are considered:

a_{tmax} – maximum depth of cut according to technological limits
a_{tmin} – minimum depth of cut according to technological limits
f_{tmax} – maximum feed rate according to technological limits
f_{tmin} – minimum feed rate according to technological limits
v_{tmax} – maximum cutting speed according to technological limits
v_{tmin} – minimum cutting speed according to technological limits.

8.1.2 Definition of part constraints

We discussed the basic part constraints to be considered when selecting the basic material removal process (Chapter 6). In this section, the dimensional and geometrical tolerances, as well as surface roughness constraints, are discussed.

It was established that the selected process can meet those specifications. However, as indicated in Chapter 6 (Table 6.4), the feasibility of the process is given as a range and not as a precise value. The selected operations and the cutting conditions determine the precise values of surface roughness or tolerances that the process will produce.

It is therefore of utmost importance to understand the effect of each decision on the outcome, and to make sure that the part will meet the drawing specifications.

In section 8.2, a detailed analysis of the parameters is made. In order to meet the specified surface roughness and tolerances, the feed rate and depth of cut should be restricted to maximum values, which are:

a_{smax} – maximum depth of cut according to part specifications
f_{smax} – maximum feed rate according to part specifications.

8.1.3 Definition of material constraints

Some researchers show that the maximum values of depth of cut and cutting speed are functions of the work material. These values will be referred to as:

a_{hmax} – maximum depth of cut depending on part material
v_{hmax} – maximum cutting speed depending on part material.

8.1.4 Definition of machine constraints

Vibrations associated with any chip formation process and affect surface finish as well as dimension accuracy. If the vibration frequency is not syncronized with the spindle revolutions, a wavy surface will be generated instead of a cylindrical one, since the relative locations of the tool and the workpiece are changing. Vibration frequencies vary between 100 Hz and 2000 Hz (hertz –

cycles per second) when using carbide tools for machining steel. In some cases it is possible to count the chatter marks on the workpiece (or rather their distance, *e*).

The following formula can be used to calculate the frequency (in Hz) of vibrations from chatter marks:

$$F_r = \frac{v}{0.06e}$$

where v is cutting speed in meters per minute and e is in millimeters.

It is customary to assume that vibration in the chipping process reduces tool life. This may occur due to instability of machine and tool, causing radial and tangential deflections as a result of the cutting forces. Under such conditions, the depth of cut and cutting speed change dynamically.

Experiments conducted indicate that cuts of very low depth of cut cause instability. This is explained by friction forces between tool and material, or by the fact that with low depth of cut, even low amplitude vibrations cause separation of contact between tool and material. This separation increases the amplitude and an unstable condition results. Some authors state 0.06 mm as the lowest limit for depth of cut. Since the machined surface might be wavy (as a result of the previous cut) and the depth of cut is not constant, a safety factor should be added.

Some experiments indicate that vibrations tend to increase with low feed rate. A critical feed rate value of 0.04 mm per revolution is set as the minimum feed rate. On the other hand, some vibration studies indicate that there is a maximum depth of cut such that above it vibration will develop. It was found that the tendency to vibrate increases with depth of cut. The critical maximum depth of cut varies from machine to machine. It is a function of machine rigidity as well as tool and workpiece rigidity.

There might be a dynamic instability when the cutting forces change. This limits the maximum cutting speed (as a function of depth of cut).

Altogether the following boundary values are to be considered:

a_{vmax} – maximum depth of cut due to chatter (7 mm)
a_{vmin} – minimum depth of cut due to chatter (0.15 mm)
f_{vmin} – minimum feed rate due to chatter (0.04 mm/rev)
v_{vmax} – maximum cutting speed due to chatter.

8.1.5 Definition of tool constraints

Research has proved that at high feed rates the width of flank tool wear is no longer the parameter that controls tool life. At high feed rates, there is plastic deformation of the cutting edge and oxidation of the secondary cutting plane. The maximum feed rate is a function of cutting speed. Recommendations are to select a maximum value range from 0.5 mm per revolution to 0.8 mm per revolution for a cutting speed of 150 m per minute.

Tools have dimensions that limit the depth of cut. It is a practical boundary when no tool with a value larger than the one used can be purchased anywhere.

Its value is considered as:

a_{kmax} – maximum depth of cut depending on tool used
f_{kmax} – maximum feed rate depending on tool used
v_{kmax} – maximum cutting speed depending on tool used.

8.1.6 Definition of user constraints

All users have their own shop practices. The tendency is to have maximum values for depth of cut, feed rate and cutting speed. The validity of such limiting values is not discussed here; the process is being generated for users whose wishes must be honored.

These values are considered as:

a_{umax} – maximum depth of cut due to user request
f_{umax} – maximum feed rate due to user request.

8.1.7 Boundary limits summary

The following boundary values have been considered:

a_{tmax} – maximum depth of cut according to technological limits
a_{vmax} – maximum depth of cut due to chatter (7 mm or 0.276 in)
a_{hmax} – maximum depth of cut depending on workpiece material
a_{smax} – maximum depth of cut according to part specifications
a_{kmax} – maximum depth of cut depending on tool used
a_{umax} – maximum depth of cut due to user request

a_{tmin} – minimum depth of cut according to technological limits
a_{vmin} – minimum depth of cut due to chatter (0.15 mm)

f_{tmax} – maximum feed rate according to technological limits
f_{smax} – maximum feed rate according to part specifications
f_{kmax} – maximum feed rate depending on tool used
f_{umax} – maximum feed rate due to user request

f_{tmin} – minimum feed rate according to technological limits
f_{vmin} – minimum feed rate due to chatter (0.04 mm/rev)

v_{tmax} – maximum cutting speed according to technological limits
v_{hmax} – maximum cutting speed depending on part material
v_{vmax} – maximum cutting speed due to chatter
v_{kmax} – maximum cutting speed depending on tool used

v_{tmin} – minimum cutting speed according to technological limits.

The minimum value of the maximum group will be considered as the working boundary limit for this group, and the maximum of the minimum group will be considered as the working boundary of the minimum group.

For example the minimum values of: a_{tmax}, a_{vmax}, a_{hmax}, a_{smax}, a_{kmax} and a_{umax} will be selected as the a_{amax}. Thus, for working boundary limits, only one maximum and one minimum will be assessed and will be designated as:

$$a_{amax}\ a_{amin} \qquad f_{amax}\ f_{amin} \qquad v_{amax}\ v_{amin}$$

8.2 ANALYSIS OF CUTTING CONDITIONS VS PART SPECIFICATIONS

In order to meet the specified surface roughness and tolerances, the feed rate and depth of cut should be restricted to maximum values. These values will be:

a_{smax} – maximum depth of cut according to part specifications
f_{smax} – maximum feed rate according to part specifications.

The part drawing specifies the required tolerances and surface roughness. These specifications must be reached in order to produce good parts. In this section, the effect of the cutting parameters on the accuracies required is discussed, the boundaries having been set in section 8.1. This section now establishes the relationship between the parameters.

8.2.1 Effect of cutting speed on surface roughness

One of the factors that control surface roughness is the primary cutting edge. Its effect is explained by the fact that, at low cutting speeds, material builds up on the tool edge (BUE – build up edge). BUE scratches the surface. As the cutting speed increases, the temperature rises, and the BUE separates from the tool. The periodic build-up and removal of the hard edge ruins the tool; it produces vibrations by lifting the tool so that it snaps back when the BUE fractures, which is detrimental to surface roughness. An additional increase in the cutting speed results in a good surface roughness. When the cutting speed increases, burn marks might appear on the machined surface.

The above is a qualitative description of the process. It is very difficult to express the above conclusions quantitatively, probably due to the fact that it is very difficult to isolate the cutting speed effect on surface roughness from other more influential factors, such as feed marks and tool geometry.

Tool material and machine rigidity are constantly improving, and the minimum and maximum values should be continuously updated. If no other values are available, it is recommended to use the value of 400 m (1312.34 ft) per minute as the upper limit and 60 m (196.85 ft) per minute as the lower limit on steel components with a carbide tool.

8.2.2 Effect of feed rate on surface roughness

Every machining process leaves its own characteristic marks on the surface. These marks are caused by many machining parameters such as true rake angle and side cutting-edge angle, cutting speed, feed rate, depth of cut, chatter, machining time, etc. Tool wear increases gradually during cut and causes a change in surface roughness, as the tool wear shape is transferred to the machined surface. Many researchers have tried to understand the surface roughness mechanism and thus to develop an equation that might be used in the prediction of surface roughness results.

Typical equations are as follows:

$$R_a = 2.95\, f^{0.7}\, r^{-0.4}\, T^{0.3} \tag{8.1}$$

where: R_a = surface roughness (μm)
f = feed rate (mm per revolution)
r = tool nose radius (mm)
T = cutting time (minutes)

or

$$R_a = 1.22 \times 10^5\, M\, f^{1.004}\, v^{-1.252} \tag{8.2}$$

where the added symbols are:
v = cutting speed (m/min)
$M = r^{-0.714}\,(\mathrm{BHN})^{-0.323}$
BHN = material hardness on the Brinell scale.

These equations are aimed at computing the surface roughness only when all other cutting parameters are known. However, the problem at hand is completely reversed. The known parameter is the surface roughness (R_a) and the need is to compute the cutting parameters. These equations are difficult to use, since they contain too many parameters with only one equation.

Most researchers agree that the main cause of surface roughness is due to feed tool marks. These marks are regularly placed and their geometry can be theoretically analyzed. There are some differences in geometry between turning, milling and hole-making operations. We address each one separately below.

Turning processes

In turning, the tool tip moves along a helical path relative to the workpiece and thus produces a thread–like surface conforming to the tool form. In the following analysis, it is assumed that the depth of cut is sufficiently large, i.e. larger than the height of the irregularities. In Fig. 8.2a, the tool is turning in the direction of the arrow, and the straight portion of the minor cutting edge is cutting on the cutting edge. Using geometric relationships and when $f/r \geqslant 2$, then the peak-to-valley roughness height h is given by:

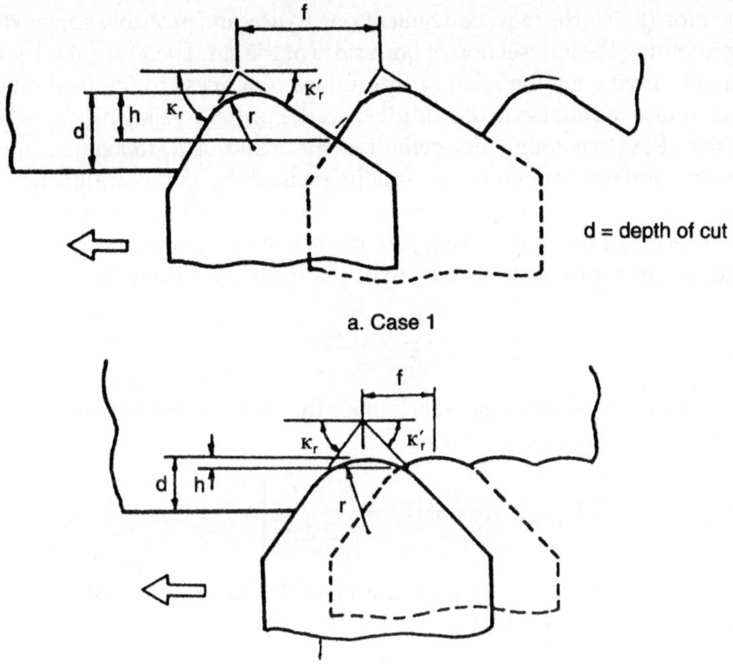

Fig. 8.2 Tool marks in turning.

$$h = f\frac{\sin\kappa_r \sin\kappa_r'}{\sin(\kappa_r+\kappa_r')} - r\left\{\frac{\cos[(\kappa_r-\kappa_r')/2]}{\cos[(\kappa_r+\kappa_r')/2]} - 1\right\} \tag{8.3}$$

where: h = peak to valley roughness height (mm)
f = feed rate (mm per revolution)
r = tool nose radius (mm)
κ_r = tool side cutting edge angle
κ_r' = tool end cutting edge angle.

In a second case, the straight portion of the minor cutting edge (the right side) is cutting on the curved portion of the cut taken by the minor cutting edge, and for $\kappa_r > \kappa_r'$ the peak-to-valley roughness height computed by geometric relationships is given by:

$$h = r(1-\cos\kappa_r') + f\sin\kappa_r'\cos\kappa_r' - \sin\kappa_r'(2rf\sin\kappa_r' - f^2\sin^2\kappa_r')^{0.5} \tag{8.4}$$

In a third case, as shown in Fig. 8.2b, when the cut is on the curvature on both sides, and when the $f \ll r$, the peak-to-valley roughness height computed by geometric relationships might be simplified in the most common case as:

$$h = \frac{f^2}{8r} \tag{8.5}$$

Equation (8.5) is the most convenient one to use, and probably comes closest to representing the intersection of both sides of the cut. Therefore, it is the most commonly used equation relating the surface roughness to the feed rate.

The surface roughness in (8.5) is given as the peak-to-valley height, which is in terms of surface roughness definition the value of R_t (maximum height). However, surface roughness is usually defined by R_a – arithmetic mean deviation (center line average). The transformation of these two values can be made by using Table 6.9 or using the mathematical computation expressions developed by approximating the arc of the circle by a parabola:

$$\frac{R_a}{R_t} = 0.256 \tag{8.6}$$

Thus to select a feed rate that will produce the specified surface roughness, the following equation should be used:

$$f = [8hr]^{0.5} = \left[\frac{8rR_a}{0.256}\right]^{0.5} \tag{8.7}$$

To simplify the use of this equation, assume that the tool nose radius will be in the range of 0.4–1.6 mm, so it will be half the value of R_a in mm or 1.6 mm. (The equation indicates that a large tool nose radius should be used. However, a large radius might cause chatter, so the radius should be limited to the above values.)

Adding a safety factor of 0.8, the following equations are recommended:

$$f = 0.8\left[\frac{8R_a}{2}\frac{R_a 10^{-3}}{0.256}\right]^{0.5} \qquad \text{for } R_a \leqslant 3.2$$

$$f = 0.8\left[\frac{8 \times 1.6 \times R_a 10^{-3}}{0.256}\right]^{0.5} \qquad \text{for } R_a > 3.2$$

or

$$f = 0.1R_a \qquad \text{for } R_a \leqslant 3.2 \tag{8.8}$$

$$f = 0.18R_a^{\,0.5} \qquad \text{for } R_a > 3.2 \tag{8.9}$$

where R_a is the value in micrometers (μm).

These equations give the value of f_{smax} for finish cuts while f_{amax} will be used for rough cuts.

It is possible to use Table 8.1 in place of these equations.

Milling processes

Milling is a very widely applied machining process. The surface roughness of face milling (Fig. 8.3) follows exactly that of turning, except that the feed rate per revolution is replaced by feed per tooth per revolution. Therefore, (8.8) and (8.9) can be used.

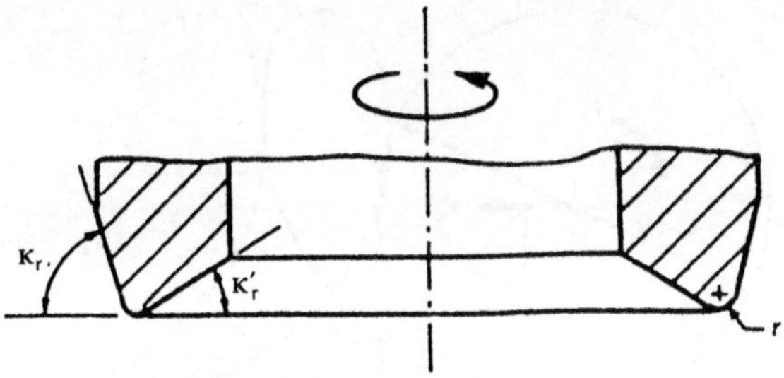

Fig. 8.3 Tool marks in face milling.

In peripheral type milling and end milling, the path of the tip of the cutter tooth is a trochoidal. Figure 8.4 shows the path for up milling and down milling. Considering the radius of curvature of the trochoid given by:

$$r = 0.5(1 \pm A)^2$$

where

$$A = \frac{v_f}{v_c} = \frac{n f_z z}{\pi D n} = \frac{f_z}{C_p}$$

v_f = feed speed of the table (mm/min)
v_c = cutting speed (mm/min)
f_z = feed rate per tooth (mm)
z = number of teeth on the tool
D = tool diameter (mm)
n = revolution per minute of the tool
C_p = circular pitch of the cutter teeth (mm)

where the (+) sign is for up milling and the (−) sign is for down milling. By geometrical relationships, a considerable amount of integration and simplification, the surface roughness is obtained by the following equation:

$$h = D\left(\frac{\pi A}{z}\right)^2 \frac{1}{8(0.5 \pm A)} \tag{8.10}$$

where h is the peak-to-valley roughness height in mm. In this case, the ratio of R_a to h is 0.227.

This equation is not convenient to use as f_z is hidden and difficult to isolate. Therefore, a simplified equation as follows is proposed:

$$f_z = \frac{0.6z}{\sqrt{D}} R_a^{0.5} \tag{8.11}$$

where R_a is the value in micrometers (μm).

This equation takes into account geometry of down milling, and some effects of cutter run-out, tool accuracy and axial displacement of individual inserts.

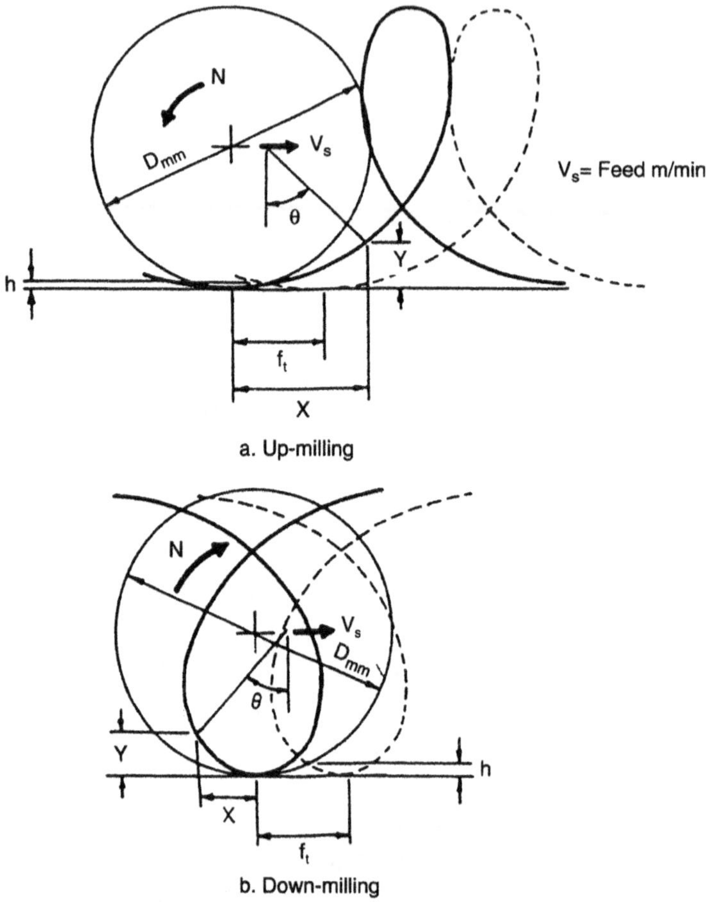

Fig. 8.4 Tool marks in peripheral type milling.

Please observe that the roughness height values in down milling are higher than those obtained in up milling. (The difference in feed rate to obtain the same surface roughness is in the range 6–12%.)

This equation gives the value of f_{smax} for finish cuts while f_{amax} will be used for rough cuts. Note that it is possible to use Table 8.2 instead of this equation. Hole making processes are discussed in Chapter 13.

8.2.3 Effect of depth of cut on surface roughness

The main cause of surface roughness comes from the feed tool marks. These marks are regularly placed and their geometry can be theoretically analyzed. This analysis has shown that the depth of cut has no effect whatsoever on surface roughness. However, this conclusion contradicts practical metal cutting observations.

Obviously, depth of cut for the finish pass should be small because of tolerances and deformation. The practical value of depth of cut varies from practitioner to practitioner and from one tool manufacture to another. The range is from 0.15 mm to 0.9 mm for a good surface finish.

There is no single equation that relates the parameters of surface roughness – depth of cut – feed rate. There are, however, some remarks and recommendations. It is recommended not to use a tool nose radius of less than 0.15 mm, to prevent breakage of the tool edge.

It is customary to increase the tool nose radius with increase of depth of cut. It is usually recommended that the radius be 10% of the depth of cut. In finish cuts, where the produced surface roughness is important, the tool nose radius should be limited to a maximum value of 1 mm in order to avoid chatter. Therefore, the depth of cut must be limited to avoid tool breakage.

The need to reduce depth of cut as a function of surface roughness is explained by the fact that as the depth of cut increases, the cutting forces increase, and therefore, deformation of the workpiece and the tool occurs, the machine structure is stressed and deflects, chatter may appear, surface integrity increases, plus many other factors that scientific analysis has not evaluated quantitatively. Therefore, to make sure that the produced workpiece will meet specifications, the depth of cut must be reduced. As it is difficult to set a rule, most users select a value, i.e. 0.5 mm as the depth of cut for the finish operation.

The need to reduce the depth of cut is mainly due to cutting forces, which are a function of workpiece material hardness. There are different degrees of surface roughness specified. Therefore, when one selects a value for finish cut, regardless of the specified surface roughness and material hardness, it seems somewhat too simplified.

It is proposed to use (8.12) to limit the depth of cut a_{smax} as a function of material hardness and surface roughness.

$$a_{smax} = \frac{32\,R_a}{BHN^{0.8}} \qquad (8.12)$$

where R_a is the value in micrometers (μm).

Equation (8.12) is a practical simplified equation (not a scientific equation). It is based on correlating the different recommended relationships of depth of cut to tool nose radius, the tool nose radius to surface roughness, the depth of cut to surface roughness and many practical recommendations of tool manufacturers and practitioners. a_{smax} is the maximum depth of cut that may be used which still produces a surface roughness as specified. A lower depth of cut (above the minimum) will probably produce a better surface roughness and is preferred as long as it does not increase manufacturing costs.

This equation gives the value of a_{smax} for finish cuts while a_{amax} will be used for rough cuts.

Tables 8.1 and 8.2 give the values of a_{smax} and feed rate as a function of the surface roughness in turning and peripheral milling.

Table 8.1 Feed and maximum depth of cut as a function of surface roughness

R_a (µm)	f_{smax} (mm/rev)	a_{smax} (mm) Brinell hardness (BHN)					
		100	200	250	300	350	400
0.8	0.08	0.64	0.37	0.31	0.26	0.24	0.22
1.0	0.10	0.80	0.46	0.39	0.33	0.30	0.27
2.0	0.20	1.61	0.92	0.77	0.67	0.59	0.53
3.0	0.30	2.41	1.38	1.16	1.00	0.89	0.80
4.0	0.36	3.22	1.85	1.54	1.34	1.18	1.06
5.0	0.40	4.02	2.31	1.93	1.67	1.48	1.33
6.0	0.44	4.82	2.77	2.32	2.00	1.77	1.59
7.0	0.48	5.63	3.23	2.70	2.34	2.07	1.86
8.0	0.51	6.43	3.69	3.09	2.67	2.36	2.12
9.0	0.54	7.23	4.15	3.48	3.00	2.66	2.39
10.0	0.57	8.04	4.62	3.86	3.34	2.9	2.65
11.0	0.60	8.84	5.08	4.25	3.67	3.2	2.92
12.0	0.62	9.65	5.54	4.63	4.01	3.5	3.18
13.0	0.65	10.45	6.00	5.02	4.34	3.8	3.45

Table 8.2 Feed vs surface roughness for peripheral milling

Feed rate (mm/rev)	Surface roughness R_a (µm)					
	D=20 z=2	D=30 z=2	D=35 z=3	D=35 z=4	D=50 z=4	D=50 z=6
0.1	—	—	—	—	—	—
0.2	0.6	0.8	0.5	—	—	—
0.3	1.2	1.9	1.0	0.6	0.8	—
0.4	2.2	3.4	1.8	1.0	1.4	0.6
0.5	3.5	5.2	2.7	1.5	2.2	1.0
0.6	5.0	7.5	3.9	2.2	3.2	1.4
0.7	6.8	10.2	5.3	3.0	4.3	1.9
0.8	8.9	13.3	6.9	4.0	5.6	2.5
0.9	11.2	16.8	8.8	5.0	7.0	3.2
1.0	14.0	20.0	10.8	6.0	8.7	3.9

Feed rate (mm/rev)	Surface roughness R_a (µm)					
	D=75 z=6	D=75 z=8	D=100 z=8	D=100 z=10	D=125 z=8	D=125 z=12
0.1	—	—	—	—	—	—
0.2	—	—	—	—	—	—
0.3	0.5	—	0.5	—	0.5	—
0.4	1.0	0.6	0.7	0.4	0.8	0.5
0.5	1.5	0.8	1.2	0.8	1.2	0.7
0.6	2.1	1.2	1.6	1.2	1.7	1.0
0.7	2.9	1.6	2.2	1.6	2.2	1.3
0.8	3.7	2.1	2.8	2.2	3.0	1.8
0.9	4.7	2.7	3.5	2.8	4.0	2.2
1.0	5.8	3.3	4.4	3.5	5.0	3.0

where: D=diameter, z=number of teeth.

8.3 OPERATIONAL AND DEPENDENT BOUNDARY LIMITS

Other types of boundary limits are those which result from process planner decisions. These decisions are based on more than one variable, e.g. depth of cut as a function of selected feed rate and depth of cut as a function of selected operation. Each boundary limit is treated separately here.

8.3.1 Depth of cut as a function of feed rate

In the previous stages, the decision of where and how to chuck the part was taken. This decision was based on the capability of holding the part while allowing tool excess and part rigidity. However, the chucking location and type may impose constraints on the allowable forces that are acting during the cutting process. These forces are:

- F_{rd} – forces that will cause movement of the part
- F_{rk} – forces that will tear the part by torsion stress
- F_{rc} – forces that will introduce compression stress in the part
- F_{rq} – forces that will exceed the allowable torsion angle
- F_{rt} – forces that will exceed the allowable torque
- F_{rb} – forces that will bend the part above the permissible value
- F_{tt} – forces that will bend the tool holder
- F_{rm} – forces that will exceed the available machine power.

Some of these forces hold true for the whole part while others are a function of the machined segment.

The above forces are reduced to three components, each one moving in the appropriate direction, the smallest one being the controlling one. They can be described by F_{rx}, F_{ry} and F_{rz}.

The cutting forces in turning can be computed (in newtons) by the following equation:

$$F_z = C_p a^u f^v \tag{8.13}$$

where: C_p = specific cutting force (for medium steel – 220)
a = depth of cut and exponent $u = 1.0$
f = feed rate and exponent $v = 0.75$.

The milling cutting forces can be computed according to:

$$F_z = K_s a B_r f_z (z/D) \tag{8.14}$$

where: F_z = total cutting force (N)
K_s = specific cutting force (N/mm²) (for medium steel ~180)
a = depth of cut (mm)
B_r = width of cut (mm)
f_z = feed rate per teeth
z = number of teeth in the tool
D = tool diameter.

The axial force (F_a), feed force (F_h) and side force (F_v) are fractions of the total cutting force.

The axial force can be estimated as 0.525 of the tangential force. The feed and side forces vary significantly with the symmetry of the tool position. It was found that the following relationships can be applied:

$$F_a = F_z \cos\left(\frac{\theta}{2}\right) \qquad \text{for } \frac{D}{B_r} > 2$$

$$F_a = F_z \cos\left(\frac{\theta}{2} + 14\right) \qquad \text{for } \frac{D}{B_r} < 2$$

$$F_v = F_z \sin\left(\frac{\theta}{2}\right) \qquad \text{for } \frac{D}{B_r} > 2$$

$$F_v = F_z \sin\left(\frac{\theta}{2} + 14\right) \qquad \text{for } \frac{D}{B_r} < 2$$

where θ is the angle of cutter engagement given in degrees (see Fig. 8.4).

The allowable exerted forces should be equated with the actual cutting forces. Thus:

$$Fx = F_{rx} \qquad Fy = F_{ry} \qquad Fz = F_{rz}$$

The cutting forces are a function of the workpiece material, which can be regarded as a constant for a given part, and the feed rate and depth of cut. These two variables are process planner decisions.

Feed rate affects machining time and cost linearly. Depth of cut effects are either step-wise or exponential.

The feed rate boundaries were set in section 8.2.2. The decision of which feed rate to use was deferred.

The unknown variable in (8.13) – similarly in (8.14) – is the depth of cut. If this equation is rearranged, it becomes:

$$a = F_z / C_p f^v \tag{8.15}$$

For the value of $f_v = f_{smax}$ for finish cut, or $f_v = f_{amax}$ for rough cut, the computed depth of cut will have its minimum value:

$$a = a_{cmin}$$

and for $f_v = f_{amin}$

$$a = a_{cmax}$$

Working with a_{cmin} will enable working with the maximum feed rate and hence the minimum machining time and cost, provided that no additional pass is necessary.

Working with a_{cmax} calls for working with minimum feed rate, resulting in a longer machining time and higher costs.

8.3.2 Depth of cut as a function of a selected operation

Any cutting operation leaves its marks on the surface of the machined workpiece. In the finish cut, care is taken that these marks will conform to part specifications (by selecting the feed). However, the limiting boundary values may not allow the finishing of the part in one operation; a rough cut or a semi-rough and semi-finishing may be needed. In the next section, an algorithm to determine the operations is presented. It is based upon the need for an intermittent operation to adjust the surface between a very heavy depth of cut and a fine low depth of cut. The basis for this algorithm is discussed in this section.

The inaccuracies in any non-finishing machining operations are as follows:

- *The dimensional tolerance obtained* Rough cut operations have economic considerations as their main concern. The tolerance used may cause more removal of material than that planned for the final pass. Hence, a scraped part will be produced.
- *The surface integrity* In any machining operation, strain and stress appear on the machined surface; this is called work hardening. Stress penetration and stress level increase with feed (and at a very low feed rate, about 0.04 mm per revolution). In a non-finishing cut, it is recommended to use as large a feed as possible, for reasons of economy. The work hardened layer must be removed from the finished part.
- *Surface roughness* In section 8.2.2, the effect of feed on surface roughness was discussed in order to propose a method of selecting a feed that will conform with part specifications. In a non-finishing operation, the maximum feed is used, having a rough surface. Its value can be computed by rearranging (8.9) and (8.11). This roughness must be removed by the finishing operation without reducing the depth of cut to a value below the minimum.
- *Geometric tolerances* A non-finishing operation is usually a heavy operation with high cutting forces which cause a deflection of the tool and the workpiece. Such deflection will reduce geometric tolerance properties and increase non-straightness, non-flatness, etc. These inaccuracies must be removed by the finishing operation without reducing the depth of cut to a value below the minimum.
- *Miscellaneous* All these inaccuracies have to be removed by the finishing cut. Since there is no established method of quantifying them, the following practical Table 8.3 covers all the inaccuracies mentioned, in turning or face milling.

Table 8.3 a_{smin} as a function of feed and depth of cut

Feed (mm/rev)	Depth of cut (mm) 1.0	1.5	2.0	2.5	3.0	3.5	4.0
0.10	0.10	0.13	0.15	0.20	—	—	—
0.20	0.15	0.17	0.19	0.23	0.28	—	—
0.30	0.17	0.20	0.25	0.30	0.39	0.43	—
0.40	—	0.28	0.30	0.38	0.46	0.54	0.60
0.50	—	0.32	0.36	0.45	0.54	0.63	0.71
0.60	—	—	0.41	0.52	0.62	0.73	0.82
0.70	—	—	0.46	0.58	0.70	0.82	0.92
0.80	—	—	0.51	0.64	0.77	0.90	1.01
0.90	—	—	—	0.81	0.97	1.13	1.30
1.00	—	—	—	1.00	1.20	1.40	1.60

Feed (mm/rev)	Depth of cut (mm) 5.0	6.0	7.0	8.0	9.0	10.0	11.0
0.10	—	—	—	—	—	—	—
0.20	—	—	—	—	—	—	—
0.30	0.61	0.74	0.86	—	—	—	—
0.40	0.76	0.92	1.06	1.24	1.37	—	—
0.50	0.90	1.08	1.26	1.46	1.61	1.80	2.00
0.60	1.03	1.24	1.45	1.65	1.85	2.06	2.27
0.70	1.16	1.39	1.62	1.85	2.08	2.31	2.55
0.80	1.28	1.54	1.79	2.05	2.30	2.56	2.82
0.90	1.62	1.98	2.27	2.59	2.92	3.24	3.56
1.00	2.00	2.40	2.80	3.20	3.60	4.00	4.40

For peripheral milling, use Table 8.3 with a correction factor (C) taken from the following:

C_p	C
1.63	0.65
1.77	0.71
2.00	0.78
2.30	0.92
2.53	1.00

where $C_p = \pi D/z$, the circular pitch of cutter teeth (in mm).

The equations and treatment of hole-producing operations are discussed in Chapter 13.

The amount of material that must be removed by the finishing cut is symbolized by a_{smin}. Its value has to be added to a_{amin} and both should be equal or smaller than a_{smax}, which is the maximum depth of cut allowed for the finishing cut. a_{smax} value can be computed by (8.12) or Table 8.1.

8.4 THE ALGORITHM FOR SELECTING CUTTING OPERATIONS

The purpose of the algorithm is to deliver the optimum operations for machining. As an axiom, the optimum is reached by minimizing the number of cutting passes while finding values of cutting conditions as high as possible, giving preference to the feed according to well-known optimization strategies.

The depth of cut boundaries developed in the previous sections are used here. They are:

- a_{amax} – maximum depth of cut allowed – see section 8.1.7;
- a_{cmax} – maximum computed depth of cut of a segment when using minimum feed rate, computed by (8.15): if $a_{cmax} > a_{amax}$ then $a_{cmax} = a_{amax}$;
- a_{amin} – absolute minimum depth of cut – see section 8.1.7;
- a_{cmin} – minimum computed depth of cut of a segment when using maximum feed rate, computed by (8.15) – if $a_{cmin} > a_{amax}$ then $a_{cmin} = a_{amax}$;
- a_{smax} – maximum depth of cut that will produce the specified surface roughness, computed by (8.12), or Table 8.1;
- a_{smin} – minimum depth of cut that must be used in order to remove inaccuracies of the previous cutting pass, or inaccuracies of the raw material – retrieve from Table 8.3; and
- a_s – previous depth of cut that results with a_{smin} value – retrieve from Table 8.3 (step 4).

Figure 8.5 shows these boundaries on a part segment.

W_G is the dimension of any intermediary cut or raw material.
W is the required dimension of the finished part.
a_n is the amount of material to be removed:

$$a_n = W_G - W$$

The most economic method of machining a part consists in using a single pass, i.e. to use a depth of cut of a_n. However, there are boundaries and a constraint to be considered. Therefore, the following steps are proposed:

Step 1. The constraint imposed by surface condition must be fulfilled. Retrieve from Table 8.1 the value of a_{smax} as a function of material hardness in Brinell and the surface roughness R_a.

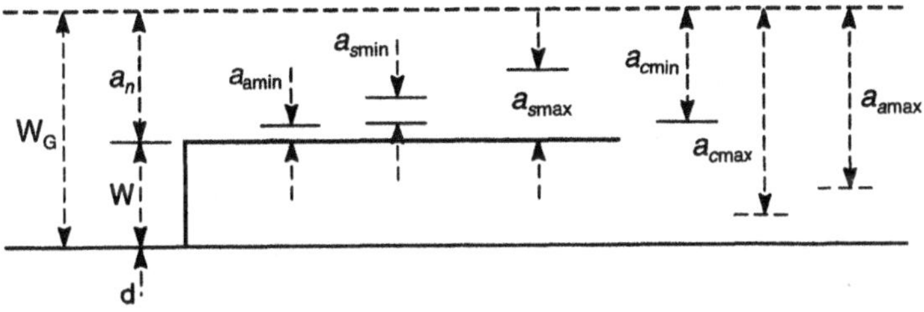

Fig. 8.5 Boundaries in part segment.

Step 2. Compute a_n. Compare a_n to a_{smax} (Fig. 8.6a).
If $a_n < a_{smax}$, proceed to step 10.
Else go to step 3.

Step 3. Compute the value of a_{sm} by: $a_{sm} = a_{smax} - a_{amin}$.

Step 4. Retrieve from Table 8.3 the values of a_s.
Scan the columns of the table, from high depth of cut to low (right to left) searching for a row with as high a feed rate as possible which has the value of a_{smin} or lower. The depth of cut of that column is a_s.
There might be several entries of depth of cut (a_s) in Table 8.3 which comply with the value of a_{smin}. In a first attempt, choose the one with the maximum a_s. If $a_s > 1.33\ a_{cmin}$ then $a_s = 1.33\ a_{cmin}$ (Fig. 8.6b).

Step 5. Compute $P_G = a_n - a_{smax}$.

Step 6. Compare P_G to a_s.
If $P_G \leqslant a_s$, proceed to step 11 (Fig. 8.6c).
If $P_G > a_s$, proceed to step 7 (Fig. 8.6d).

Step 7. Compute B_i.
In this case at least three cutting passes are required, as an intermediary cut is needed between the rough cut and the finish cut. The number of cuts will be determined by computing:

$$B = \frac{a_n - a_s}{1.33 \times a_{cmin}}$$

and rounding it up to the closer upper integer (B_i).
(i.e. 2.01 is rounded to 3 and 2.99 is rounded to 3).
The number of cutting passes will be B_i + semi-finish + finish cut.

Step 8. To determine depth of cut distribution, compute:

$$a_p = (a_n - a_s)/B_i$$

Step 9. If $a_p \leqslant a_{cmin}$, proceed to step 12 (Fig. 8.6e).
If $a_p > a_{cmin}$, proceed to step 13 (Fig. 8.6f).

Step 10. $a = a_n$; f is taken from Table 8.1 (Fig. 8.6a).
One pass may reduce the dimension to the one required. A check should be made in the case of raw material (cast or forged) to assure that $a_n > a_{smin}$. If not, it is impossible to produce a good part.
Exit the algorithm.

Step 11. See Fig. 8.6c.
In this case two cutting passes can produce the segment.
Select from Table 8.3 the column with a depth of cut value of P and a maximum feed rate to comply with the value of a_{smin}. Mark the new a_{smin} value by Q.
Finish cut – $a_0 = a_{amin} + Q$. f is taken from Table 8.1.
Rough cut $a_1 = a_n - a_0$. f is taken from Table 8.3.
Exit the algorithm.

Step 12. See Fig. 8.6e.
The $B_i - 1$ rough cut will be with depth of cut $a = a_{cmin}$, $f = f_{amax}$.

The last rough cut will be $a_l = a_n - 1.37$, $f = f_{amax}$.
Semi-finish cut: $a = 1$ mm, f is taken from Table 8.3.
Finish cut: $a = 0.37$ mm, f is taken from Table 8.1.
Exit the algorithm.

Step 13. See Fig. 8.6f.
The $B_i - 1$ rough cut will be of depth of cut $a = a_p$.
The feed rate will be

$$f_{amax} \times \left(\frac{a_{cmin}}{a_p} \right)^{0.75}$$

Compute $P_G = a_n - B_i \times a_p - a_{smax}$.
Proceed to step 11.

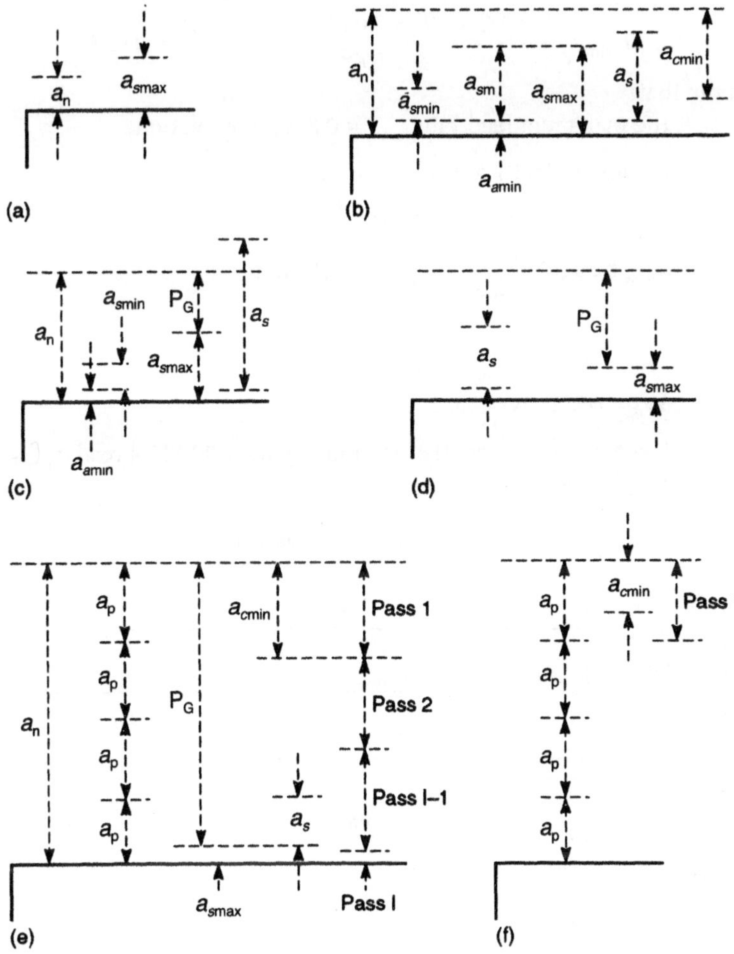

Fig. 8.6 Cases in depth of cut decisions.

8.5 EXAMPLES OF USING THE ALGORITHM

Example 1

A part as shown in Fig. 8.7 has to be machined. Let us assume that the allowed deflection in the rough cut is $\delta = 0.1$ mm, and in the finish cut, 20% of the tolerance $\delta = 0.08 \times 0.2 = 0.016$ mm.

The allowable acting forces on the part during machining will be computed as a beam at one end and loaded at the other.

$$\delta = \frac{FL^3}{3EI} \quad \text{or} \quad F = \frac{3EI\,\delta}{L^3}$$

$$F_{rough} = \frac{3 \times 2.2 \times 10^4 \times 0.049 \times 45^4 \times 0.1}{100^3} = 1326 \text{ N}$$

$$F_{finish} = \frac{3 \times 2.2 \times 10^4 \times 0.049 \times 45^4 \times 0.016}{100^3} = 212 \text{ N}$$

Assume that

the maximum feed rate $f_{amax} = 0.8$ mm/revolution

minimum depth of cut $a_{amin} = 0.2$ mm

$a_{amax} = 8.0$ mm.

$$a_{cmin}\ (\text{rough}) = \frac{1326}{220 \times 0.8^{0.75}} = 7.12 \text{ mm}$$

$$a_{cmin}\ (\text{finish}) = \frac{212}{220 \times 0.8^{0.75}} = 1.14 \text{ mm}$$

$W_G = 65/2$; $W = 45/2$.

Step 1. From Table 8.1 row 0.8 μm and column 200 BHN read $a_{smax} = 0.37$; $f_{smax} = 0.08$

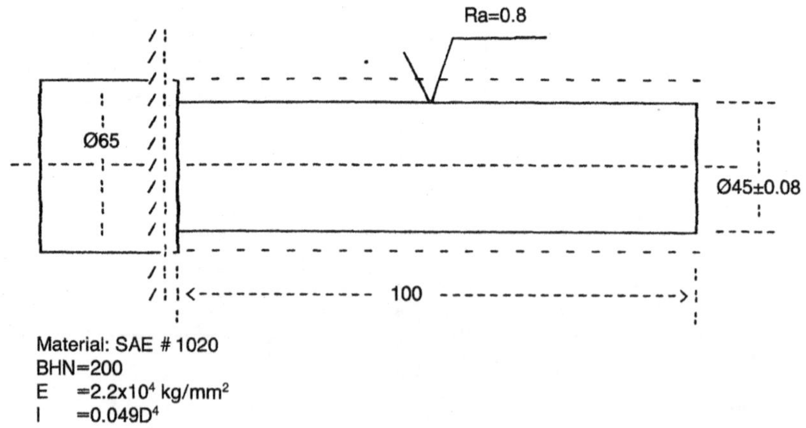

Fig. 8.7 Example of part to be machined.

Step 2. $a_n = (65-45)/2 = 10$
In this case $a_n > a_{smax}$ $(10 > 0.37)$

Step 3. $a_{smin} = a_{smax} - a_{amin} = 0.37 - 0.2 = 0.17$

Step 4. From Table 8.3 in the third column, a value of $a_{smin} = 0.15$ and thus $a_s = 2.0$ mm.

$$1.33\ a_{cmin} = 1.33 \times 7.12 > a_s = 2.0$$

Step 5. $P_G = a_n - a_{smin} - a_{amin} = 10.0 - 0.17 - 0.2 = 9.65$ mm

Step 6. Compare $P_G - a_s$: $9.65 > 2$; proceed to step 7.

Step 7.

$$B = \frac{a_n - a_s}{1.33 \times a_{cmin}} = \frac{10-2}{1.33 \times 7.12}$$

$B = 0.84$
Round to integer: $B_i = 1$

Step 8.

$$a_p = \frac{a_n - a_s}{B_i} = 8$$

Step 9. Compare $a_p > a_{cmin}$: $8 > 7.12$; proceed to step 13.

Step 13. The decisions are:
Rough cut: $a = a_p = 8.00$ mm

$$f = f_{amax}\left(\frac{a_{cmin}}{a_p}\right)^{0.75}$$

$$= 0.8\left(\frac{7.13}{8}\right)^{0.75} = 0.73 \text{ mm/rev}$$

$P_G = 10.0 - 1 \times 8 - 0.17 - 0.2 = 1.63$ mm
New $a_n = a_n - B_i \times a_p = 10 - 1 \times 8 = 3$
Proceed to step 11.

Step 11. From Table 8.3 (interpolation) $Q = 1.63$; $f = 0.2$
Finish cut: $a_0 = a_{amin} + Q = 0.37$ mm; $f = 0.10$ mm/rev
Semi-finish cut: $a_1 = a_n - Q = 1.63$ mm; $f = 0.20$ mm/rev
Decisions summary:
op. 010: depth of cut = 8.00: feed rate = 0.73
op. 020: depth of cut = 1.63: feed rate = 0.20
op. 030: depth of cut = 0.37: feed rate = 0.10.

Example 2

A part as shown in Fig. 8.8 has to be machined. Let us assume that the allowed deflection in the rough cut is $\delta = 0.1$ mm, and in the finish cut, 20% of the tolerance $\delta = 0.08 \times 0.2 = 0.016$ mm.

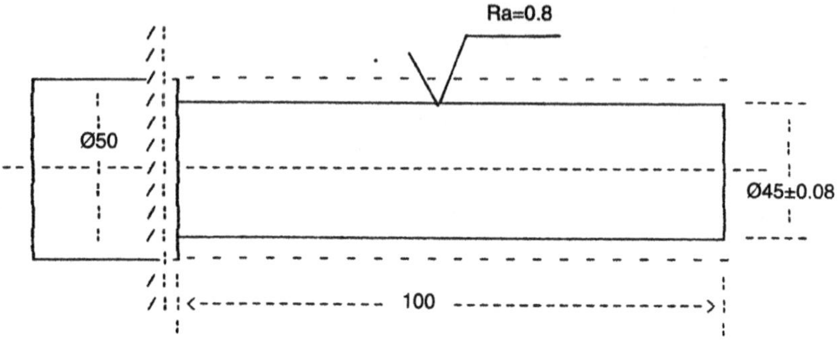

Fig. 8.8 Example of part to be machined.

The allowable acting forces on the part during machining will be computed as a beam fixed at one end, loaded at the other end.

$$\delta = \frac{FL^3}{3EI} \quad \text{or} \quad F = \frac{3EI\,\delta}{L^3}$$

$$F_{rough} = \frac{3 \times 2.2 \times 10^4 \times 0.049 \times 45^4 \times 0.1}{100^3} = 1326 \text{ N}$$

$$F_{finish} = \frac{3 \times 2.2 \times 10^4 \times 0.049 \times 45^4 \times 0.016}{100^3} = 212 \text{ N}$$

Assume that

the maximum feed rate is $f_{amax} = 0.8$ mm/revolution
minimum depth of cut $a_{amin} = 0.2$ mm
$a_{amax} = 8.0$ mm
a_{cmin} (rough) $= 1326/(220 \times 0.8^{0.75}) = 7.12$ mm
a_{cmin} (finish) $= 212/(220 \times 0.8^{0.75}) = 1.14$ mm
$W_G = 50/2$; $W = 45/2$.

Step 1. From Table 8.1, row 0.8 μm and column 200 BHN
$a_{smax} = 0.37$; $f_{smax} = 0.08$ mm/rev.
Step 2. $a_n = (50 - 45)/2 = 2.5$
$a_n > a_{smax}$.
Step 3. $a_{smin} = a_{smax} - a_{amin} = 0.37 - 0.2 = 0.17$
Step 4. From Table 8.3 in the third column a value of $a_{smin} = 0.15$ and thus $a_s = 2.0$ mm.
Step 5. $P_G = a_n - a_{smin} - a_{amin} = 2.5 - 0.15 - 0.2 = 2.15$ mm
Step 6. Compare $P_G > a_s$, $2.15 > 2$

Step 7.

$$B = \frac{a_n - a_s}{1.33 \times a_{cmin}} = \frac{(2.5 - 2)}{(1.33 \times 7.12)}$$

$B = 0.053.$
$B_i = 1.$

Step 8.

$$a_p = \frac{(a_n - a_s)}{B_i} = 0.5.$$

Step 9. $a_p < a_{cmin}$; $0.5 < 7.12$; proceed to step 12.
Step 12. Rough cut $a = a_n - 1.37$; $a = 2.5 - 1.37 = 1.13$ mm
$f = f_{amax} = 0.80$ mm/rev
Semi-finish cut: $a = 1.0$ mm; $f = 0.30$ mm/rev
(From Table 8.3 for $a = 1$ and $a_{smin} = 0.17$, $f = 0.3$)
Finish cut: $a = 0.37$ mm; $f = 0.10$ mm/rev.

Decisions summary:
op. 010: depth of cut = 1.13: feed rate = 0.80
op. 020: depth of cut = 1.00: feed rate = 0.30
op. 030: depth of cut = 0.37: feed rate = 0.10.

Example 3

The height of a block of SAE 1020 steel of hardness of 200 Brinell has to be machined from a dimension of 32 mm to 30 ± 0.15 with R_a 6.4 μm. Recommend the necessary operations.

Assume that
the maximum feed rate is $f_{amax} = 0.6$ mm/rev
minimum depth of cut: $a_{amin} = 0.2$ mm
$a_{amax} = 5.0$ mm
$a_{cmin} = 5.0$ mm

Step 1. Retrieve from Table 8.1, row of 6.0 μm and 200 BHN
$a_{smax} = 2.77$ mm; $f_{smax} = 0.44$ mm/rev.
Step 2. $a_n = 2.0$ (given). Compare a_n and a_{smax}: $2.0 < 2.77$ so proceed to step 10.
Step 10. Make one cutting pass with:
$a = 2.0$ mm; $f_z = 0.44$ mm/teeth.

CHAPTER 9

How to select cutting speed

In Chapter 8, operation planning was analyzed. It was shown that selecting an operation plan is not an independent decision; it depends on the selected depth of cut. However, the decision of selecting a depth of cut was partially based on the feed rate decision. Moreover, it will be shown that selecting a cutting speed depends on the depth of cut and feed rate.

At this point, the unknown variables of selecting cutting speed are the tool grade and machine power. Tool selection is discussed in Chapter 11 after selecting cutting speed and a machine. The logic of this sequence of decisions is that there is a huge variety of tools; once the characteristics of the tool requirements are known, an appropriate tool will no doubt be found. Selecting a tool before all parameters are known will impose artificial constraints on the decisions which follow.

The selection of a machine, with its power constraint, is discussed in Chapter 10. The logic of this sequence of decisions is based on the fact that, before deciding the cutting conditions, the required power of the machine, its torque value and the speed range are unknown. Taking a decision without this data will introduce an artificial constraint. At this stage, the only known parameters for selecting a machine are its physical sizes.

The decision on the cutting speed selection has a direct bearing on the economics of machining, as can be seen from (9.1).

$$T = \frac{L}{z \times n \times f_z} = \frac{\pi D \times L}{v_c \times f_z} = \frac{L}{v_f} \tag{9.1}$$

where: T = machining time (min)
L = tool travel length (mm)
n = number of revolutions per minute
f_z = feed rate (mm/tooth)
z = number of teeth per tool
D = tool or part diameter (mm)
v_c = cutting speed (mm/min)
v_f = feed speed in mm/min ($= z \times n \times f_z$).

Naturally, the higher the cutting speed, the shorter the machining time and thus the machining cost.

There are several sources the process planner may use in selecting cutting speed.

Machining data handbooks

One of the major handbooks is the *Machining Data Handbook* published by Machinability Data Center, Metcut Research Associate Inc., Cincinnati, Ohio. This organization, as its name implies, has devoted its research to establishing recommended machinability data. Its purpose is to provide starting data recommendations for important machining situations. In the preface, it states that:

> The data herein were collected from many industry sources and from extensive searches of technical literature. Data were then evaluated by technical personnel in order to provide realistic machining parameters for general application in large and small shops.

The handbook is arranged by machining types such as, turning (single point and box tool, ceramic tools), milling operations (face milling and end milling) – slots. Under each type of operation is a list of 38 material groups, including their hardness and condition. Then, for a given depth of cut, a tool material, feed rate and cutting speed are recommended. In a separate chapter, cutting fluids and tool geometry are recommended.

A similar handbook was published in France by Centre Technique des Industries Mécaniques (CETIM). There is a separate volume for each type of machining operation, including an illustration of the tool and the workpiece under operation. Extra data is given for tool diameter selection in milling, the number of teeth per tool, computed feed speed per minute and estimated power requirements.

In Germany, the Technical University of Aachen has developed a machining database called INFOS, which includes collected and tested data from industry. It can be retrieved by computer inquiry, or by addressing the institute and receiving an hardcopy with the (very detailed) recommendations. Extra information as to tool life and wear land size VB (a definition of tool life) is added in the tables, as well as parameters to compute machining power.

Machinability ratings

In this methodology, materials are classified in terms of their relative machining speeds. Various metals are first classified in major groups numbered 1 through 10. Materials in group 1 are considered the most difficult to machine; materials in each successive group are easier to machine than the previous group. Group 8 (B1112 steel) was used as a percentage base for the first eight groups and was assigned a machinability rating of 100%. Groups 9 and 10 were assigned machinability percentages of 130 and 225, respectively.

The basic cutting conditions recommendations vary for different types of metal removal operations.

Technical books

Cutting conditions recommendations can be found in many technical books and in general handbooks, such as the *Tool and Manufacturing Engineers' Handbook* and *The Machinery Handbook.*

Tool manufacturers

Usually tool manufacturers supply technical data for their products. The given data varies from supplier to supplier and from one catalog to another. In some cases, one finds recommendations given with very wide ranges, such as tool steel, hardness range 200–330 BHN; surface speed for finishing cut 45–250 m/min. Others might provide very detailed information, such as ISCAR data.

Machinability computerized systems

Several attempts have been made to computerize the machining data. A computerized system has the following advantages over a book type databank:

- It may store data from different sources.
- It may use shop parameters instead of theoretical and general data.
- It concentrates accumulated experience in an easily accessible form.
- The database may be kept up to date.
- Fast retrieval of selected data is possible.
- Rapid optimization computations are possible.
- Comparison of alternative cutting conditions is easy.
- Calculation of standard machining and geometrical formulae is provided.
- It displays recommendations in a concentrated form.

The machinability computerized systems allow for user specification of material data, machine group speeds and feeds and operation factors. This provides results which are consistent with user machining practices. Figure 9.1 shows a data sheet for collecting data for end mill operation.

These different sources of cutting speed information are nominal recommendations and should be considered only as good starting points, as they do not – and cannot – consider variables such as part configuration, specifications of the machine, type of fixturing, dimensional tolerance, surface finish requirements, etc.

In the following approach, some of these variables are known, or have been decided beforehand, and therefore an optimum performance may be sought. This is discussed below.

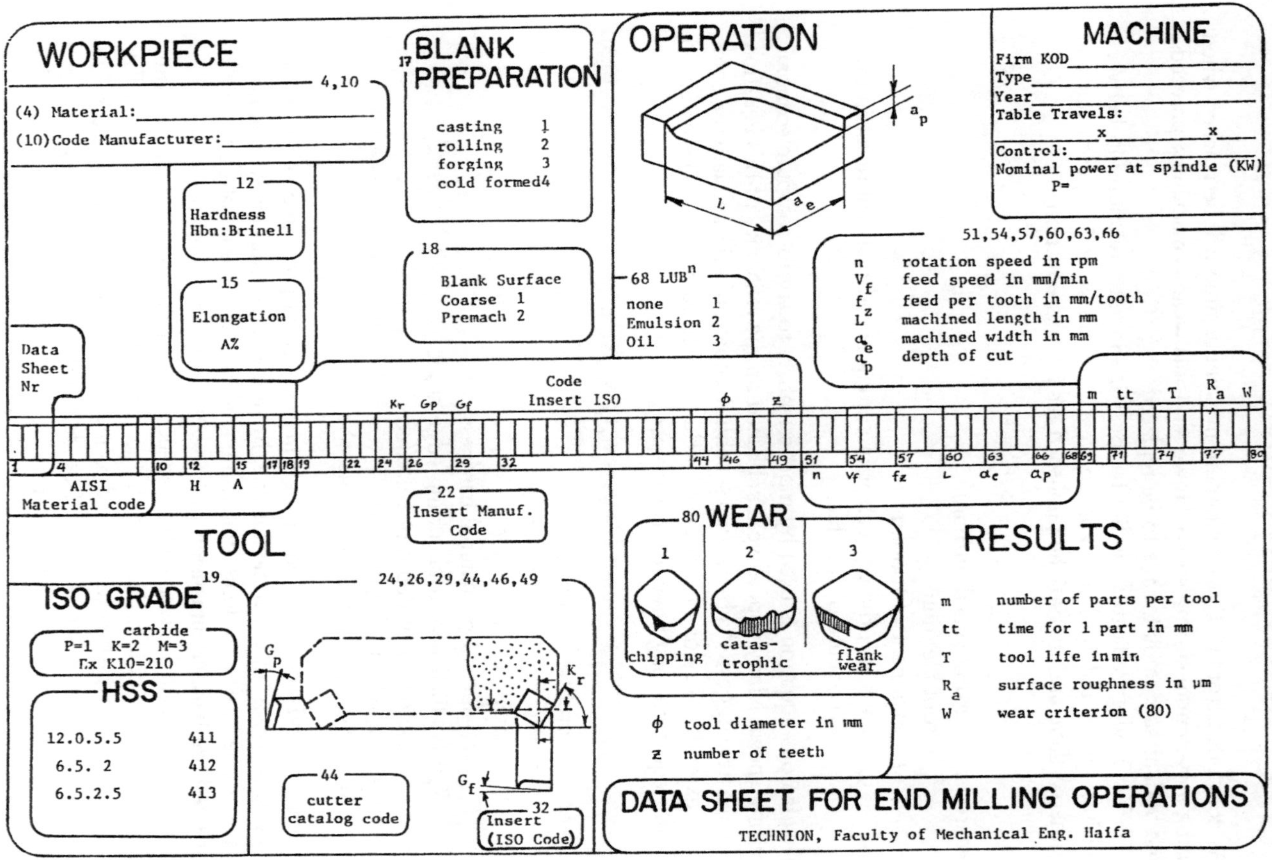

Fig. 9.1 Data sheet for end milling operation.

9.1 CUTTING SPEED OPTIMIZATION

As concluded from (9.1), the minimum machining time will be reached when using a maximum cutting speed. However, tools wear as machining proceeds, and will, at some point, have to be replaced.

F. W. Taylor dedicated his life to investigating the relationship between cutting speed and tool life. For 26 years, he conducted experiments and gathered data, and in 1907 he published his famous equation:

$$v_c T_L^n = C_v \tag{9.2}$$

where: v_c = cutting speed (m/min)
T_L = tool life (min)
n = exponent of tool life
C_v = constant; the equivalent of cutting speed that will result in one (1) minute of tool life.

This famous equation is used by most scientists today to compute economic cutting speed. Many papers were published on the reliability of this equation. No one has suggested a better or a substitute equation, although latterly, some have extended Taylor's equation to include other parameters. The extended Taylor equation may have forms such as:

$$v_c T_L^n a^q f^p = C_v V_B^m K_{mv} K_\tau \tag{9.3}$$

where a = depth of cut (mm)
f = feed rate (mm/revolution)
V_B = flank wear value (mm)
K_{mv} = coefficient of material hardness
K_τ = coefficient of tool cutting edge angle
q = exponent of depth of cut effect
p = exponent of feed effect
m = exponent of tool wear effect.

The right-hand side of the equation is in many cases presented as a constant. This equation is an experimental one, therefore dimension analysis is not in place. If one insists on doing dimension analysis, then the constant C_v will be assigned the balance dimension.

This relationship between cutting speed and tool life indicates that, as cutting speed increases, tool life decreases. In fact, any cutting speed may be used. The decision is not which cutting speed to use, but which tool life to select. The machining direct cost or time will be decreased while the tooling cost and change time will increase. This effect can be seen in Fig. 9.2 which indicates that there is an optimum tool life (and thus an optimum cutting speed). An increase or decrease of cutting speed from the optimum will increase the cost or machining time.

The common criteria of optimization are minimum cost or maximum production. They can be computed by using Taylor's equation. It is interesting

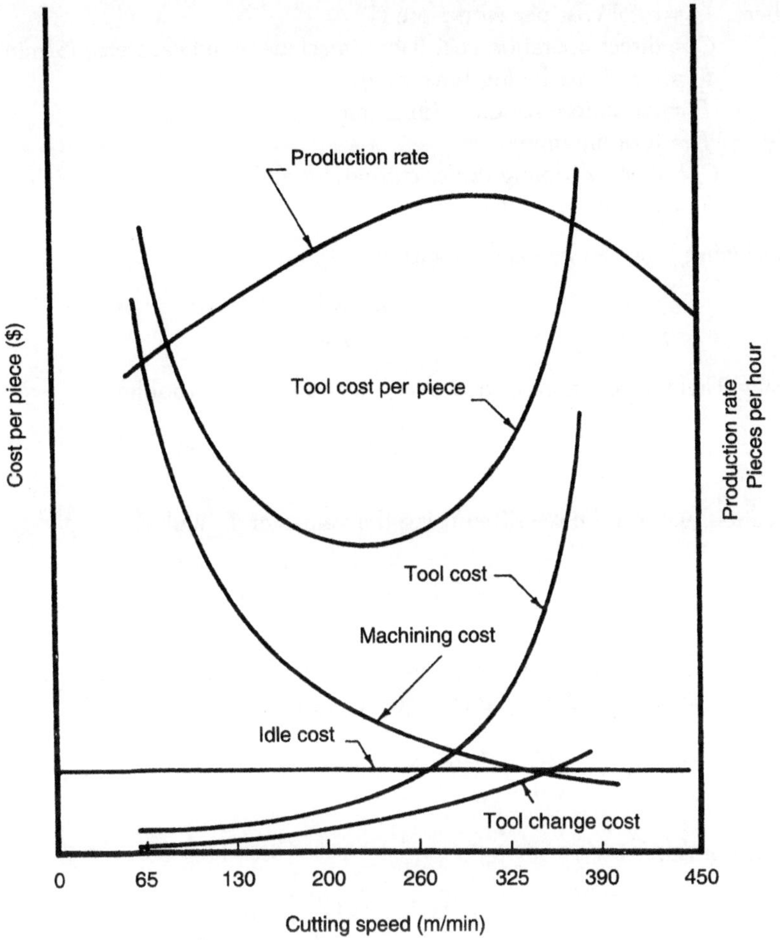

Fig. 9.2 Optimization cutting speed curve.

to note, that whether the original equation or the extended equation is used as the basis for optimization, the same results for optimum tool life will be obtained. The optimization is for tool life and not for cutting speed. The optimum tool life will be introduced to the preferred cutting speed equation, (9.2) or (9.3) and the optimum cutting speed will be computed.

The optimization (as shown in Fig. 9.2) can be obtained by standard mathematical optimization procedures. For the criterion of minimum cost, an equation of total cost is set as in (9.4)

$$C_t = C_d T_m + C_d T_h + \frac{T_m}{T_L}(C_w + C_d T_c) \tag{9.4}$$

where: C_t = total cost per workpiece (\$)
C_d = direct operation cost (labor, machine, overheads etc.) (\$/min)
T_m = direct machining time (min)
T_h = workpiece handling time (min)
T_L = tool life (min)
C_w = cost of tooling (\$ per cutting edge)
T_c = tool change time (min)

Machining time can be expressed as in (9.1).

$$T_m = \frac{\pi D \times L}{v_c \times f}$$

Using Taylor's equation as in (9.2) and rearranging for tool life T_L

$$T_L = \frac{C_v^{1/n}}{v_c^{1/n}}$$

equation (9.4) can be rewritten using the values of T_m and T_L:

$$C_t = C_d \frac{\pi DL}{v_c f} + C_d T_h + \left(\frac{\pi DL}{v_c f} \frac{v_c^{1/n}}{C_v^{1/n}} \right) (C_w + C_d T_c) \tag{9.5}$$

In order to arrive at a minimum cost, differentiation of (9.5) with respect to cutting speed (v_c) is done and set equal to zero:

$$\frac{dC_t}{dv_c} = -C_d \frac{\pi DL}{v_c^2 f} + \frac{\pi DL}{C_v^{1/n} f} (C_w + C_d T_c) \left\{ \left(\frac{1}{n} - 1 \right) v_c^{(1/n-2)} \right\} = 0$$

Rewritten as:

$$\frac{\pi DL}{f v_c^2} \left[-C_d + \frac{C_w + C_d T_c}{C_v^{1/n}} \left(\frac{1}{n} - 1 \right) v_c \frac{1}{n} \right] = 0$$

$$-C_d + \frac{C_w + C_d T_c}{C_v \frac{1}{n} \Big/ v_c \frac{1}{n}} \left(\frac{1}{n} - 1 \right) = 0 = -C_d + \frac{C_w + C_d T_c}{T_L} \left(\frac{1}{n-1} \right)$$

Solving for economic tool life for minimum cost:

$$T_{Lopt} = \left(\frac{1}{n-1} \right) \frac{C_w + C_d T_c}{C_d} \tag{9.6}$$

The economic cutting speed for minimum cost is therefore:

$$V_{min} = \frac{C_v}{\left[\left(\frac{1}{n-1} \right) \frac{(C_w + C_d T_c)}{C_d} \right]^n} \tag{9.7}$$

In a similar way, the cutting speed that yields maximum production rate can be derived:

$$V_{max} = \frac{Cv}{\left\{\left(\frac{1}{n-1}\right)Tc\right\}^{n}} \tag{9.8}$$

Equations (9.7) and (9.8) are based upon Taylor's equation. It should be remembered that this basic equation is an experimental one derived from experimental data that contains random errors. These random variations tend to distort its reliability and accuracy. Also, there is a danger in extrapolating the Taylor equation beyond the range over which the experimental data was collected, thus obtaining wrong values.

9.2 HOW EFFECTIVE IS CUTTING SPEED OPTIMIZATION?

Cutting speed is a major determinant of machining time and cost. For practical purposes, it must be selected in order to instruct the operator to change tools after a predetermined number of pieces have been machined. The number of parts to be machined per tool should be set conservatively to ensure that the cutter will not fail. The number of parts is computed by dividing the tool life time by the machining time. But these two variables are dependent. Moreover, the same tool may be used in several machining operations, each one with different cutting conditions.

The selection of cutting speed usually relies on the process planner's experience. The optimization method described in section 9.1 is mathematically true. However, it should be remembered that cutting speed selection is an economic decision rather than a technological one. The selected cutting speed should be regarded as a variable that may be modified (within the boundary limits) in order to achieve economic machining.

Several arguments to support this statement are given in the following sections.

9.2.1 Tool life definition

Taylor's equation is based on the term 'tool life': how is tool life measured? Can it be a constant value?

For sure, a clear end of the tool life is a catastrophic failure of the tool, such as breakage. However, nobody wants to work with this definition. Before the era of carbide tools, it was easier to determine tool life because the failure of the tool could readily be seen from the appearance of a burnished ring on the workpiece when the tool was dull (the Schlesinger criterion). A burnished ring usually cannot be observed on a workpiece when machined with carbide tools, so this method is of no use today.

Hence, tool life is generally defined as the cutting time in minutes taken to produce a given wear land for a set of machining conditions. This means that tool life depends on an assumed maximum of the wear of the tool and is therefore a matter of agreement and standardization. The ISO standard 3865 defines the values of tool wear (ISO 3685 for turning and ISO 8688 for milling).

The criteria most commonly used for sintered carbide tools is that the average width of the flank wear land should be $V_B = 0.3$ mm. However, the flank wear land is usually not of uniform width, but erratic, as shown in Fig. 9.3.

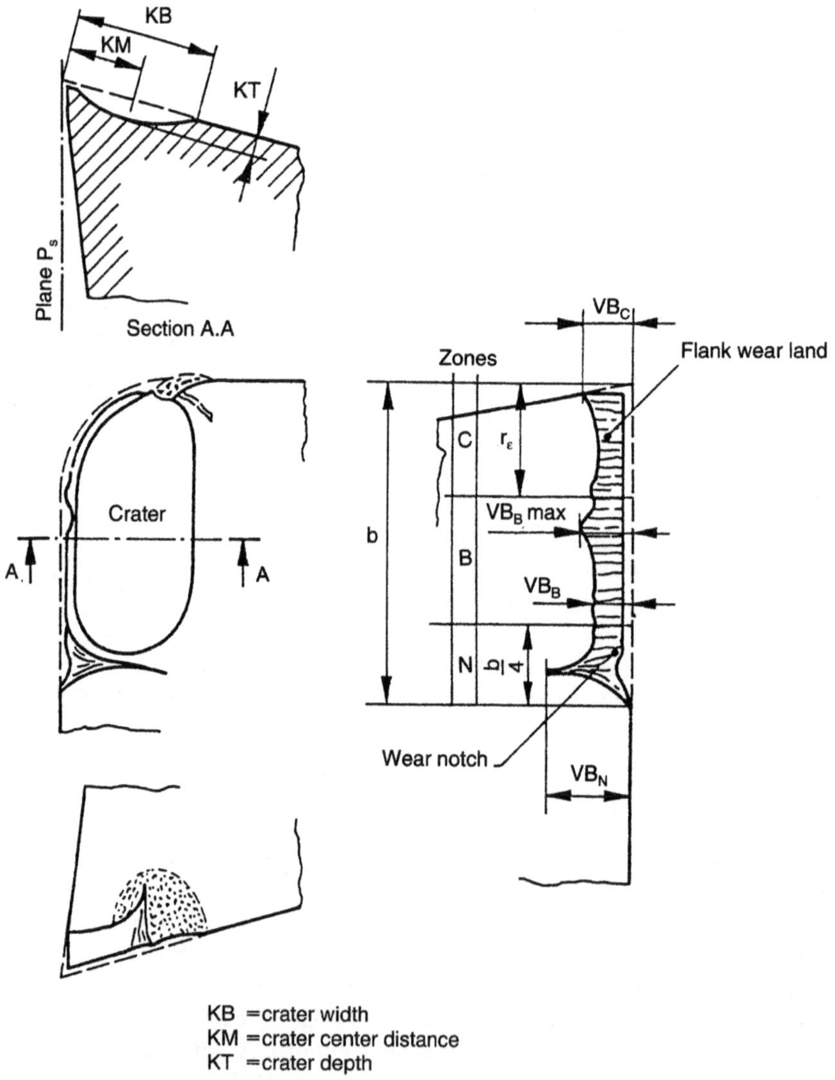

Fig. 9.3 Some types of tool wear.

Thus the question arises as to whether the maximum or the mean of the wear land should be taken as the basis for tool life. The definition of tool life was therefore broadened: the width of the tool is divided into three zones C, B, N as shown in Fig. 9.3. If the flank wear land is considered to be regularly worn in zone B the tool life limit is when $V_{B_B} = 0.3$ mm. If the flank wear is not regularly worn in zone B, then the tool life limit is when $V_{B_B} = 0.6$ mm.

Cratering at the tool face complicates the situation considerably as it depends on the feed rate. The standard agreement is that the tool life limit is $K_T = 0.06 + 0.3 \times$ feed rate (where $K_T =$ crater depth).

In conclusion, a precise tool life definition is important for comparing metal cutting experiments, or tool data of different suppliers. For a process planner, tool life should be selected as a function of the cut (rough or finish), the required surface finish, the machine stability, etc. It is good practice to use the mathematical approach, but the process planner should apply common sense in the final decision on the selected tool life and the resultant cutting speed.

9.2.2 Lot size effect

Cutting speed affects tool life and the number of parts to be machined per tool. Lot size or order size are established by production management personnel and are not of a technological nature. In cases where plant practice is to start each batch with a new tool, it is reasonable to adjust the tool life to the batch conditions.

For example, if the cutting speed decision calls for changing a tool after 68 parts and the production batch size is 70 parts, flexibility should be applied either by changing the instruction to change the tool after 70 parts rather than 68, or reducing the cutting speed slightly to increase tool life.

9.2.3 Machining a part

The optimum cutting speed is computed for each operation separately. When tool life is over, the tool has to be changed. The tool changing procedure in many machines is as follows: the machine has to be stopped, the tool holder has to be moved to the changing location, the tool is changed and the tool holder is moved back to its working position. The tool changing time comprises two elements, setup time for tool changing and tool changing time.

Usually, several tools are used to machine a part. The total machining time and cost is a summation of the time and cost of each tool operation. This may lead to an uneconomic situation. The following will demonstrate this phenomena:

Example

1000 parts are needed. The economical process plan calls for three tools. The economic cutting speed was computed for each operation separately. The results are:

Tool number	*Tool type*	*Machining time in minutes*	*Parts per insert*	*Inserts per batch*
1	MCLNR	0.4100	68	15
2	I TD	0.6127	97	11
3	CCLPR	0.2289	92	11

In this case, machining will be stopped for tool changing after the following parts have been machined:

Stop number	*Incremental no. of parts*	*Accumulated no. of parts*
1	68	68
2	24	92
3	5	97
4	39	136 (68 × 2)
5	48	184 (92 × 2)
6	13	197 (97 × 2)
7	7	204 (68 × 3)
8	68	272 (68 × 4)
9	4	276 (92 × 3)

This procedure suffers from two drawbacks:

1. it is difficult to instruct the operator when to change tools; and
2. time is wasted because the setup time for each tool change is counted separately.

It is proposed that the economic cutting speed and economic tool life be modified in order to result in an overall part machining optimum. This can be achieved by one of two means:

1. replace a tool before its nominal life has expired; or
2. reduce cutting speed, thereby increasing tool life and reducing the number of tool change setups.

These proposals assume that the speed cannot be increased due to some other constraints.

The logic of the proposal is that the number of parts between tool changes must be a constant. To arrive at this number, a simulation is used. In this example, the computed optimization advises a tool change always after 72 parts, i.e to reduce cutting speed for tool number one and replace tools 2 and 3 before their life is over. The results are shown in the following table:

Tool number	*Tool type*	*Machining time in minutes*	*Parts per insert*	*Inserts per batch*
1	MCLNR	(0.4100) 0.4179	(68) 72	(15) 14
2	I TD	(0.6127) 0.6127	(97) 72	(11) 14
3	CCLPR	(0.2289) 0.2289	(92) 72	(11) 14

The computation results depend on the machine's hourly rate, the tool cost, the tool change time, the tool change setup time and the criterion of optimization.

9.3 DATA FOR THE EXTENDED TAYLOR EQUATION

Using Taylor's equation poses two types of problems:

1. Material definition
   Different countries have different material standards such as:

   - SAE/AISI — USA
   - BS and EN — Great Britain
   - DIN — Germany
   - AFNOR — France
   - JIS — Japan
   - SS — Sweden

   Altogether, there are thousands of material standard symbols, so it is impossible to assign constants and exponent values to each of them. Therefore, for metal cutting purposes, the materials are grouped into a smaller number of material groups, where the constants and exponents are assigned to the group and not to a specific material code. In the *Machining Data Handbook*, there are 38 material groups. Other sources might have a different number of material groups.

   The problem is that the different groups and the different material standards are not in unanimous agreement, i.e. a material that according to one source belongs to one group might be said by another source to belong to another group.
2. Retrieving data for the constants and the exponents
   The few sources of this data are to be found in many research papers and other scattered sources.

   To assist the user, data was collected from different sources and compared with each other and the results evaluated. It is published here as a guide, not as an exact data source. It is mostly experimental data which requires evaluation by the user.

Table 9.1 Values of constants and exponents of the Taylor equation

*Number***		*Name*	C_v	*n*	K_{mv}
		Carbon steel			
1	11	– free machining	565	0.2	$4.3-3.3\ (HBN/150)^{0.13}$
2	12	– wrought/cast low carbon	419	0.2	$4.0-0.03\ (HBN/105)^{0.1}$
3	13	– wrought/cast medium carbon	351	0.2	$4.5-3.5\ (HBN/150)^{0.12}$
4	14	– wrought/cast high carbon	302	0.2	$4.0-3.0\ (HBN/200)^{0.18}$
		Alloy steel			
5	21	– free machining	360	0.2	$3.2-2.2\ (HBN/175)^{0.26}$
6	22	– wrought/cast low carbon	331	0.2	$3.5-2.5\ (HBN/150)^{0.18}$
7	23	– wrought/cast medium carbon	292	0.2	$2.8-1.8\ (HBN/200)^{0.35}$
8	24	– wrought/cast high carbon	323	0.25	$1.85-0.85\ (HBN/200)^{0.63}$
9	25	High strength steel wrought/cast	301	0.25	$2.3-(HBN/263)^{0.9}$
10	26	Maraging steel wrought	357	0.25	$1.56-0.56\ (HBN/300)^{1.5}$
11	27	Tool steel wrought/cast	290	0.25	$1.6-0.6\ (HBN/175)^{0.73}$
		Free machining			
12	31	– S.S. wrought ferritic	457	0.25	$(160/HBN)^{0.35}$
13	32	– S.S. wrought austenitic	379	0.25	$(160/HBN)^{0.35}$
14	33	– S.S. wrought martensitic	457	0.25	$4.0-3.0\ (HBN/160)^{0.2}$
		Stainless steel			
15	34	– wrought/cast ferritic	511	0.3	$(160/HBN)^{0.4}$
16	35	– wrought/cast austenitic	351	0.35	$2.0-(HBN/160)^{0.2}$
17	36	– wrought/cast martensitic	345	0.3	$2.0-(HBN/250)^{0.72}$
18	37	PH* stainless steel wrought/cast	294	0.35	$2.1-(HBN/200)^{0.5}$
19	41	Gray cast iron	533	0.25	$2.0-(HBN/135)^{0.75}$
20	42	Compacted graphite cast iron	268	0.25	200/HBN
21	43	Ductile cast iron	512	0.25	$3.4-2.4\ (HBN/190)^{0.35}$
22	44	Malleable cast iron	553	0.25	$2.0-(HBN/135)^{0.75}$
23	51	High temper alloys wrought/cast cobalt and nickel base	78	0.4	$2.0-(HBN/205)^{0.8}$
24	52	High temper alloys wrought/cast iron base	142	0.4	$(205/HBN)^{0.4}$
		Refractory alloys			
25	53	– wrought/cast columbium	266	0.4	200/HBN
26	54	– wrought/cast molybdenum	296	0.4	255/HBN
27	55	– wrought/cast tantalum	100	0.4	225/HBN
28	56	– wrought/cast tungsten	133	0.4	$(300/HBN)^{0.5}$
29	61	Nickel alloys wrought/cast	434	0.4	$2.0-(HBN/125)^{0.55}$
30	62	Titanium alloys wrought/cast	503	0.4	$2.0-(HBN/140)^{0.53}$
31	71	Aluminum alloys wrought/cast	837	0.15	1
32	72	Aluminum – titanium	560	0.15	1
33	73	Aluminum – bronze	390	0.15	1
34	74	Copper alloys free machining	710	0.2	1
35	75	Copper alloys	545	0.2	1
36	76	Copper nickel	330	0.2	1
37	77	Phosphor – bronze	350	0.2	1

* PH = Precipitation hardening.
** = Machinability groups.

The extended Taylor equation for computing cutting speed in turning with a single point, uncoated tool is:

$$v_c T_L^n a^q f^p = C_v V_B^m K_{mv} K_\tau = K_V \tag{9.9}$$

where: v_c = cutting speed (m/min)
T_L = tool life (min)
a = depth of cut (mm)
f = feed rate (mm/revolution)
C_v = constant; the equivalent of cutting speed that will result in one minute of tool life
V_B = flank wear value (mm)
K_{mv} = coefficient of material hardness
K_τ = coefficient of tool cutting edge angle $(\kappa_r) = (45/\kappa_r)^{0.3}$
K_V = constant
n = exponent of tool life
q = exponent of depth of cut effect = 0.10
p = exponent of feed effect = 0.25
m = exponent of tool wear effect = 0.46

For milling operations, a similar equation is proposed as

$$v_c T_L^n a^q f^p z^x . B_r^y = C_v D^g V_B^m K_{mv} K_\tau K_n \tag{9.10}$$

where: z = number of teeth per cutter
x = number of teeth exponent (= 0 for carbide)
B_r = width of cut (mm)
y = width of cut exponent = 0.2
D = cutter diameter (mm)
g = cutter diameter exponent = 0.2
K_n = factor of workpiece material condition
without skin = 1.0
forging with skin = 0.9
casting with skin = 0.8.

In practice, the number of teeth does not have any effect on the cutting speed.

There is a correlation between cutter diameter and width of cut. It is recommended to have a ratio of 1.3. Hence their effect is: $1.3^{0.2} = 1.0538$, which can be regarded as negligible.

The material condition is an important factor and may be added to the turning equation. It may be concluded that the same data may serve turning and face milling operations.

Selecting cutting speed and its data for hole-making operations such as drilling, reaming, boring, milling, etc. is discussed in Chapter 13.

Table 9.1 gives the values of the constants and exponents of the extended Taylor equation, for 37 groups of materials for cutting speed in turning with a single point tool, uncoated carbide tool and face milling.

CHAPTER 10

How to select a machine for the job

When selecting a machine for a job, one has to consider many parameters, such as the machine's physical size, power, torque at the spindle, machine accuracy, available speeds and feeds, number of tools, tool change times, hourly rate, batch quantity, etc.

One method of selecting a machine is to consider the size of the workpiece and its accuracy demands and estimate the required parameters, especially power. This is the most common method. Unfortunately, by this method, a machine is selected before the exact operations it has to perform are known so that the machine specifications act as constraints on the operation selection. This is an artificial constraint. Moreover, a machine has to perform several operations, each operation requiring different accuracy and power. Selecting a machine on the basis of the most extreme demands will result in a waste of resources, such as using a machine of 35 kW to perform an operation that requires 0.75 kW, or using an accurate machine to perform a rough operation.

Selecting a machine should be a compromise between the separate operation requirements; in other words, an optimization should be made on the part level and not on the individual operation level. Such an optimization method is described below.

The aim of the process planner is to generate the most economical process plan; therefore, no artificial constraints should affect the decisions. To accomplish this, it is recommended that the process activities be separated into two phases.

The first phase handles the engineering stage and is limited only by engineering and technological considerations. Thus, a theoretical process is generated, theoretical, that is, from a specific shop point of view (e.g. such a machine may not be available), although practical from an engineering point of view, since all technical constraints have been taken into account.

For each individual operation, the parameters have been defined: depth of cut, feed rate and cutting speed. These values guarantee that the part will be produced economically and meet all specifications as defined in the drawing. However, it does not consider the available machines in a given industrial shop, nor the other operations.

The problem at the second phase is to transform the theoretical operations into practical ones from a shop point of view, i.e. to adjust the process operations to the available facilities. This is basically a combinatorial problem where all the alternatives available in terms of facilities have to be generated and computed. The only technical knowledge required is how to make the specific operation comply with a particular machine specification. These adjustments can easily be constructed, as shown in section 10.3. Thus the problems of machine selection and process planning are transformed from an engineering problem into a mathematical one. The problem is to search for the best sequence of operations and the best sequence of machine utilization. This is basically a combinatorial problem which leads to an untractable number of alternatives which have to be compared in terms of economic efficiency. The following sections will suggest solutions to this problem.

10.1 DEFINITION OF THE COMBINATORIAL PROBLEM

The individual operations have been optimized as explained in Chapters 8 and 9. The optimization problem, at this stage, consists of choosing for every operation the 'best' available machine, taking into consideration time for part transfer from machine to machine.

To illustrate this problem, let us assume that, through operation optimization, four operations are required to produce a part. Their requirements are shown below:

Operation number	*Power (kW)*	*Accuracy (mm)*		*Machining time (min)*
		location	*geometric*	
010	28	0.3	0.2	0.27
020	12	0.1	0.08	0.89
030	0.6	0.06	0.02	2.56
040	25	0.2	0.15	0.55

The decision to make is to select the 'best' available machines to produce this part. Operations that require less power than the machine is capable of may be machined without increasing machining time (but **not** decreasing it). Operations that require higher power than is available may be machined, but the machining time will be increased. With accuracy, there are no compromises. An inaccurate machine cannot produce an accurate part, even with added machining time.

If the 'best' machine for an individual operation is selected, additional time has to be added covering transfer time (and cost) between machines. If the sequence of operations can be altered, and machining operation 040 is allowed to be produced after operation 010, savings in transfer time may be realized.

It is clear that transfer time and cost are functions of the required quantity. For low quantities, the best compromise might be to machine the part with only one machine. This will increase the direct machining time, but reduce the transfer time. For very high quantities, however, the best compromise might be to machine each operation with the best machine, which will increase transfer time and reduce direct machining time. For moderate quantities, the best compromise is not so clear. Calculations should be made to find the point when direct machining time and transfer time should be increased, to arrive at the minimum total production time (or cost).

If the order of precedence of the operations allows some changes they can be introduced in the sequence of the optimal path. These variants can ultimately permit a better optimization of the whole sequence. For example, for the part represented in Fig. 10.1 and taking in account anteriority rules such as 'roughing before finishing' or 'tool access restrictions', a feasible sequence of operations could be 1,2,3,4,5 or 1,2,3,5,4 or 1,2,5,3,4 for roughing operations and 6,7,8 for the finishing operations, although they can be put in any order e.g. 1,2,3,5,4,8,6,7 or 1,2,5,8,3,6,4,7.

According to the sequence chosen, extra cost has to be added of course, when parts are transferred from machine to machine, because of transfer time, setup time, chucking time and additional inspection time. When considering transfer time, it should be remembered that chucking time is added for every part, whereas setup time is only added once for the whole batch. Therefore, the optimal result will also depend on the quantity of parts in a batch. It is also very important at this stage to realize that the constraints of capacity planning of the shop and of job routing will seriously influence the strategy to find the optimal path of operations. In the following analysis, though, we do not take this factor into consideration.

In conclusion, the optimization problem at hand can be defined in the following way: given M machines and N operations, and taking in account the cost or time of machining of every operation on every machine, as well as the sequence of operations, what is the best combination of operations and machine which will result in the optimum cost or time?

The criterion of optimization is normally the minimum cost or the minimum time of the whole series of operations to be performed.

The following section presents a solution to this type of problem on the basis of an application of the theory of dynamic programming.

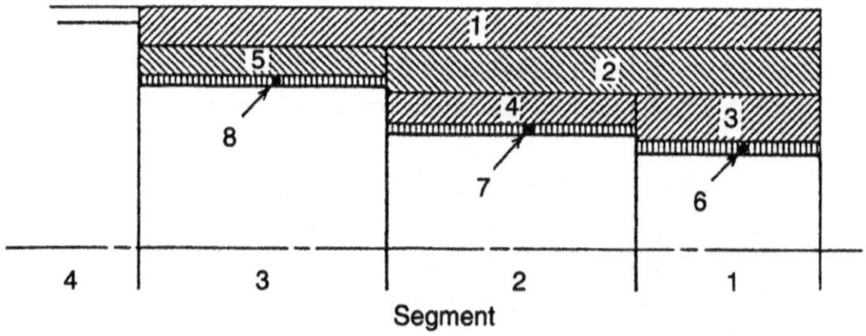

Fig. 10.1 Operation sequence as specified in first stage.

10.2 DYNAMIC PROGRAMMING OF THE SEQUENCE OF OPERATIONS AND MACHINES

Given a list of operations to be performed and a list of available facilities, a decision is required as to which machine (or machines) to use, which operation(s) to perform on each machine, what their sequence should be and what cutting conditions to employ.

The operations in the list are the absolute optimum, that is, theoretical in relation to the existing facilities. They are not arranged according to any reasonable machining sequence.

It is proposed to present the problem in a matrix format, where the rows of the matrix represent operations, while the columns represent resources (Fig. 10.2). The elements of the matrix are the costs of each operation with respect to each resource.

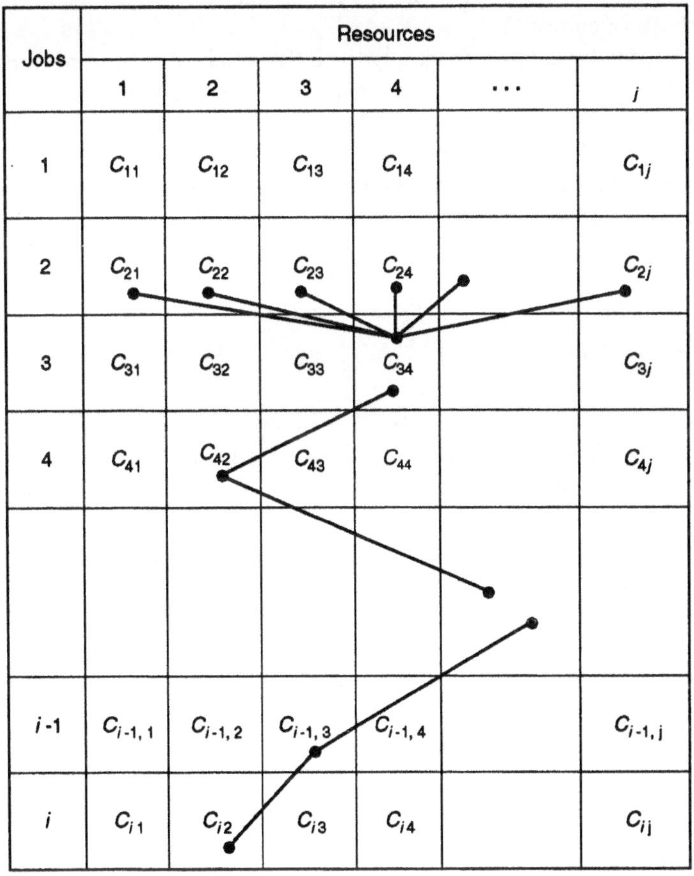

Fig. 10.2 Dynamic programming procedure.

If no restrictions existed on the sequence of operations, the number of combinations to consider would be equal to $N!M^N$, i.e. the product of $N!$ arrangements of operations with M eventualities of machines for each operation. This number rapidly becomes very large when the number of operations and machines is growing and therefore practically intractable. However, if we use R. Bellman's theory from the dynamic programming field, which states that, 'if the payoffs from each decision be additive and that no matter how a state arose, then the consequences for the future are the same', then the number of useful combinations to arrive at the desired optimum may be drastically reduced.

The following example demonstrates this logic. At an intermittent point of the series of decisions it was decided that operation 3 is to be performed on resource 4 (Fig. 10.2). At this stage we can ignore how and why we reached this decision. The accumulated cost or time of all operations up to operation 3 is $T_{3,4}$. The time (or cost) of operation $i=2$, on machine j is $C_{i,j}$. The problem at hand is: to where should we proceed from this point, that is, on which resource should we perform operation 2?

This is done by comparing simply the differential paths from operation 3 to operation 2 and choosing the optimal one. The following step begins at operation 2 to look for the best path to operation 1 which will define the total optimal path of the whole job.

This problem can be solved by finding which of the following expressions has the minimum values:

$$T_{2,1}=T_{3,4}+C_{2,1}$$

$$T_{2,2}=T_{3,4}+C_{2,2}$$

$$\dots$$

$$T_{2,j}=T_{3,4}+C_{2,j}$$

This is a finite problem and can be solved easily and fast. It says that if the pay-offs at every stage are additive, and because, at any stage of the optimal path, the partial path to reach this point is also an optimal path, it is only necessary to find a number of combinations equal to $N \times M$ (when the sequence of operations is fixed), a number which is substantially smaller.

Thus, the solution is a series of decisions, starting from the last row (i) of the matrix and working all the way up to the first row. The minimum value of $T_{1,i}$ is the resource for the first operation and it points to the resource of the second operation, and so on.

This method solves the general problem but it does not consider the sequence priority, i.e. the allowable change in sequence of operations. The total solution will be described in the following section.

10.3 CONSTRUCTING THE OPERATION–MACHINE MATRIX

The machines (j) will be listed horizontally, and the operations (i) will be listed vertically in the matrix. The cost, or time of each operation on each machine $C_{i,j}$ ($T_{i,j}$) will be entered accordingly.

The decisions regarding the basic operations, i.e. theoretical operations were computed as discussed in the previous chapters. At this stage, the theoretical operations are known and machine specifications and constraints are known. The adjustments of the theoretical operations to the practical machines are discussed in the following sections.

10.3.1 Power and force adjustment

Power is a linear function of cutting speed and cutting forces.

$$P = K_1 v_c F \tag{10.1}$$

where: P = power (kW)
v = cutting speed (m/min)
F = cutting force (N).

The cutting forces constraints have been acted on when selecting depth of cut and feed rate, as discussed in Chapter 8. However, there were eight different sources of force limiting factors. One of them (F_{rm}, section 8.3.1) is due to a machine constraint. This factor was taken as a maximum value and at this stage it has to be adjusted for each separate machine. This is done as described in Chapter 8, i.e. by adjusting the depth of cut or feed rate.

If the resulting power (of the theoretical operation) is equal to or less than the machine power, the selected cutting conditions are in effect. However, if the required power is above that available, a further adjustment must be made – the power must be reduced. There are several courses of action. Rules must be made to keep the process at the optimum level for that machine. The rules are formulated as follows:

The cutting speed can be expressed as:

$$v_c = \pi D n = K_2 n \tag{10.2}$$

where: D = tool or workpiece diameter (mm)
n = revolutions per minute

The cutting force can be expressed as:

$$F = K_3 a^x f^y = K_4 f^y \tag{10.3}$$

where f is the feed rate and a is the depth of cut.

Substituting (10.2) and (10.3) into (10.1), gives

$$P = K_1 K_2 n K_4 f^y = K n f^y \tag{10.4}$$

This can be rewritten as

$$P=Knf^{y}=K\frac{nf}{f}f^{y}=K\frac{nf}{f^{1-y}};\quad nf=\frac{Pf^{1-y}}{K} \tag{10.5}$$

The machining time can thus be expressed as:

$$T=\frac{L}{nf}=\frac{KL}{Pf^{1-y}} \tag{10.6}$$

where: T = machining time (min)
L = machining length (mm)

Equation (10.6) tells us that as the power (P) increases, machining time decreases, and vice versa. However, the tool and strength of the workpiece limit the maximum power that can be exerted in any single operation. This maximum power is the one specified by the theoretical cutting conditions.

This means that there is no advantage, from the machining time standpoint, in using a machine that is more powerful than necessary.

When machining power is less than that required, the cutting speed or the feed rate or the depth of cut has to be reduced. If P_o designates the available machine power, then the machining time as expressed in (10.6) becomes

$$T_1=\frac{KL}{P_o f^{1-y}} \tag{10.7}$$

In this case, it is assumed that the reduction of machine power is achieved by reducing the cutting speed.

Another approach is to maintain the cutting speed and reduce the feed rate to a value f_o. In such a case, the machining time is

$$T_2=\frac{KL}{P_o f_o^{1-y}} \tag{10.8}$$

If we take the ratio of the machining times that result from the above two methods, we obtain

$$\frac{T_2}{T_1}=\frac{KL}{P_o f_o^{1-y}}\,\frac{P_o f^{1-y}}{KL}=\left\{\frac{f}{f_o}\right\}^{1-y} \tag{10.9}$$

Since by definition, f is always greater than f_o, T_2 will always be greater than T_1. This means that it is more profitable to reduce the cutting speed than to reduce the feed rate. Reducing cutting speed will also result in an increase of tool life. A practical conclusion is that power adjustment should be made according to the following priorities:

- If the power has to be reduced by less than 50%:
  1. reduce the cutting speed down to its lower value;
  2. reduce the feed rate down to its lower limit value; and

3. reduce the depth of cut, i.e. split the cut into more than one pass and adjust cutting speed and feed rate.

- If the power has to be reduced by more than 50%, split the depth of cut to more than one pass and then use the priorities above.

A more sophisticated measure may include a change in tool cutting edge angles and in the number of teeth in a milling tool also. The rules in such cases are:

- down by 85% – reduce cutting edge angle;
- down by 70% – as above, plus a reduction in cutting speed;
- down by 60% – as above plus a reduction in feed rate;
- down by 51% – split depth of cut to two passes;
- down by 40% – split and reduce cutting speed;
  – split depth of cut.

Note: a reduction of cutting forces is not effected by reducing the cutting speed. Use only the reduction of feed rate as a first attempt and if this is not enough, split the depth of cut.

Remember that if the cutting forces **and** the power have to be reduced, handle first the reduction of cutting forces, and check to ensure that the power problem still exists.

10.3.2 Maximum depth of cut constraints

Each machine has its own maximum depth of cut value. Chatter might appear if depth of cut is above this value. If the operation depth of cut is greater than the maximum value, it must be reduced. The reduction can be made by changing the cut distribution (between rough and finish, or between several unequal rough cuts), or by splitting the operation into several passes. In either case, the machining time is affected and must be adjusted accordingly.

10.3.3 Maximum torque constraints

Some machines are defined by power and maximum allowed torque on the spindle. If the operation torque is greater than the allowed torque, it must be reduced. Torque is the multiplication of cutting force by radius (of part or tool). Therefore in milling, an attempt should be made to reduce the torque by reducing tool diameter, and if not, reducing the cutting forces as recommended earlier. The machining time is affected and must be adjusted accordingly.

10.3.4 Machine accuracy constraints

Some machines might be worn out and thus their accuracy capabilities are below standard. If the part tolerances cannot be maintained by the specific machine, the machine cannot perform the operation that calls for this tolerance.

However, the machine might be more suitable for other operations such as rough cut, due to its low hourly rate. Therefore the machine remains in the matrix. As a common practice, for all incapabilities a value of $C_{i,j} = T_{i,j} = 9999$ is inserted. This will eliminate the selection of such a machine for the performance of an operation of which it is incapable.

10.3.5 Spindle bore constraint

If the spindle bore diameter of a machine (lathe) is less than the part diameter, the free length of the part in chucking is not controlled by the chucking location, but rather by the machine spindle bore and the hole through the chuck and chuck width. In such cases the allowable bending forces should be adjusted to the new conditions and the cutting conditions have to be recomputed. This will affect machining time and thus the selection of a machine for the operation.

10.3.6 Time and cost conversion

The optimization criterion can be either maximum production or minimum cost. Maximum production actually means minimum machining time. Thus both criteria call for a path that results in a minimum value. In order to use one solution method, the content of the matrix is altered from time to cost. Time is an engineering value and is determined by the cutting conditions. Cost is the multiplication of time by the hourly rate. Each machine has its hourly rate to be used for cost computations.

10.4 PRELIMINARY MACHINE SELECTION

In order to lessen the number of alternatives and thus reduce solution time, a preselection of machines will be made. This preselection should be made so as not to affect the optimum solution.

10.4.1 First step in machine selection

The first step in machine selection is based on the type of machine and its physical dimensions. For round symmetrical parts, for example, only a lathe machine group is considered. A lathe machine is partially defined by the swing over bed, swing over cross slide, total cross slide travel, and so on. These are the physical dimensions of the machine. A part with a radius greater than swing over bed cannot by gripped on the machine. A part whose length is greater than the machine bed cannot be turned on that machine.

In the first step of machine selection, only those machines that absolutely cannot perform any one of the operations are excluded from further consideration.

10.4.2 Second step in machine selection

A machine whose power is lower than the minimum required does not stand a chance of being selected for the operations. Therefore, it can be excluded from further consideration without affecting the optimum process plan, unless there are no other machines.

A machine with more power than required has no advantage over lower-power machines. Therefore, such a machine can be excluded from further consideration, unless its hourly rate is lower than the low-power machine or it has a spindle speed higher than that of the lower-power machine and this spindle speed is required by one of the operations.

10.4.3 Third step in machine selection

This third step is employed only if the number of remaining machines is too high. An estimation of a machine's chance of being chosen serves as the basis of this step. The chances are estimated by comparing the machine time or cost for the given machine with those of other machines.

- Select the three machines that have the lowest total machining time (single machine solution for the job).
- Select machines that have a minimum value in any single operation.
- If there are still too many machines, select those which have a minimum value in the higher number of operations. Decide when to stop.

10.5 MATRIX SOLUTION

The matrix solution should specify machine selection and the sequence of operations. The solution is a result of generating all alternatives and selecting the optimum one. The theoretical operations are the 'absolute optimum' that probably cannot be reached, but can be used as a yardstick. This theoretical optimum can be used to determine whether a single machine exists that gives an optimum process and thus whether there is any sense in generating and evaluating all other alternatives. If such a machine does not exist, the general matrix solution is employed.

10.5.1 Single-machine solution

An imaginary machine is added to the matrix (T_{j+1}). It is assumed that this is an ideal machine, that is, it has the properties most desired in any machine present in the matrix. Thus, this machine has a value of $T_{i,j+1}$ equal to that of the machine with the lowest value. The minimum value of each operation is thus inserted into this machine column (as shown in Fig. 10.3). The vertical sum

$$T_M = \Sigma \min(T_i) = \Sigma(T_{i,j+1})$$

Operation (i)	Machine (j)					Imaginary machine
	1	2	3	....	J	$J+1$
1	$T_{1,1}$	$T_{1,2}$	$T_{1,3}$		$T_{1,J}$	$T_{1,\min}$
2	$T_{2,1}$	$T_{2,2}$	$T_{2,3}$		$T_{2,J}$	$T_{2,\min}$
3	$T_{3,1}$	$T_{3,2}$	$T_{3,3}$		$T_{3,J}$	$T_{3,\min}$
I	$T_{I,1}$	$T_{I,2}$	$T_{I,3}$		$T_{I,J}$	$T_{I,\min}$
Total value of a single machine	$(\Sigma I T_i)_{,1}$	$(\Sigma I T_i)_{,2}$	$(\Sigma I T_i)_{,3}$		$(\Sigma I T_i)_{,J}$	$(\Sigma I T_i)_{,\min}$
	T_1	T_2	T_3		T_J	T_M

Fig. 10.3 Imaginary machine concept.

is thus the practical theoretical optimum, i.e. the minimum value that can be expected for this part.

The total part value for each separate machine $T_j = \Sigma T_i$ is computed for each machine. Within these total values there exists a minimum $(T_j)_{\min}$ (Fig. 10.3). This minimum points to a single machine solution.

The general solution attempts to arrive at a lower value than $(T_j)_{\min}$. However, for each additional machine, transfer time must be added. Therefore, the condition that must prevail in order for a single machine solution to be optimum is

$$T_M + T_{RN} \geqslant (T_j)_{\min}$$

or

$$(T_j)_{\min} - T_M \leqslant T_{RN}$$

where T_{RN} is the transfer time or cost value.

10.5.2 General matrix solution

The general solution is based (mathematically) on a dynamic programming technique. The basic feature of dynamic programming is that the optimum is reached stepwise, proceeding from one stage to the next. An optimum solution set is determined, given any conditions in the first stage. This optimum solution set from the first stage is then integrated with the second stage to obtain a new optimum solution, given any conditions. Then, in a sense ignoring the first and second stages as such, this new optimum solution is integrated into the third stage to obtain still further optimum solutions and so on until the last stage. It is the optimum solution that is carried forward rather than the previous stage.

In the problem at hand the stages are referred to as operations, and decisions are made by choosing the optimum path between any two operations. However, since the sequence of operations listed in the matrix is not fixed, this sequence can be changed. One of the problems to be solved is which sequence of operations will result in an optimum solution. Therefore, the general dynamic programming solution procedure has to be modified in order to handle the problem at hand.

The proposed solution is divided into two phases. The first phase is from the bottom up, that is, from the last operation up to the first. It will proceed operation by operation, determining the optimum path (machine selection) for each stage independently of the previous operation. However, at each operation a review of all previous optimum decisions is made in order to examine the effect of the sequence of operations. The sequence that results in a total path optimum is selected.

The second phase is from top down, that is from the first operation down to the last. It reviews the optimum achieved by examining the effect of the sequence of operations from any operation up to the first operation. The sequence that results in a total path optimum will be used.

The following describes how this technique is applied. For convenience, two auxiliary matrices are constructed (or the same matrix is used with additional values in each box). In this example the total solution uses three machine (j) – operation (i) matrices:

1. element value $T_{i,j}$
2. total downward value $C_{i,j}$ $(i \neq 1)$
3. path pointers $P_{i,j}$

The computation starts with operation $I-1$ and machine 1. The problem is as follows: to which machine should we proceed from this point in order to arrive at a minimum value? Since this is the last operation, the transfer time $R_{i,j}$ should be added when the machine changes. Thus the alternatives are:

$$S_1 = T_{I-1,1} + T_{I,1} (= C_{I,1})$$

$$S_2 = T_{I-1,1} + R_{1,2} + T_{I,2} (= C_{I,2})$$

$$S_3 = T_{I-1,1} + R_{1,3} + T_{I,3} (= C_{I,3})$$

$$\ldots$$

$$S_J = T_{I-1,1} + R_{1,J} + T_{I,J} (= C_{I,J})$$

The chosen path will be where S_j is the minimum value. This minimum value is placed in the total matrix as $C_{I-1,1}$. The path matrix lists the machine number of operation I that results in the above minimum value. Thus $P_{I-1,1} = K$. These values are shown in Fig. 10.4.

This process is repeated for operation $I-1$ and machine 2 and the resulting values are placed in $C_{I-1,2}$ and $P_{I-1,2}$ and so on until machine J and all values of $C_{I-1,J}$ and $P_{I-1,J}$ are computed. Covering all junction points of operation $I-1$,

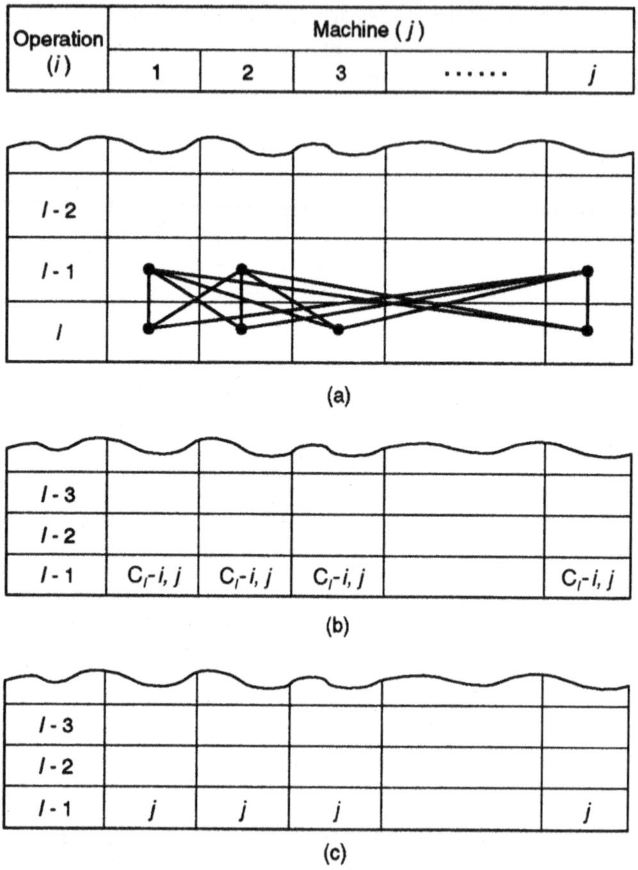

Fig. 10.4 General solution matrices: (a) element value matrix; (b) total downward value matrix; (c) path pointer matrix.

the solution proceeds upward to handle all junction points of operation $I-2$, $I-3$ and so on until the first operation.

The general junction alternatives to be evaluated can be expressed as

$$S_j = T_{i,j} + C_{i+1,k}\Big| + R_{j,k}\Big|_{k=1}^{k=j}$$

The junction to be evaluated is operation i on machine j. Its time or cost is $T_{i,j}$. From this point it is possible to proceed downward to perform operation $i+1$ with one of the available machines. The optimum solution for each machine in operation $i+1$ is the total $C_{i+1,k}$ and is independent of the path by which it was reached.

The term $R_{j,k}$ is the transfer time covering the expenses caused by shifting the work from machine j to machine k. The value of the transfer time is path-dependent. It is possible that by changing the sequence of operations, no transfer time should be added. This case occurs either when the path from the current operation down to the last operation passes through the current machine and the operation that uses that machine can be shifted upward, i.e. to be performed right after the current operation, or when the current operation can be shifted downward to be performed right before the other operations. In such cases, transfer time has already been added to the total and no extra is required.

The information on whether an operation can be shifted upward or downward is made available by the priority code. Figure 10.5 demonstrates such a case (upward phase). The optimum path from operation 10, machine 6 is (*i,j*) 10,6; 11,7; 12,4; 13,4, as shown by the heavy line in Fig. 10.5(a). Transfer time has been added twice so far. The junction of operation 9 machine 4 is evaluated. One of the alternatives is to proceed to operation 10 machine 6. This calls for adding

$$T_{9,4} + C_{10,6} + R_{4,6}$$

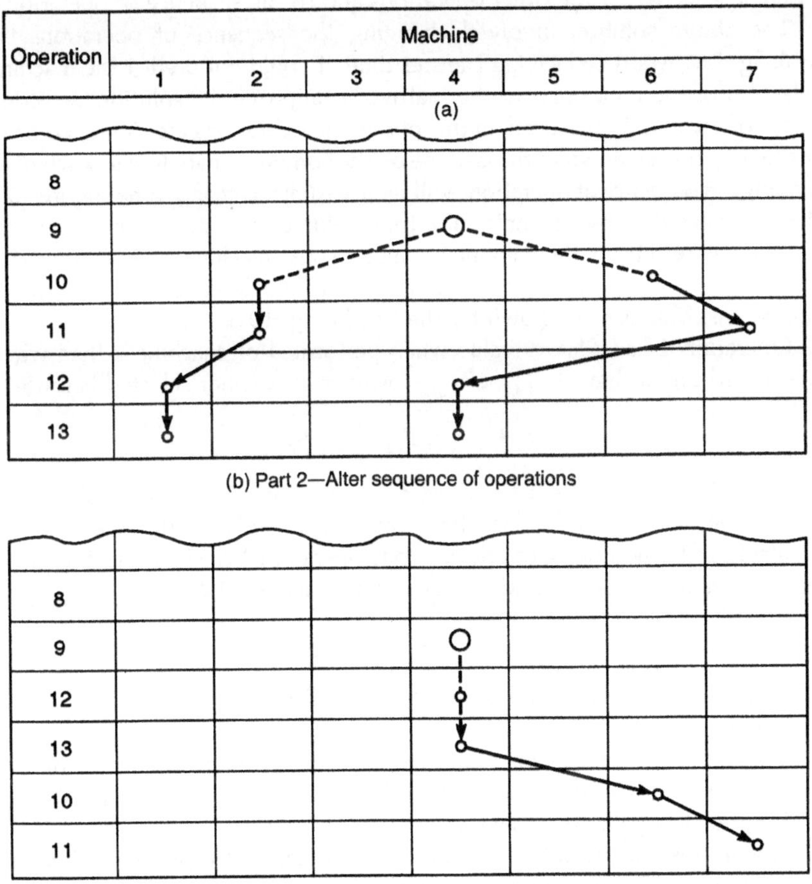

Fig. 10.5 Selection of the sequence of operations – upward phase.

Scanning down the path reveals that machine 4 is selected for operations 12 and 13. If the sequence of operations can be altered to read 9, 12, 13, 10, 11 (as shown in Fig. 10.5(b)), only two machine transfer times occur in this path and no extra transfer time should be added. Thus, the total value should be $T_{9,4}+C_{10,6}$.

If this is the best alternative for the junction of operation 9 machine 4, the operation sequence will be altered.

This computation and the path checking are performed for any alternative. Thus, when evaluating the alternative of proceeding from operation 9 machine 4 to operation 10 on machine 2, transfer time must be added (Fig. 10.5), since machine 4 does not participate in the above path.

The priority number indicates whether the sequence of operations can or cannot be changed. The value in the first column of the total matrix ($C_{i,j}$) represents the total cost or time to produce the part when starting with any one of the available machines. The machine chosen for the first operation is the one with the minimum value of $C_{i,j}$. The path matrix will then lead through the machine selected for the other operations and to the sequence of operations.

The above solution involved changing the sequence of operations by looking downward and saving transfer time. It could not predict the machine selection of the upper part of the matrix. To improve the solution, a second phase of computation is used. In this phase, the operations are examined from the first operation down to the last one on the computed path to check whether a change in sequence of operations will reduce total machining time. Figure 10.6 demonstrates this case. Scanning the total value of operation 1 indicates that machine 4 results in the minimum value. Thus, machine 4 is selected for operation 1. The path matrix leads to machine selection for the other operations. This path is shown by the line in Fig. 10.6.

Operation 4 has a lower value when performed on machine 4 than when performed on machine 2: $T_{4,4}<T_{4,2}$. However, it was not selected because

$$T_{3,2}+T_{4,2}<T_{3,2}+T_{4,4}+R_{2,4}$$

The transfer time $R_{2,4}$ must be added since machine 4 is not available on the lower side of the path. Looking from top down, we know that machine 4 is available, and if according to priority code operation 4 can be moved forward, no transfer time should be added. Examining junction 3,2 results in

$$T_{3,2}+T_{4,2}>T_{3,2}+T_{4,4}+0$$

Thus the sequence of operations should be modified to read: 1, 2, 4, 3, 5, 6, 7, . . . (Fig. 10.6).

If operation 4 cannot be processed before operation 3, this change of sequence is not allowed. However, it might be possible to machine operations 3 and 5 prior to operations 1 and 2.

This means that operation 4 is not going to be pulled up, but rather that operations 1 and 2 are going to be pulled down. This will result in a machine selection and sequence of operations as shown in the bottom part of Fig. 10.6.

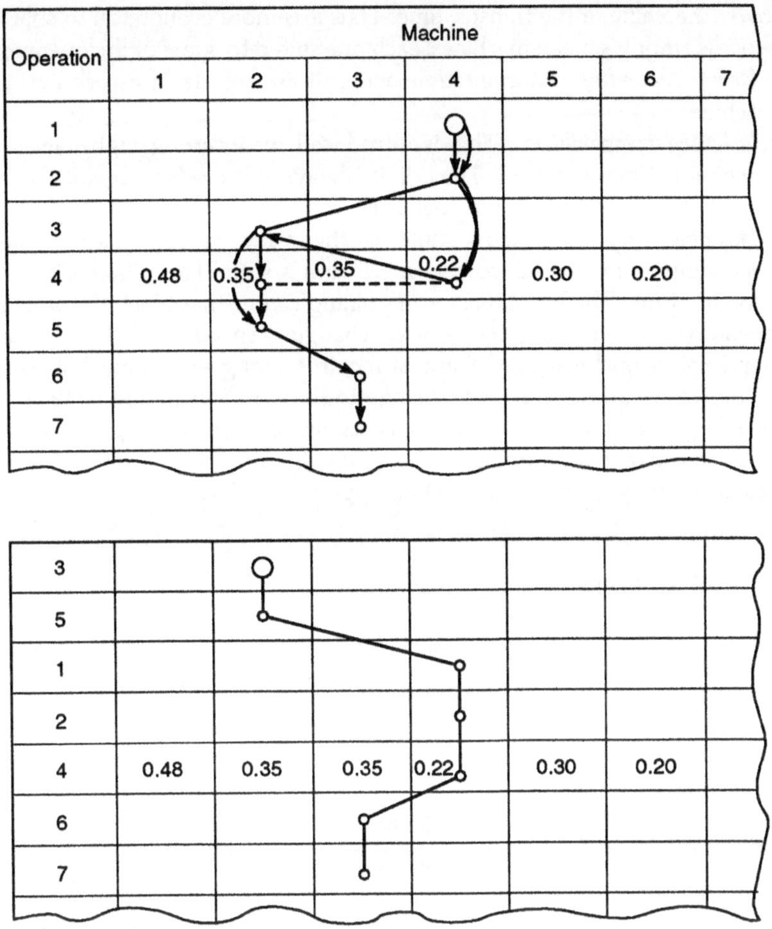

Fig. 10.6 Selection of the sequence of operations – downward phase.

10.6 CONCLUSION

The proposed method results in an optimum selection of machines and sequence of operations. The selection and decision process is purely mathematical and is not affected by intuition or rule of thumb as used today. An example of rule of thumb is to use old, inaccuracte machines for rough cuts and accurate machines for finish cuts. Such a decision will, when true (for large quantities) automatically be reached by the proposed method. However, for small quantities, a different decision might result. The decisions and thus the process planning takes the following into consideration:

- *Quantity* The quantity affects the transfer time. The larger the quantity, the lower the value of the transfer time. Thus it is more economical to split the process among several machines, each one suited to some of the operations.

- *Machine capability* Machine time and cost are adjusted for each available machine.
- *Machining incapabilities* This is introduced by inserting high values for machining time and cost. Thus, it is always more advantageous to add transfer time and bypass this machine.
- *Machining type* Machining time is the same in any machine type, depending only on the power, speed and so on. The difference lies in handling time. The NC, DNC or machining centers machines perform such operations as adjusting the tool or changing speed much faster than a universal manual machine. Automat requires a long setup time, but has the ability to do a fast return and change tools in a short period of time. The handling timetable can have many columns, one for each type of machine. Thus, the general matrix solution can handle different machine types and select the most economical machine type for the job.

FURTHER READING

Ackoff, R.L. and Sasieni, M.W. (1968) *Fundamentals of Operation Research*, John Wiley and Sons.

CHAPTER 11

How to select tools for a job

The selection of tools for a job involves many parameters, such as, insert shape, insert grade, tool holder type, method of holding the insert in the holder, etc. Furthermore, there are many tool manufacturers offering a large variety of tools. Only standard tools will be considered. It is not the purpose of this book to promote the use of special tools that may perform an operation in a more efficient way.

The International Standard Organization (ISO) and the American National Standard Institute (ANSI) have published Toolholders and Inserts Identification Systems which are discussed in this chapter. Following the standards and a discussion of the meaning of each digit and how to define it, we end up with a universal nomenclature of tool designation. This nomenclature is known to all tool manufacturers, who can convert it to their catalog number. They can also suggest an improvement of some types of tools.

The ISO and ANSI standards relate to different types of machining processes. Each one handles the general features and the specifications for that particular machining process (turning, boring, threading, milling, drilling, etc.). In this book, we concentrate on turning, which can serve as an example to all other standards.

11.1 SELECTING INSERT SHAPE AND TOOLHOLDER TYPE

In order for the tool to remove material by cutting, the cutting edge angle must have an inclination of a minimum of 3° towards the machined surface. The larger the angle, the better the machining result. At the minor cutting edge, the tool should not scratch or rub the machined surface. Therefore, a clearance angle of a minimum 3° should be used (Fig. 11.1). The angles are always measured from the center line of the machine. The actual angles on the selected tools are dependent on the tool type and the insert shape. Tool manufacturers do not offer an infinite number of tool holder types for, say, cutting edge angles or insert shapes. Therefore, the tool with the best approximation of the needs should be selected.

On a machine, there may be constraints regarding the number of tools that can be accommodated. Moreover, tool change adds handling time and cost to the fabrication process. Therefore, it is recommended, from an economic point

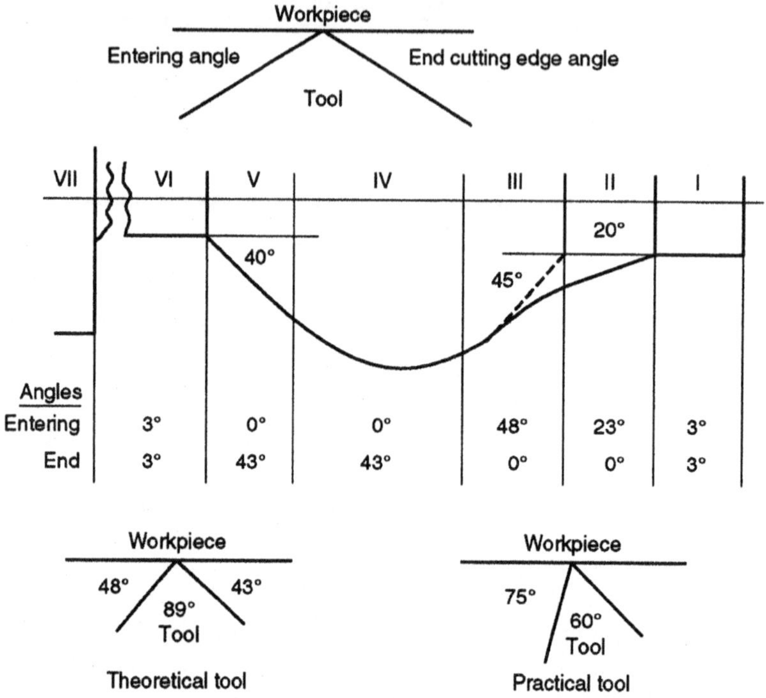

Fig. 11.1 Clearance angles and tool shape selection.

of view, to use a minimum number of tools, provided that it is not imposing a penalty.

At this stage of process planning, the machines have been selected and the operations to be made on each machine have been determined. The problem at hand is to select the tools for the job.

The first step in selecting tools consists of making decisions about operations which will use the same tools. To arrive at this decision, the operations are separated into internal and external operations and into segments of the parts that require special tooling, such as threading tools, grooving or parting tools.

The second step will list the minimum cutting edge (κ_r) and minor cutting edge (κ_r') angles for each segment. This can be seen in Fig. 11.1. In this figure there are six segments:

1. The first segment is a cylinder, therefore $\kappa_r = \kappa_r' = 3°$.
2. The second segment is a cone of 20°, therefore $\kappa_r = (20 + 3 =)23°$ and $\kappa_r' = 0°$, as the machined surface is of $-20°$ and minus does not count.
3. The third segment is a curve, where the largest tangent to it is 45°. Therefore $\kappa_r = (45 + 3 =)48°$; $\kappa_r' = 0$ as before.
4. The fourth segment is a curve around the center of the segment. $\kappa_r = 48°$ (as in the conjunction curve of segment 3). This curve is a tangent to a

cone of 40°, which is the maximum tangent angle to the curve, hence $\kappa_r'=(40+3=)43°$.
5. The fifth segment is a cone with a minimum slope of 40°. Hence $\kappa_r=0°$ and $\kappa_r'=(40+3=)43°$.
6. The sixth segment is a cylinder, therefore $\kappa_r=\kappa_r'=3°$. The maximum value of κ_r along this path is 48° and the maximum value of κ_r' is 43°.

   The maximum body of the insert is 180 degrees minus κ_r and κ_r', which in this case is $(180-48-43=)89°$. This means that a square insert cannot be used.

   However, the practical cutting edge angle (κ_r), is determined by the available cutting edge angles on the commercial toolholders. The closest toolholder cutting edge angles are 45° and 60°, therefore a toolholder with 60° can be used (instead of 48°), which leaves for the insert body $(180-60-43=)77°$. From the available standard economical inserts, the closest one is the triangle insert with 60°.

If the part had a 7th segment (Fig. 11.1) the minimum value of segment 6 would have been 90° (or more, as discussed later), thus the insert minimum angle can be $(180-90-43=)47°$. There exists a diamond tool with an angle of 35° that can be used. However, as this tool cannot withstand heavy loads, it is used mainly as a copying or profiling tool for finishing passes, where the forces are low. Hence, in such a case, one tool cannot be used to machine the whole path, and use of a second tool to machine segment 6 is necessary.

The above procedure is our first proposed selection. It may be changed due to other factors, as discussed in the following sections.

11.2 SELECTING THE INSERT GRADE

The international standards for toolholder and inserts do not include the definition of tool grades, i.e. the material the insert is made of. There is an independent standard for this.

The environment that a cutting edge experiences is very rough. When forming a chip, a new surface is generated that is chemically very active and that reacts with the insert edge material. In processes that intermittently generate chips (such as in milling or turning a slotted part), the edge undergoes cyclic mechanical and thermal loadings. The severity of the loading depends partially on the cutting conditions selected by the process planner. The conditions that affect the cutting edge are:

- elevated temperature generated by the cutting process;
- cyclic mechanical and thermal loading;
- chemical reaction with the surface in contact; and
- high normal and tangential stress.

The design requirements of the insert are:

- hardness greater than that of the workpiece;
- reduction in wear of the insert;
- ability to retain hardness at elevated temperatures;
- toughness to resist impact cyclic loading; and
- chemically inert to the workpiece material and the cutting fluid.

The problem is that it is very difficult to find one material that possesses all the above requirements. Materials can be found that have high hardness and are chemically inert at room temperature, but they lose their hardness at higher temperatures. Tool manufacturers are constantly developing new materials that improve performance, such as cemented carbide, coated tools, ceramics, cubic boron nitride, TiC coated sintered carbide, TiN coated sintered carbide, etc. So far however, no insert material has been developed to include all the above requirements. A tool may have maximum toughness or maximum hardness (wear resistance), but not both at their maximum values. That means that tools appropriate for tough jobs will use lower cutting speeds than tools for light jobs.

ISO and ASA have set up a standard of tool material properties showing relative hardness and toughness. The hardness curve can be seen in Fig. 11.2 and the values are in Table 11.1.

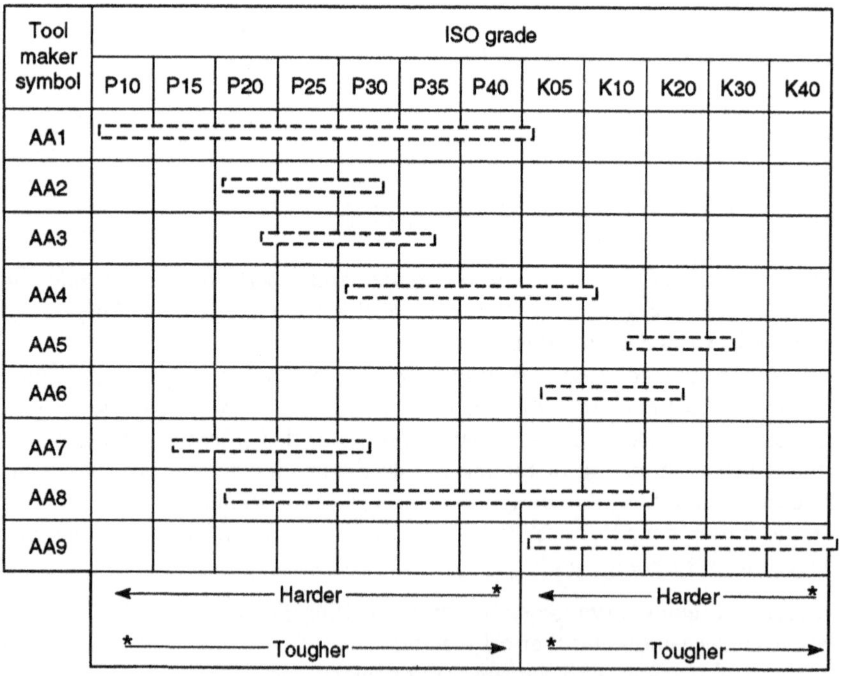

Fig. 11.2 Insert material grade of commercial inserts.

Table 11.1 Standard designation and properties of insert grades

Designation		*Transverse rupture strength*		*Hardness Rockwell A*
ISO	*ASA**	*TRS psi*	*N/mm²*	HR_A
P05	C8	220 000	1450	92.8
P10	C7–C8	260 000	1800	92.5
P20	C7	280 000	2000	92.5
P25	C6	340 000	2400	91.7
P40	C5	370 000	2600	90.9
K01	C4	280 000	1950	93.0
K10	C4	330 000	2250	92.5

* ASA = American Standards Association.

Commercial tools usually have the manufacturer's code name for the insert material, a material that usually covers the range of several ISO or ASA classes. For example, a tool that covers the range of P10 to P35 is highly suitable for a range of operations.

The definitions of rough cut, semi-finish cut or finishing cut are somewhat cumbersome and arbitrary. However, this definition will affect the selection of insert grades. For rough, tough passes, a P50 grade will be recommended and for finishing passes P01 will be recommended. A finishing pass, according to surface roughness definition is a pass that brings the workpiece to its specified dimension, regardless of feed rate and depth of cut.

The acting forces in the cutting operation are a function of depth of cut (a) and feed rate (f), which has bearing on the toughness and the wear resistance and hardness and affects cutting speed (v_c). Table 11.2 gives recommendations for the ISO insert grade to be used as a function of workpiece material, depth of cut (a) and feed rate (f).

The actual insert to be used must be a commercial one. Tool manufacturers publish their tool grades and the range covered by them (Fig. 11.2). The process planner has to check that the tool used for several segments (Fig. 11.1) can be a commercial tool that covers the range. Otherwise, an additional tool must be used. It will probably use the same toolholder and have the same insert shape, but have a different insert grade.

The use of the recommendations in Table 11.2 does away with the need for a rule of thumb for using different tools for roughing cuts and finishing cuts. Such a decision will be reached automatically. Moreover, the recommended cutting speed for tools with higher toughness, to withstand high forces, is lower than for tools with high wear resistance. The above equations and tables take care of these considerations.

Table 11.2 Recommendation for ISO insert grades as a function of workpiece material, depth of cut (a) and feed rate (f)

Number		*Name*	$a \times f = AF^*$	ISO grade
1	11	Carbon steel – free machining	$AF \leqslant 0.25$	P10–P20
2	12	– wrought/cast low carbon	$0.25 < AF \leqslant 3.0$	P25–P35
3	13	– wrought/cast medium carbon	$AF > 3.0$	P40–P50
4	14	– wrought/cast high carbon		
5	21	Alloy steel – free machining	$HBN > 450$	
6	22	– wrought/cast low carbon	$AF \leqslant 0.25$	P01–P05
7	23	– wrought/cast medium carbon	$AF > 0.25$	P10–P20
8	24	– wrought/cast high carbon	$HBN \leqslant 450$	
9	25	High strength steel wrought/cast	$AF \leqslant 0.25$	P10–P20
10	26	Maraging steel wrought	$0.25 < AF \leqslant 3.0$	P25–P35
11	27	Tool steel wrought/cast	$AF > 3.0$	P40–P50
12	31	Free machining – SS* wrought ferritic	$AF \leqslant 0.25$	P10–P20
13	32	– SS wrought austenitic	$0.25 < AF \leqslant 3.0$	P25–P35
14	33	– SS wrought martensitic	$AF > 3.0$	P40–P50
15	34	Stainless steel – wrought/cast ferritic	$HBN > 450$	
16	35	– wrought/cast austenitic	$AF \leqslant 0.25$	P01–P05
17	36	– wrought/cast martensitic	$AF > 0.25$	P10–P20
			$HBN \leqslant 450$	
			$AF \leqslant 0.25$	P10–P20
			$0.25 < AF \leqslant 3.0$	P25–P35
			$AF > 3.0$	P40–P50
18	37	P.H. Stainless steel wrought/cast	$AF \leqslant 0.25$	P10–P20
			$0.25 < AF \leqslant 3.00$	P25–P35
			$AF > 3.0$	P40–P50
19	41	Gray cast iron	$AF \leqslant 0.25$	K10–K15
20	42	Compacted graphite cast iron	$AF > 0.25$	K20–K25
21	43	Ductile cast iron	$HBN > 240$	
22	44	Malleable cast iron	$AF \leqslant 0.25$	P01–P05
			$AF > 0.25$	P10–P20
			$HBN \leqslant 250$	
			$AF \leqslant 0.25$	P10–P20
			$AF > 0.25$	P25–P35
23	51	High temper alloys wrought/cast cobalt and nickel base	$AF \leqslant 0.1$	K01–K05
			$0.1 < AF \leqslant 0.5$	K10–K15
24	52	High temper alloys wrought/cast iron base	$AF > 0.5$	K20–K25
25	53	Refractory alloys wrought/cast columbium		
26	54	Refractory alloys wrought/cast molybdenum	$AF \leqslant 0.25$	K01–K05
27	55	Refractory alloys wrought/cast tantalum	$AF > 0.25$	K10–K25
28	56	Refractory alloys wrought/cast tungsten		

Table 11.2—*continued*

Number		Name	$a \times f = AF^*$	ISO grade
29	61	Nickel alloys wrought/cast	$AF \leqslant 0.25$	P01–P05
			$0.25 < AF \leqslant 3.0$	P10–P20
			$AF > 3.0$	P25–P35
30	62	Titanium alloys wrought/cast	$AF \leqslant 0.25$	K01–K05
			$0.25 < AF \leqslant 3.0$	K10–K15
			$AF > 3.0$	K20–K25
31	71	Aluminum alloys wrought/cast		
32	72	Aluminum – titanium		
33	73	Aluminum – bronze	$AF \leqslant 0.25$	K01–K05
34	74	Copper alloys free machining	$AF > 0.25$	K10–K25
35	75	Copper alloys		
36	76	Copper nickel		
37	77	Phosphor – bronze		

* SS = Stainless steel.

11.3 STANDARDS FOR INDEXABLE INSERTS

Each digit in the ISO code is independent of the others and is assigned a value.

Toolholders and inserts selection have to be made jointly, as they consist of one working tool. Some decisions should be made based on the insert constraints, while others should be based on the toolholder constraints. This section concentrates on inserts, and the next section elaborates on toolholders.

The first digit

The first digit represents insert shape. The shape of an insert largely determines its strength and cost per cutting. There are six insert shapes commonly used in turning and boring, as shown in Table 11.3. A shape is obviously stronger than

Table 11.3 Available cutting edges for tool inserts

Insert shape	*Code*		*Available cutting edges*	
			Rake attitude	
	ISO	*ANSI*	*Positive/neutral*	*Negative*
Round	R	R	4–10	8–20
Square	S	S	4	8
80°/100° diamond	C	C	4	8
Triangle	T	T	3	6
55° diamond	D	D	2	4
35° diamond	V	V	2	4

another if it has a larger included corner angle. Thus, it is recommended that the strongest insert shape be selected that will produce the part configuration, as demonstrated in section 11.1. Moreover, stronger inserts cost less per cutting edge and dissipate heat better, which allows an increase in cutting speed without a corresponding loss in tool life.

The 35° diamond tool should be used only for low strength operations such as finishing.

The round insert is a high strength tool. However, the direction of the feed forces is highly dependent on the depth of cut. The higher the depth of cut, the greater the component of feed force which acts in the radial direction. Therefore, on operations where chatter or inability to hold tolerances due to deflection is a problem, round inserts are not recommended.

The second digit

The second digit represents clearance angles. The clearance angle decreases friction between the major flank surface and the workpiece. This reduces heating at the clearance surface and consequently reduces wear. The selection of the clearance angle depends on the properties of the material being machined and the feed rate.

A large clearance angle is generally used for soft and ductile materials; hard and brittle materials require less clearance. The clearance angle can improve tool life by reducing friction, but, as the clearance angle increases, the strength of the tool decreases, which decreases tool life. A decrease in feed rate increases the wear on the flank of the tool, requiring a larger clearance angle for optimum performance. Therefore, tools for finishing operations, which employ low feed, should have a larger clearance angle than tools used for roughing operations which need improved strength. Table 11.4 might serve as a recommendation for selecting clearance angles for carbide tools.

In practice, the most popular clearance angles on the market are: $N=0°$; $C=7°$; and $P=11°$. The clearance angle is also the fourth digit of the toolholder ISO coding system.

The actual clearance angle is a combination of the toolholder and the insert. The advantage of selecting a toolholder that results in the clearance angle is that an insert with zero (0°) clearance angle can be used on both sides, i.e. double the number of cutting edges per insert. Inserts with a larger clearance angle can be used only from one side.

The third digit

The third digit represents tolerances. Standard inserts are available in varying tolerance classes. The tolerances refer to thickness of the insert, theoretical diameter of the insert inscribed circle and maximum height or the difference between maximum height and the inscribed circle (depending on insert shape). Moreover, in some shapes the tolerance is also a function of size.

Table 11.4 Recommendations for selecting clearance angles

Workpiece material	*Clearance angle for carbide tools* Feed > 0.3 mm	Feed < 0.3 mm
Carbon and alloy steel		
$\sigma < 450$ N/mm^2	—	—
450–770	8	12
> 770	—	12
Ductile brass	—	—
Malleable cast iron		
< 160 BHN	8	12
160–220 BHN	—	—
Cast iron		
160–220 BHN	—	—
> 220 BHN	6	10
Steel casting	—	12

Unground inserts usually have code 'M' (molded) tolerance, while ground inserts usually have code 'G' (ground) tolerance. The biggest reason for preferring an unground insert is that it costs less and is somewhat stronger. It is usually offered with a center locking hole and a formed chip groove, which means that either a pin or a finger clamp can be used to lock an unground insert. Mechanical chipbreaker plates are not required if they have pressed-in chip control grooves.

Ground inserts should only be used whenever they can eliminate manual adjustment of cutting tool position (operation is initially set up). They can eliminate manual tool adjustment whenever tolerances on the operation are not very close nor very wide, i.e. in the range 0.07–0.15 mm. The exact value of the endpoint of the range depends on maximum allowable flankwear and other factors.

Ground inserts are not helpful on close tolerance work (less than approximately 0.07 mm). Such tolerances are obtained by manual adjustments and thus precision inserts are simply not necessary. Neither are they needed when tolerances are wide (more than approximately 0.15 mm), because unground inserts can hold wide tolerances without adjustment.

The fourth digit

The fourth digit represents the type of insert. It depends mainly on two parameters: clamping system and workpiece material.

The clamping system specifies if the insert has to have a clamping hole in it or not. This is the first digit in the ISO toolholder coding system, and is discussed in the appropriate section.

The workpiece material affects the types of chip that will be produced. If the rake surface of the insert is a flat plane, the chip may flow as a continuous ribbon. It may then wrap around the tool post, take a turn out toward the operator, or wind back to the workpiece and possibly be recut. In short, it is out of control. To bring it under control, grooves are added to the surface of the insert, sometimes called a chip breaker. If the geometry of the groove is suited to the material and cutting conditions, the chip will normally hit the uncut workpiece or the side of the toolholder below the insert and break off after one turn.

Use of pressed-in groove generally increases true rake angle, thus improving cutting tool performance, reducing cutting forces, increasing tool life or increasing cutting speed without loss of tool life.

The shape of the groove determines whether a cutting edge can be provided on both sides of the insert. High feed groove geometries normally are deeper and wider than conventional grooves. The increased size of these grooves weakens the insert and prevents putting them on both sides of the insert. Tool manufacturers may supply charts or tables for selection of groove geometry and selection of the type of insert.

The fifth digit

The fifth digit represents the cutting edge length and should be selected according to the maximum depth of cut to be taken. In order to support the forces of the cut adequately, the available cutting edge length of the insert used should be at least twice the maximum cutting edge engagement. The smallest insert with a cutting edge length that meets the above rule should be selected in order to minimize tool cost.

There are only a limited number of cutting edge lengths on the market and a standard size is always recommended.

This digit should be preceded by a zero. i.e. 12.7 will be '12'; 9.525 ($\frac{3}{8}$ in) will be '09'. In the ANSI insert identification system, the size is:

- number of $\frac{1}{32}$nds on inserts less than $\frac{1}{4}$ in, inscribed circle (IC);
- number of $\frac{1}{8}$ths on inserts $\frac{1}{4}$ in IC and over; and
- for rectangle and parallelograms inserts, two digits
  1st position – number of $\frac{1}{8}$ths in width
  2nd position – number of $\frac{1}{4}$ths in length.

The sixth digit

The sixth digit represents the insert thickness. The thickness of the insert should be capable of withstanding the intended feed rate and depth of cut, i.e. the cutting forces. Harder materials generally require thicker inserts than softer materials. The user may use tool manufacturer's nomographs or tables, which are based on feed rate and cutting edge length engagement, or use the following equation rounding the answer to the closest commercial equivalent available.

$$T = 3 \times L^3 \times f^{0.6}$$

where: T = insert thickness (mm)
L = cutting edge length engagement (mm)
f = feed rate (mm/revolution)

This digit should be preceded by a zero, i.e. 4.76 will be '04'; 6.35 will be '06'.

The seventh digit

The seventh digit represents the tool nose radius. All inserts have a nose radius because it improves surface finish and is much stronger than a sharp corner. The nose radius provides increased resistance at the point where the major cutting edge and minor cutting edge join.

However, when turning slender workpieces of insufficient stiffness using carbide tools with a very large nose radius, chatter may occur resulting in chipping of the cutting edge. Thus, vibration or chatter and the resulting rapid failure of the cutting edge sets the limits of tool nose radius. Reducing the nose radius lowers the radial component of the cutting force and the interface temperature as well. A large nose radius is recommended for machining brittle materials which produce discontinuous chips, and is preferred with a larger depth of cut. Table 11.5 gives recommended nose radii as a function of depth of cut.

The ISO designation includes only the first seven symbols. The eighth and ninth digits are supplementary, where each tool manufacturer may add additional symbols.

Table 11.5 Recommended nose radii as a function of depth of cut

Depth of cut (mm)	*Nose radius (mm)*
0.5 or less	0.2 to 0.4
0.5 to 1	0.5 to 0.8
1 to 3	0.8 to 1
3 to 10	1 to 1.5
10 to 20	1.5 to 2
20 to 30	2 to 3

11.4 STANDARDS FOR TOOLHOLDERS

Each digit in the ISO code is independent of the others.

Toolholders and inserts selection have to be made jointly, as they consist of one working tool. Some decisions should be made based on the insert constraints while others should be based on the toolholder constraints. This section elaborates on toolholders.

The first digit

The first digit designates clamping systems. The heart of a toolholder is the pocket that holds the insert and the mechanism that locks it in place (Fig. 11.3).

The insert is located on the back of the pocket wall. This poses accuracy problems of two kinds. Firstly, the wall is never at a perfect 90° angle. Machining pressure forces the cutting edge back against the pocket wall and the edge can be chipped. Secondly, after indexing the insert, a wear land of the insert is the locating wall, which has an erratic shape. Therefore, manufacturers are now using different pocket designs, mainly to ensure a line contact instead of a surface contact, thereby eliminating the accuracy problem.

One very common insert locking method uses an unthreaded cam pin as a primary locking member. The pin is located in a hole in the insert pocket. The top end of the pin fits into the insert hole. When the pin is rotated, the insert is forced back into the pocket. Friction between pin and insert prevents the pin from rotating after it is locked. The problem with this type of locking system is that a locking mechanism works whether the pin is rotated clockwise or counterclockwise, yet the location of the insert in the pocket will not be the

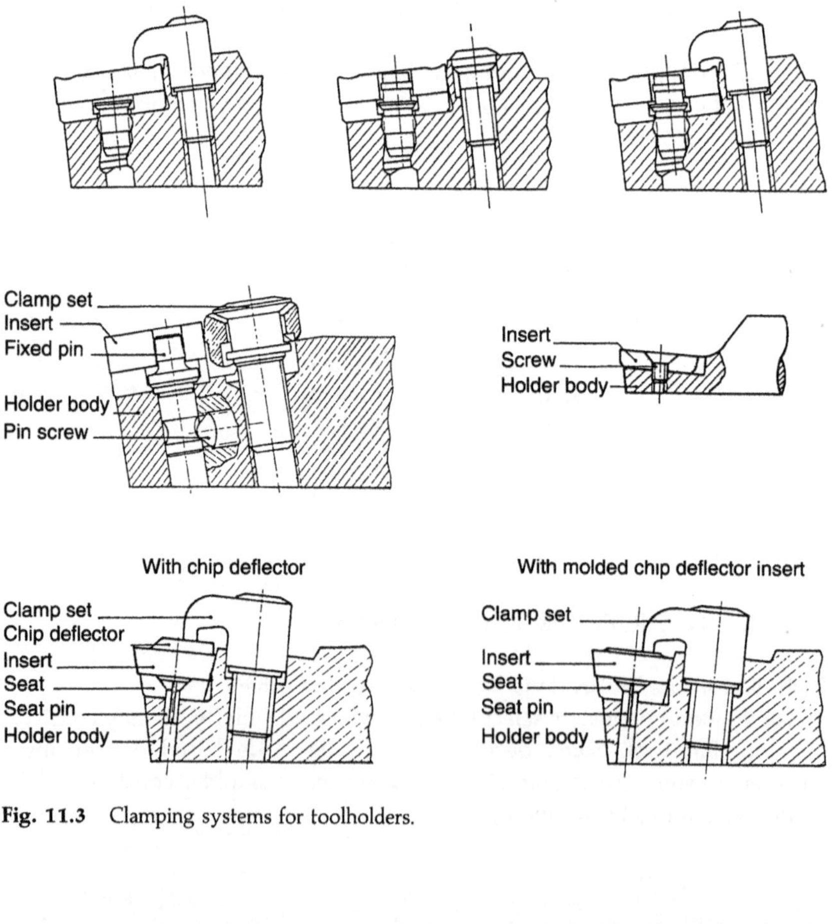

Fig. 11.3 Clamping systems for toolholders.

same. Tool manufacturers are aware of this phenomenon and have developed several patents to overcome it.

The top clampings are primarily intended for external finishing operations which demand a positive cutting geometry and high precision tolerances. It is a simple design with few parts and allows the use of small shank dimensions.

The hole clampings are used for external rough and finishing operations in standard tolerances. The absence of a protruding top clamp provides unhampered chip flow. It is a low cost and quick change and indexing operation. The design allows the use of a shorter toolholder.

The second digit

The second digit represents the insert shape.

The third digit

The third digit represents toolholder type in relation to entering angles. Both are linked together and were discussed in section 11.1.

The toolholder type is based on entering angles (also referred to as cutting edge angle, or lead angle) and shank offset (Fig. 11.4a). Usually, the user cannot choose the specific type required, but has to select a toolholder from the limited variety offered by tool manufacturers. Smaller cutting edge angles are preferred, for the following reasons:

- the cut is spread over a broader section of the cutting edge (Fig. 11.4b);
- it has a greater heat dissipating capacity; and
- it protects the tool nose radius area (Fig. 11.4c).

There are nevertheless some limitations on selecting tools with a small cutting edge angle (Fig. 11.4d):

- part specifications, see section 11.1;
- an increasing tendency to chatter (because of cutting forces directions);
- a decrease in the chucking stiffness (because of cutting force directions); and
- the insert size must be increased.

Selecting a toolholder with or without shank offset (Fig. 11.4a) depends on part specifications and the operations to be performed. Toolholders with shaft offset can work nearer to the chuck jaws without turret carriage interference.

The fourth digit

The fourth digit represents rake angle attitudes (clearance angle). The clearance angle was discussed in section 11.3 under *Second digit*.

The working clearance angle is a combination of the insert and the toolholder rake attitude (Fig. 11.4e). In ISO standards, the insert clearance angle is given, whereas the working clearance angle is not specified. In ANSI standards, this digit represents the rake attitude as follows:

N = negative rake
O = neutral rake
P = positive rake.

If an insert with flat rake surface is used, the rake angle is determined solely by the toolholder. In other cases, the rake angle is measured on the toolholder with the insert attached to it (Fig. 11.4f).

The primary advantage of negative rakes is that the insert will have twice as many cutting edges as the one designed for neutral or positive rakes. Positive and neutral rakes require relief under the cutting edge to provide clearance for

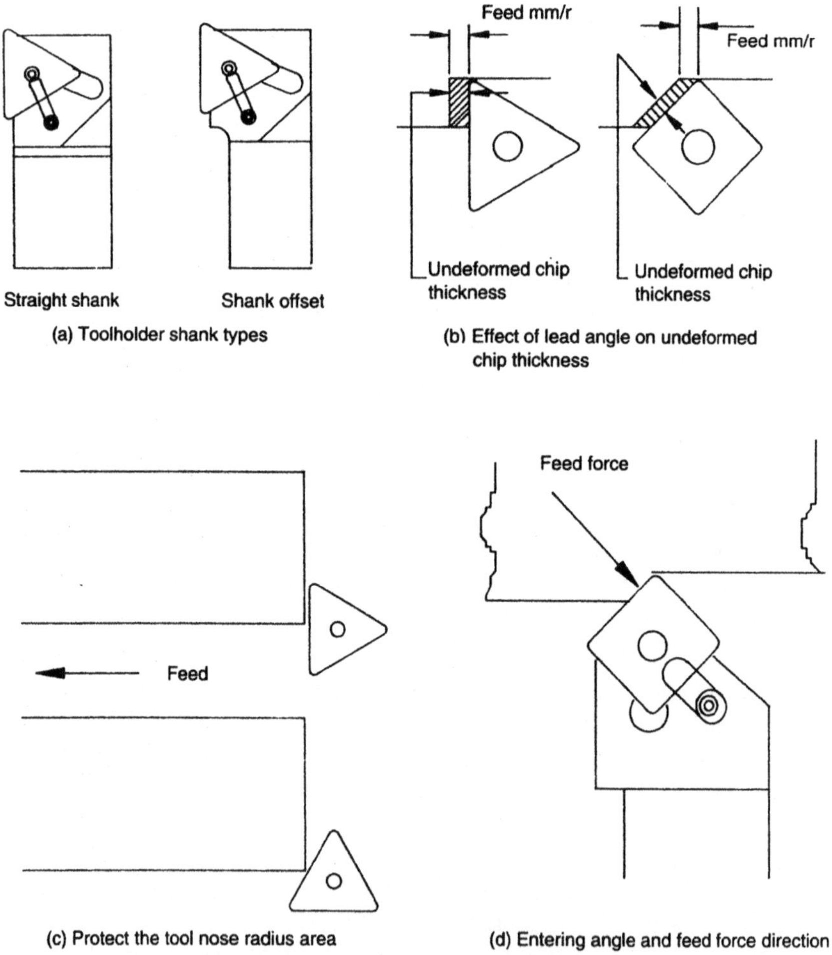

Fig. 11.4 Toolholder angles.

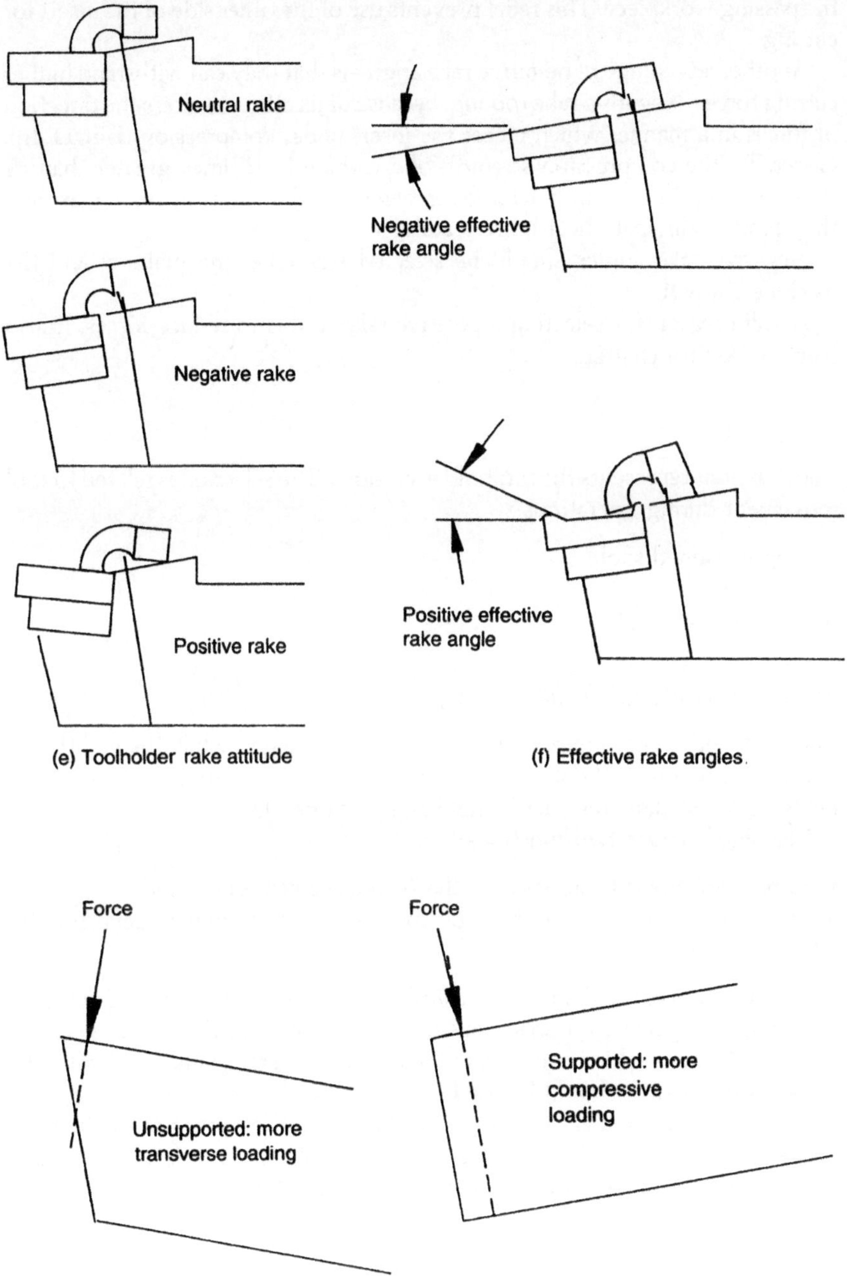

(e) Toolholder rake attitude

(f) Effective rake angles

(g) Rake angle and cutting forces

Fig. 11.4—*continued*

the passing workpiece. This relief prevents use of the other side of the insert for cutting.

Another advantage of negative rake angles is that they can withstand higher cutting forces. Negative rake tooling, because of its attitude, alters the direction of forces in a manner which places the insert under compression (Fig. 11.4g). Generally, the compression strength of a carbide is $2\frac{1}{2}$ times greater than its transverse rupture strength. Moreover, negative rake inserts increase mass and they tend to dissipate heat more readily.

Negative rake angles should be selected whenever the material and the machine allow it.

A valid reason for selecting a positive rake tool is to reduce forces. It also helps to control chatter.

The fifth digit

The fifth digit represents the toolholder versions. The selection is related to tool movement during operations:

R = right handed tool
N = neutral
L = left handed tool.

The sixth, seventh, eighth and ninth digits

The sixth and seventh digits represent the shank height, while the eighth and ninth digits represent the shank width. Both of these digits are two-digit integer fields with a leading zero, i.e. 8 mm height will be '08'.

The shank serves two functions:

- to position the cutting edge on the workpiece centerline; and
- to provide enough support to the cutting edge to minimize deflection by cutting forces.

The selection of the shank size is usually determined by the selected machine. Each machine specifies the shank size that may be used, so there is no selection problem. If the machine tool can accommodate several shank sizes, then the largest size possible should be used.

A standard toolholder is designed so that the cutting point of the insert is located in the same plane as the top of the shank, and is automatically positioned in the longitudinal centerline of the machine.

The tenth digit

The tenth digit represents the tool length. To provide enough support to the cutting edge to minimize deflections by cutting forces, the length should be as short as possible, while being able to perform the operations without toolholder interference in the machine or the workpiece.

11.5 CONCLUSION

For each type of machining process, there are specific ISO standards covering tools. They all are structured in a similar manner.

The purpose of this chapter was to show the logic of the standard and the selection of the digit for each parameter. The same considerations can be employed in selecting digits in the standards for types of processes other than turning.

CHAPTER 12

SPC – statistical process control

Statistical process control (SPC) is a technique for error prevention rather than error detection. SPC products will be of the required quality because they are manufactured properly rather than because they are inspected. Thus, it increases productivity by reducing scrap and rework and provides continuous process improvement. Other methods – such as flexible manufacturing systems (FMS), computer integrated manufcturing (CIM) and just-in-time (JIT) – which have aimed at increasing productivity, concentrate on hardware flexibility, integration of information flow or reduction of inventory. Seldom does a system use technology flexibility to produce the correct product.

SPC is accomplished by technological means, using statistics for detection and technology for prevention.

12.1 INTRODUCTION TO SPC

SPC is statistically based and logically built around the phenomenon that variation in a product is always present.

There is a natural variation inherent in any process due to wear of tools, material hardness, spindle clearance, jigs and fixtures, clamping, machine resolution, repeatability, machine accuracy, tool holder accuracy, accumulation of tolerances, operator skill, etc. Variation will exist within the processes. Parts that conform to specifications are acceptable; parts that do not conform are not acceptable. However, to control the process, reduce variation and ensure that the output continues to meet the expressed requirements, the cause of variation must be identified in the collected data or in the scatter of data. Collection of this data is characterized by a mathematical model called 'distributions' which is used to predict overall performance.

Certain factors may cause variations that cannot be adequately explained by the process distribution. Unless these factors, also called 'assignable causes', are identified and removed, they will continue to affect the process in an unpredictable manner.

A process is said to be in statistical control when the only source of variation is the natural process variation and 'assignable causes' have been removed.

SPC identifies changes between items being produced over a given period, and distinguishes between variations due to natural causes and assignable causes. Corrective action may therefore be applied before defective products are produced. A properly conducted SPC program recognizes the importance of quality and the need for a never-ending search to improve it by reducing variations in process output. Parts will be of the required quality because they are manufactured properly, not because they are inspected. SPC is basically opposed to methods involving part sorting, such as the sorting of conforming parts from noncomforming ones.

Variations that are outside the desired process distribution can usually be corrected by someone directly connected with the process. For example, a machine set improperly may produce defective parts. The responsibility for corrective or preventive action in this case will belong to the operator, who can adjust the machine to prevent recurring defects. Natural variation will establish process capability. The process must be in control in order to apply SPC. A process in control has its upper and lower control limits, which establish the suitability of the process to the task and the anticipated scrap and rework percentages. Inherent capability of process factor (C_p) will indicate if:

(a) the process is capable;
(b) the process is capable but should be monitored; or
(c) the process is not capable (Fig. 12.5).

Natural process variations may only be corrected by redesigning the part and the process plan.

Successful SPC control requires action in the form of a monitoring system and a feedback loop, in a corrective and preventive action plan. A control chart may be in place to record the average fraction defective at a work station, but it is only of marginal value unless the people responsible for the process know what action to take when the process moves out of control.

SPC eliminates subjectivity and provides a means of comparing performance to clearly defined objectives. The control chart used to identify variability and the existence of assignable causes is also used to track process improvements.

Through application of statistical techniques, problems are identified, quantified and solved at source in an optimum time. Out-of-control conditions become evident quickly, as does the magnitude of the problem. With this information, action can be taken before the condition becomes a crisis.

Immediate feedback is the key to the success of any SPC system. SPC is not solely a quality department function. The responsibility for control is in the hands of the producer. This provides the dual advantage of giving the operator a better understanding of what is expected, as well as providing a means of detecting undesirable conditions before it is too late.

12.2 GOALS AND BENEFITS OF SPC

The goals of an SPC program are consistent with typical company goals of:

- improving quality
- increasing profit
- enhancement of a competitive advantage.

SPC analyzes and controls the performance of the activities performed on given inputs to produce resultant outputs, i.e. manufacturing processes. A controlled process offers many advantages to both the producer and the consumer.

The producer will attain lower production costs and lower rework and scrap costs. In addition to these economic considerations, effective process control may justify a reduction in the amount of inspection and test performed on final products.

Specific goals of SPC are as follows:

- *To improve the quality and reliability of products without increasing cost.* This objective is not simply an intrinsically 'good' thing to do, but is a necessity for an organization that wants to remain competitive. Steps taken to improve a process will result in fewer defects and therefore a better quality product delivered to the consumer.
- *To increase productivity and reduce costs.* Application of SPC can produce immediate improvements in yield, reduce defects and increase efficiency, all of which are directly related to cost reduction.
- *To provide a practical working tool for directing and controlling an operation or process.* Implementation of SPC creates a high degree of visibility of process performance. The same statistical technique used to control the process can be used to determine its capabilities.
- *To establish an ongoing measurement and verification system.* Measurements will provide a comparison of performance to target objectives and will assess the effectiveness of problem solutions.
- *To prioritize problem-solving activities* and help with decisions on allocation of resources for the best return on investment. SPC directs efforts in a systematic and disciplined approach in identifying real problems. Less time and effort will be spent trying to correct non-existent or irrelevant conditions.
- *To improve customer satisfaction* through better quality and reliability and better performance to schedule.

Effective process control will enable the producer to fulfill his responsibility of delivering only conforming products and on time. When an out-of-control process is detected, the producer should attempt to identify and correct the cause since the nonconforming products are probably not being produced by chance.

The consumer receives products of the required quality and on schedule.

Additional benefits are, improved quality which will result in lower maintenance, repair, and replacement cost; a smaller inventory of spare parts; higher reliability; better performance and reduced time lost due to defective products.

Benefits of SPC include defect or error prevention rather than merely detection. This means more machine up-time, lower warranty costs, the avoidance of unnecessary capital expenditure on new machines, an increased ability to meet cost targets and production schedules and an increase in productivity and quality.

Additionally, SPC has been used as a basis for product and process design. With detailed knowledge obtained from SPC on product variability with process changes, designers have the capability to design and produce items of the required quality from the first piece. Therefore, SPC control not only helps with design, but results in reduced start-up and debugging effort and cost.

12.3 BASIC STATISTICAL CONCEPTS

Statistics has been defined as the science of organizing and analyzing numerical data for the purpose of making decisions in the presence of uncertainty. In SPC, statistics is employed to identify changes between items being produced over a given period, and to distinguish between variations due to both natural and assignable causes.

Natural causes (common or constant causes) are the distribution of outcomes over a long run, and can be expressed by probabilities. Manufacturing processes behave like a natural causes system; if left to produce parts continually without change, variations would remain, and cannot be altered without changing the process itself.

Assignable causes are events that disturb a process without an immediate known cause, with an outcome that seems unnatural. A change of material, excessive tooling wear, a new operator or accidents may be at the root of assignable causes.

SPC aims for the separation of natural causes from assignable causes and the elimination of assignable causes of variation. The only reliable method of achieving this separation is through the use of control charts, which provide a graphic comparison of a measured characteristic against computed control limits. They plot variation over time, and help to distinguish between the two causes of variation through the use of control limits (Fig. 12.7).

In statistics, the terms population and sample are used quite often. 'Population' refers to the whole collection of units of interest. More precisely, population refers to all the cases or situations to which statistical conclusions, estimates or inferences can be applied. It may be logically impossible or impractical to study all members of the population and thus one has to select a smaller portion, known as a sample, to represent the population. A sample is a subset of the units or individuals from the population. Samples are usually pulled into rational groups called subgroups. Groups of samples that are pulled together in a manner that show little variation between parts within the group, such as consecutive parts taken on a manufacturing line, are considered rational subgroups. The average value of the subgroup members ($\bar{x}$) will be referred to as the mean. It is computed by:

$$\bar{x} = \frac{(x_1 + x_2 + x_3 + \cdots + x_n)}{n} = \frac{\Sigma x_i}{n} \tag{12.1}$$

where n is the size of the subgroup.

Once data of several subgroups has been collected, the overall average can be computed. It will be referred to as $\bar{\bar{x}}$ or the grand average.

$$\bar{\bar{x}} = \frac{(\bar{x}_1 + \bar{x}_2 + \cdots + \bar{x}_N)}{N} \tag{12.2}$$

where N is the number of subgroups.

It is very difficult and time consuming to measure every part manufactured, so the use of a subgroup and statistical analysis will give an idea of what all the parts in the population look like. The statistical concept used to draw conclusions about the population is a measure of the central tendency.

Many processes are set up to aim at a target dimension. The parts that come off the process vary of course, but they are close to the nominal and very few fall outside the high and low specifications. Parts made in this way exhibit what is called a central tendency. The most useful measure of central tendency is the mean or average. The diagram in Fig. 12.1 displays the frequency distribution. The average value should not necessarily be among the high frequency values, i.e. the average does not provide enough information about the subgroup consistency.

Two measures of dispersion are used in statistics, the range and the standard deviation:

1. The range (R) is a measure of the difference between the highest value and

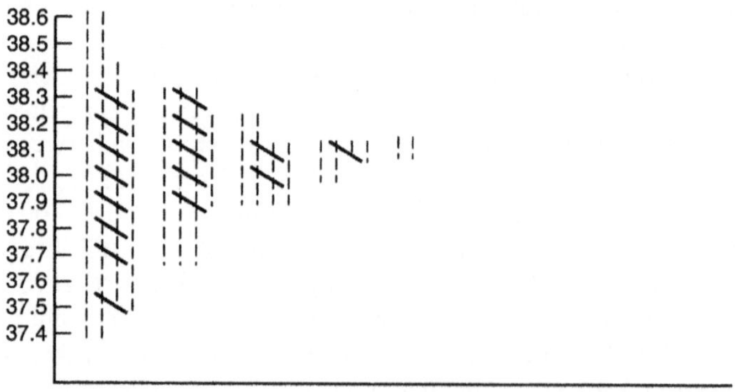

Fig. 12.1 Frequency distribution diagram.

the lowest value of the observation. It gives the overall spread of the data, and is computed by

$$R = x_{max} - x_{min} \tag{12.3}$$

2. The standard deviation (σ) of a sample describes how the points are dispersed around the sample mean ($\bar{x}$). The defining formula for the standard deviation of a population of data is:

$$\sigma = \sqrt{\frac{\Sigma(x-\bar{x})^2}{(n-1)}} \tag{12.4}$$

Note: when the sample size n is large (>30) the denominator $(n-1)$ will be replaced by n.

To make the calculation easier, the following equation has been developed:

$$\sigma_x = \frac{\bar{R}}{d_2} \tag{12.5}$$

where d_2 is a constant taken from Table 12.1.

This equation uses the sample range (R), where the sample size (subgroup) is usually of three to five samples and gives a good estimate of the population standard deviation. However, the sample range is very sensitive to sample size. The range value for larger samples will generally be larger than those for small samples. The d_2 constant, depending on n, is designed to give consistent values of σ regardless of the sample size. The value of $\bar{R}$ is computed by:

$$\bar{R} = \frac{(R_1 + R_2 + R_3 + \cdots + R_k)}{k} \tag{12.6}$$

where k is the number of samples, (usually on 20 to 25 small samples).

Table 12.1 Constants for statistical equations

Sample size	d_2	A_2	D_3	D_4
2	1.128	1.880	0	3.267
3	1.693	1.023	0	2.574
4	2.059	0.729	0	2.282
5	2.326	0.577	0	2.114
6	2.536	0.483	0	2.004
7	2.704	0.419	0.076	1.924
8	2.847	0.373	0.136	1.864
9	2.970	0.337	0.184	1.816
10	3.078	0.308	0.223	1.777

Probability of distribution

There are two basic types of distribution: discrete and continuous. In a discrete distribution, observations are limited to specific values. Typical discrete distributions are the binomial, Poisson and hypergeometric. Observations that can take any value are a continuous or normal distribution, which is the most useful in statistical quality control.

Some of the characteristics of the normal distribution curve are as follows:

- It is represented by a symmetrical 'bell shape' curve, centered about the mean $\bar{x}$.
- The two extremes of the curve are asymptomatic, i.e. as the observations values move away from the mean, the curve gets closer and closer to the horizontal axis, but never reaches it.
- The total area under the curve is equal to 1, and so the area between any two points along the horizontal scale represents the probability or relative frequency of observed value between these two points, (Fig. 12.2). Thus, it is sometimes referred to as the normal probability distribution.

A technique for finding the area between any two points uses values from the normal distribution probability table (Fig. 12.3). Any normal distribution can be converted to standard normal distribution by changing the variable x to a variable z by the formula:

$$z = \frac{(x - \bar{x})}{\sigma} \tag{12.7}$$

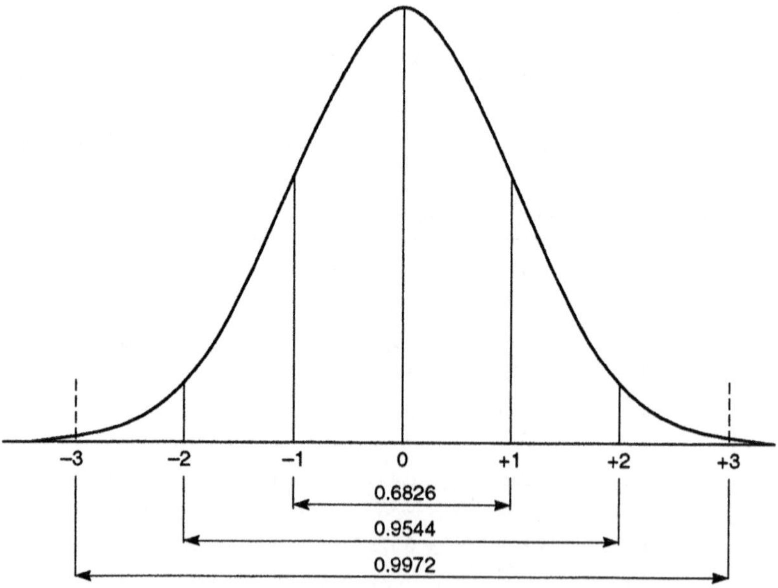

Fig. 12.2 The normal curve.

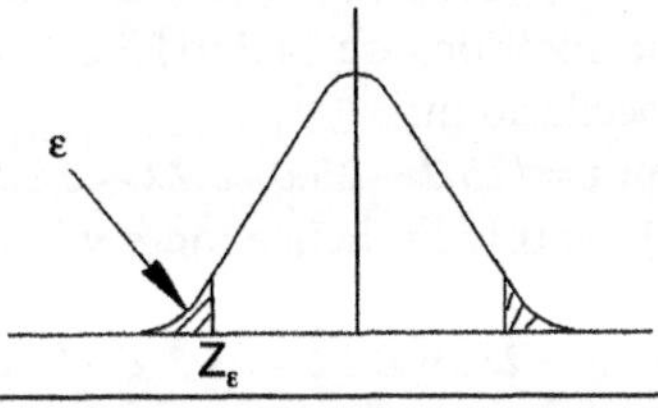

Z_ε	=0	1	2	3	4	5	6	7	8	9
0.0	0.5000	0.4960	0.4920	0.4880	0.4840	0.4801	0.4761	0.4721	0.4681	0.4641
0.1	0.4602	0.4562	0.4522	0.4483	0.4443	0.4404	0.4364	0.4325	0.4286	0.4247
0.2	0.4207	0.4168	0.4129	0.4090	0.4052	0.4013	0.3974	0.3936	0.3897	0.3859
0.3	0.3821	0.3783	0.3745	0.3707	0.3669	0.3632	0.3594	0.3557	0.3520	0.3483
0.4	0.3446	0.3409	0.3372	0.3336	0.3300	0.3264	0.3228	0.3192	0.3156	0.3121
0.5	0.3085	0.3050	0.3015	0.2981	0.2946	0.2912	0.2877	0.2843	0.2810	0.2776
0.6	0.2743	0.2709	0.2676	0.2643	0.2611	0.2578	0.2546	0.2514	0.2483	0.2451
0.7	0.2420	0.2389	0.2358	0.2327	0.2296	0.2266	0.2236	0.2206	0.2177	0.2148
0.8	0.2119	0.2090	0.2061	0.2033	0.2005	0.1977	0.1949	0.1922	0.1894	0.1867
0.9	0.1841	0.1814	0.1788	0.1762	0.1736	0.1711	0.1685	0.1660	0.1635	0.1611
1.0	0.1587	0.1562	0.1539	0.1515	0.1492	0.1469	0.1446	0.1423	0.1401	0.1379
1.1	0.1357	0.1335	0.1314	0.1292	0.1271	0.1251	0.1230	0.1210	0.1190	0.1170
1.2	0.1151	0.1131	0.1112	0.1093	0.1075	0.1056	0.1036	0.1020	0.1003	0.0985
1.3	0.0968	0.0951	0.0934	0.0918	0.0901	0.0885	0.0869	0.0853	0.0838	0.0823
1.4	0.0808	0.0793	0.0778	0.0764	0.0749	0.0735	0.0721	0.0708	0.0694	0.0681
1.5	0.0668	0.0655	0.0643	0.0630	0.0618	0.0606	0.0594	0.0582	0.0571	0.0559
1.6	0.0548	0.0537	0.0526	0.0516	0.0505	0.0495	0.0485	0.0475	0.0465	0.0455
1.7	0.0446	0.0436	0.0427	0.0418	0.0409	0.0401	0.0392	0.0384	0.0375	0.0367
1.8	0.0359	0.0351	0.0344	0.0336	0.0329	0.0322	0.0314	0.0307	0.0301	0.0294
1.9	0.0287	0.0281	0.0274	0.0268	0.0262	0.0256	0.0250	0.0244	0.0239	0.0233
2.0	0.0228	0.0222	0.0217	0.0212	0.0207	0.0202	0.0197	0.0192	0.0188	0.0183
2.1	0.0179	0.0174	0.0170	0.0166	0.0162	0.0158	0.0154	0.0150	0.0147	0.0143
2.2	0.0139	0.0136	0.0132	0.0129	0.0125	0.0122	0.0119	0.0116	0.0113	0.0110
2.3	0.0107	0.0104	0.0102	0.0099	0.0096	0.0094	0.0091	0.0089	0.0087	0.0084
2.4	0.0082	0.0080	0.0078	0.0075	0.0073	0.0071	0.0069	0.0068	0.0066	0.0064
2.5	0.0062	0.0060	0.0059	0.0057	0.0055	0.0054	0.0052	0.0051	0.0049	0.0048
2.6	0.0047	0.0045	0.0044	0.0043	0.0041	0.0040	0.0039	0.0038	0.0037	0.0036
2.7	0.0035	0.0034	0.0033	0.0032	0.0031	0.0030	0.0029	0.0028	0.0027	0.0026
2.8	0.0026	0.0025	0.0024	0.0023	0.0023	0.0022	0.0021	0.0021	0.0020	0.0019
2.9	0.0019	0.0018	0.0018	0.0017	0.0016	0.0016	0.0015	0.0015	0.0014	0.0014
3.0	0.0013	0.0013	0.0013	0.0012	0.0012	0.0012	0.0011	0.0011	0.0010	0.0010

Fig. 12.3 The normal distribution probability table.

Example

A process has a product mean $\bar{x}=25.0$ and the standard deviation is $\sigma=0.22$. Assuming that the product specifications are 24.5 to 25.2, find the percentage of products which are out of specifications.

Solution: for the upper limit $z=(25.2-25.0)/0.22=0.909$ from the table in Fig. 12.3, the area for $z=0.91$ is 0.1814. hence there will be 18.14% product oversize.

For the lower limit, $z=(24.5-25.0)/0.22=-2.27$. From the above table the area is 0.0116, i.e. there will be 1.16% of products undersize.

Total nonconforming products will be $18.14+1.16=19.3\%$.

12.4 PREREQUISITES FOR SPC – PROCESS CAPABILITY

Process capability is the measure of a process performance. Capability refers to how capable a process is of producing parts that are well within engineering specifications.

A capability study is done to find out if the process is capable of making the required parts, how good it is, or if improvements are needed. It should be done on selected critical dimensions.

To employ SPC and the process capability, the process must be under control and have a normal distribution. If the process is not under control, normal capability indices are not valid, even if they indicate that the process is capable.

There are three statistical tools to use in order to determine that the process is under control and follows a normal distribution.

1. Control charts
2. Visual analysis of a histogram
3. Mathematical analysis tests.

The control charts (which are explained in the next section) are used to identify assignable causes. The capability study should be done only for random variations data.

A histogram is a graphic representation of a frequency distribution. The range of the variable is divided into a number of intervals (usually, for convenience, of equal size) and a calculation is made of the number of observations falling into each interval. It is essentially a bar graph of the results. In many cases a statistical curve is fitted and displayed on top of the histogram.

It is possible to obtain useful information about the state of a population by looking at the shape of the histogram. Figure 12.4 shows typical shapes; they can be used as clues for analyzing a process. For example:

- case (a): general type, normal distribution; the mean value is in the middle of the range of data and the shape is symmetrical.
- case (c): positive skew type; the process capability may be excellent, i.e. it uses only part of the tolerance. However, problems of excessive variation

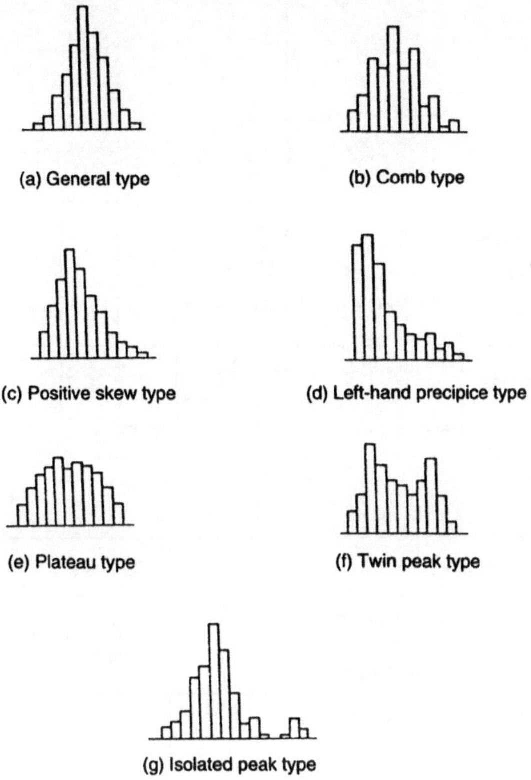

Fig. 12.4 Types of histograms.

caused by shift and an out-of-control process may appear. The cause may be attributed to the machine, the operator or the gauges.

The $\pm 3\sigma$ of a normal distribution curve is regarded as a reasonable process capability and can be computed from (12.5). The capability of the process to meet engineering specifications is measured by comparing the $\pm 3\sigma$ with the tolerance. Figure 12.5 shows the production tolerance versus $\pm 3\sigma$ of process capability. If the tolerances are within the $\pm 3\sigma$, it means that there will be rejected parts. The probability of the percentage of rejects, rework and scrap can be computed by the method shown in the example given in section 12.3. If the tolerances are much wider than the process capability, no production problems of size are encountered, and inspection and SPC are probably not needed.

The most commonly used capability indices are C_p and C_{pk}. C_p, standing for capability of process, is the ratio of tolerance to 6σ. It is computed by:

$$C_p = \frac{\text{tolerance}}{6\sigma} \tag{12.8}$$

As can be seen, the greater the C_p number the better the process. $C_p = 1$ means

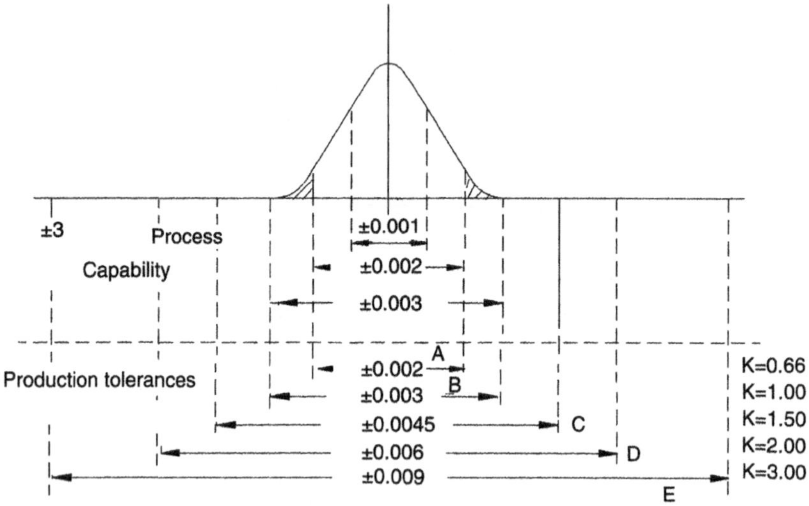

Case	Remarks	Samples
A	High production risk; any shift in average will increase failure.	100% inspection is a must
B	No room for process, average shift, accurate setup	1:1 1:3 1:5 depending on part value
C	Standard requirement of system	1:5 1:10 1:15 depending on part value
D	Improve type C	1:15 1:20 1:30
E	Wide open tolerance, no production problems anticipated	Once a day or batch

Fig. 12.5 Production tolerances vs process capability.

that 99.73% of the parts will be within engineering tolerances. However, any minute deviation from the mean will produce more rejected parts. Therefore, it is usual to aim for $C_p = 1.33$. Nowadays, a target of $\pm 6\sigma$ or $C_p = 2$ becomes a dominant figure.

C_p is only a measure of the spread of the distribution; it is not a measure of centering. The distribution midpoint may not coincide with the nominal dimension, and thus, even when the C_p shows good capability, parts may be produced out of specification. Therefore, the C_{pk} index is introduced.

C_{pk} is a measure of both dispersion and centeredness. It computes the capability index once for the upper side and then for the lower side and selects as the index the lower of both. It is computed by:

$$A_1 = \frac{(\text{upper specification limit} - \text{mean})}{3\sigma}$$

$$A_2 = \frac{(\text{mean} - \text{lower specification limit})}{3\sigma}$$

$$C_{pk} = \min(A_1 \text{ and } A_2) \qquad (12.9)$$

Sometimes, a C_r index is used as a substitute for C_p. It is simply the reciprocal of C_p.

12.5 CONTROL CHARTS

Control charts are the tools for statistical process control. Statistics and parameters by themselves are hard to interpret and visualize. The control chart however is a pictorial method which enables the operator to tell at a glance how well the process is controlling the quality of items being produced.

There are essentially two kinds of control charts, one for variable data (quantitative measurements) and one for attribute data (qualitative data or count). The variable control charts are more sensitive to changes and therefore are better for process control. The $\bar{x}$ and R chart is the most common form of control chart and one of the most powerful for tracking and identifying causes and variations. In this book, only this control chart will be discussed.

Since parts come off a process in large numbers, we need a way to establish and monitor the process without having to measure every part. This is done by taking samples (subgroups) of two to 10 parts (five as the most common size) and plotting the measurements on a chart (Fig. 12.6). The side-way distribution curve (histogram) represents the mean and the range of each subgroup. The $\bar{x}$ and R chart (Fig. 12.7) is easier to make. Instead of calculating and graphing small histograms of data, separate graphs for the mean ($\bar{x}$) and the range (R) are used.

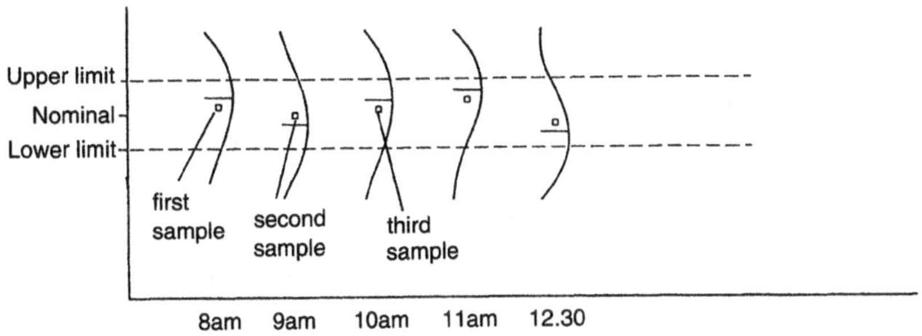

Fig. 12.6 Subgroup and samples.

Sample number	1	2	3	4	5	6	7	8	9	10	11	12	13	14	15	16	17	18	19	20	21	22	23	24	25
Sample measurements	933	911	889	882	903	890	892	908	895	916	901	908	909	895	893	909	885	897	912	882	896	912	926	917	884
	897	898	915	913	930	940	912	920	920	890	892	895	904	902	906	907	892	904	896	894	912	909	903	917	889
	885	900	905	930	890	895	895	896	922	891	892	896	906	902	917	904	942	916	932	941	907	913	908	918	912
	900	905	902	900	890	909	896	894	928	920	895	925	872	932	910	923	911	912	936	934	928	915	910	914	919
	879	862	873	871	900	915	902	906	926	915	898	933	927	932	925	888	916	920	913	917	926	928	885	925	898
Median	897	900	902	900	900	909	896	906	922	915	895	908	906	902	910	907	911	912	913	917	912	913	908	917	898
Range	54	49	42	59	40	50	20	26	33	30	9	38	55	37	32	35	57	23	40	59	32	19	41	11	35

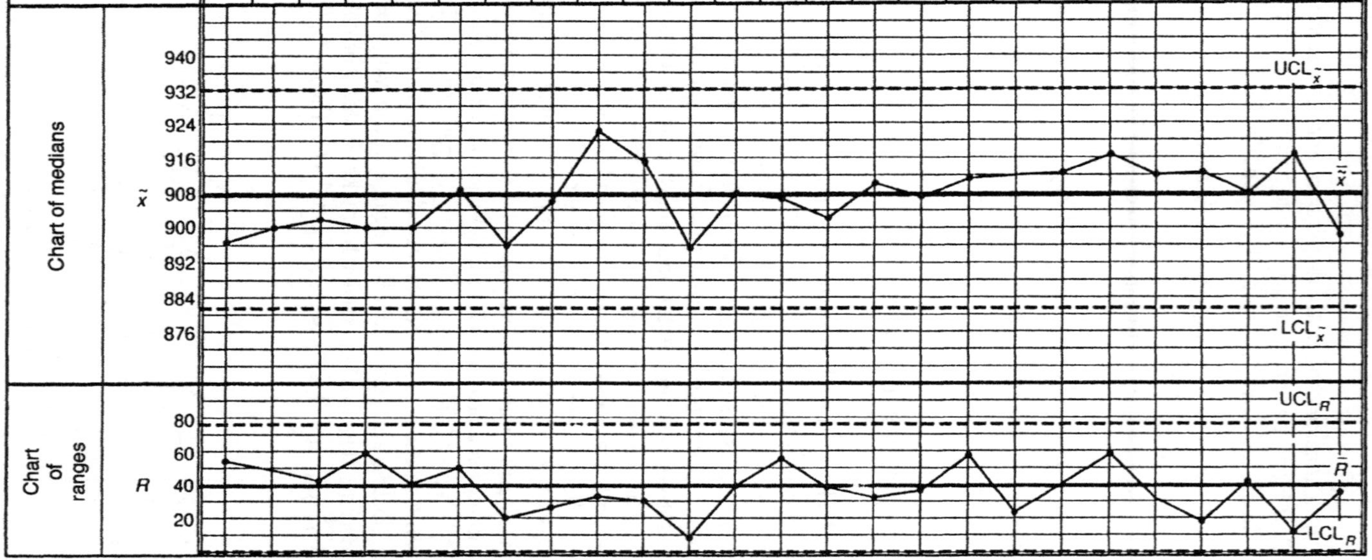

Fig. 12.7 The $\tilde{x}$ and R chart.

To interpret the chart at a glance, the centerline and control limits are drawn on the chart.

The centerline is marked as $\bar{\bar{x}}$ and is the average of the $\bar{x}$ values and computed using equation (12.2).

The purpose of the control limits on the chart is to indicate if the process is under control, i.e. that 99.73% of all the average of subgroups $\bar{x}_i$ will be within these limits. More accurately, under-control customarily means that all $\bar{x}_i$ are within the estimated $\pm 3\sigma\bar{x}$ limits of the process.

According to statistics theorems, the sample mean from a normally distributed population is exactly distributed as a normal distribution, and even if the distribution of a population is not normal, the sample mean is approximately a normal distribution. The approximation holds best for large samples (n), but is adequate for a value of n as low as 5. The formula to calculate the estimated sample deviation is:

$$s = \frac{\sigma}{\sqrt{n}} \tag{12.10}$$

Just as the σ is a measure of variation in a sample, the s is a measure of variation which may be expected when obtaining one observation ($\bar{x}_i$) from the distribution mean. Figure 12.8 shows the relationship between σ and s.

An easier method to compute the control limit is to use the following equations:

$$\begin{aligned} \bar{x} \text{ upper control limit, } U_{\bar{x}} &= \bar{\bar{x}} + A_2 \times \bar{R} \\ \bar{x} \text{ lower control limit, } L_{\bar{x}} &= \bar{\bar{x}} - A_2 \times \bar{R} \\ \text{Range upper control limit, } U_R &= D_4 \times \bar{R} \\ \text{Range lower control limit, } L_R &= D_3 \times \bar{R} \end{aligned} \tag{12.11}$$

where the constants A_2, D_3, D_4 are given in Table 12.1, $\bar{\bar{x}}$ is computed using equation (12.2) and $\bar{R}$ is computed using equation (12.6).

The general rule is that at least 20 points on the control chart (20 subgroups) representing 100 measurements are needed before control limits can be calculated.

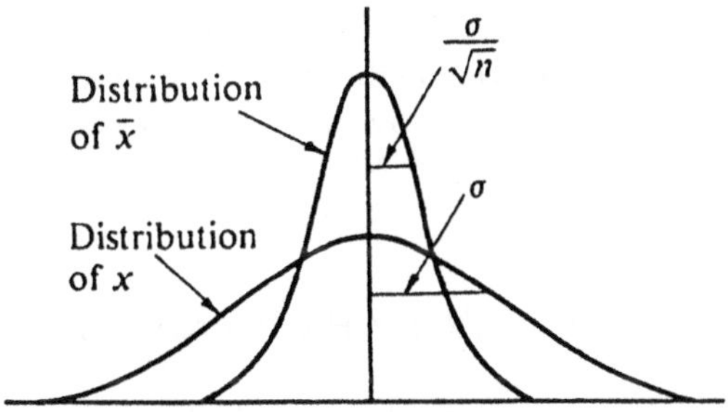

Fig. 12.8 Distribution of $\bar{x}$ and R.

12.5.1 Control chart parameters selection

The points on a control chart are the mean and the range of the sample subgroup. In this section, the subgroup size and frequency of taking measurements is discussed.

A rational subgroup is one where there is a very low probability of assignable causes creating variations measurements within the subgroup itself. If a subgroup has five measurements, then the opportunities for variation among those measurements must be made deliberately small. This usually means the subgroup should be taken from a batch of pieces made when the process operates under the same setting – one operator and no tooling or material changes.

Five consecutive pieces might be the easiest to collect. The logic behind rational subgrouping is that if the variability between pieces within a subgroup is entirely due to common causes, then the differences in subgrouping averages and range will be due to assignable causes. The effect of assignable causes will not be buried within a subgroup and dampened by averaging. They will appear on the chart in the form of a point that exceeds the control limits, or have an identifiable pattern.

If the time of day may contribute to variations between pieces, then one subgroup of five consecutive pieces from the process should be collected at selected times throughout the day. The time interval between subgroups reflects the expected time of variations in the process and the cost and ease of taking the measurements. In stable processes, every few hours might be satisfactory. In processes where tools wear rapidly, or other changes in short periods of time exist, short intervals should be used.

If in a multiple spindle machine some of the spindles are less scattered than others, the subgroup should be five pieces from one spindle and five from the next rather than one piece away from each spindle averaged together.

An $\bar{x}$ and R chart is used to plot one set of causes. If several different machines contribute to a single lot of parts, a chart of samples taken from the lot will not reveal nearly as much as separate charts from each machine. On the other hand, if one machine with the same setup produces different parts, one chart may be sufficient to control all the parts, as we control the process and not the parts. To do it, a chart of deviations from the nominal is used instead of charting the nominal itself.

In cases of small lots produced on the same machine with the same tooling, the technique of charting 'moving averages' or nominals or σ might be the answer.

12.6 INTERPRETING CONTROL CHART ANALYSIS

Analysis is accomplished by the use of control charts, mainly the R chart and the $\bar{x}$ chart. The most common feature of a process showing stability is the absence of any recognizable pattern. The points on the charts are randomly distributed between the control limits. A rare point out of limits on a process that has shown stability over a long period of time can probably be ignored. The characteristics of a stable process are:

- most points near the centerline;
- some points spread out and approaching the limits; and
- no points beyond the control limits.

The characteristics of a non-stable process are:

- points outside the control limits;
- four out of five successive points outside $\pm\sigma$ limit;
- points crowded near centerline;
- two out of three successive points falling outside $\pm 2\sigma$ limits on the same half of the chart;
- a trend of increasing or decreasing trend of seven points;
- 12 out of 14 successive points on the same side of the center line;
- a cycle or pattern that repeats itself; and/or
- patterns that may appear which are unnatural.

Figure 12.9 shows several such cases.

It is recommended to review the *R* chart first, as it is more sensitive to changes. If defective parts start appearing in a process, they will affect the *R* chart. The variation will increase, so some points will be higher than normal. The lower the points in the *R* chart, the more uniform the process. Two machines producing the same part may produce different forms of the *R* chart. Any changes in the process, such as operator inexperience, poor material, tool wear or lack of maintenance will tend to shift points upwards.

When the *R* chart is unstable, the $\bar{x}$ chart can be very misleading. With a stable *R* chart the variations in the $\bar{x}$ chart might be due to a change of material, temperature change, a new tool, machine setup, gradual tool wear, etc.

SPC controls the process and not the parts. Therefore any pattern on the charts that is not normal from a statistical point of view should be an alarm, i.e. something has changed in the process. Action must be taken to correct it. There is a 50% chance of having one point above the centerline and one below it. There is also a chance that 68.25% of the reading will be around the centerline, and so on. Any deviation from such criteria should alert the user.

When a point is out of the control limits, the process shoud be stopped and immediate action should be taken to remove the cause before more incorrect pieces are produced.

When there is a trend, the pieces are still within control; however, in a short period of time, they will run out of control. Therefore the process might keep on producing parts, but action should be taken to determine the cause and eliminate it.

When all points on the $\bar{x}$ chart are within $\pm\sigma$ and very low on the *R* chart, it means that all pieces have been produced within control. Statistically, however,

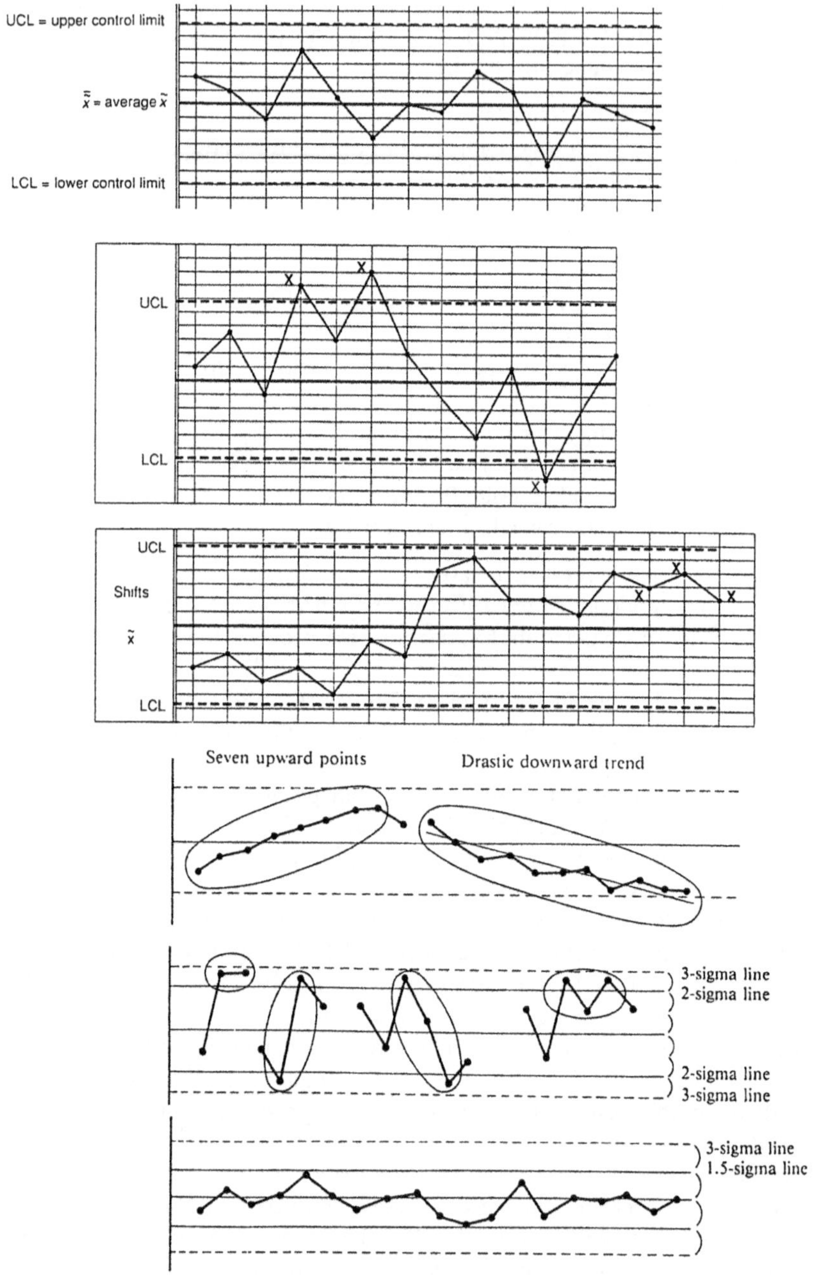

Fig. 12.9 Control chart interpretation.

the result is too good and is therefore impossible! The process may continue to produce pieces, but action should be taken to learn the cause of this shift in the process. A sticking gauge might be the cause, or a new material or a tool.

Action should be taken whenever the process does not behave according to statistical laws, even if it improves the process.

12.7 CAUSE AND EFFECT ANALYSIS – TROUBLESHOOTING

The statistics role in SPC is to spot variations in the process and to alert the operator to take action when needed. However, it does not tell the operator what action to take. The action must be handled by a cause and effect study or troubleshooting technique.

The Pareto diagram is used to highlight the few most important causes and the results of improvements, and assists in determining the relative importance of cause. It can be the first useful document after data collection. Most of the defective items are usually due to a very small number of causes. Thus, if the causes of these vital few defects are identified, most losses can be eliminated by concentrating on these particular ones, leaving aside the other trivial defects for the time being. Figure 12.10 shows a Pareto diagram.

Once the problem is defined, a cause and effective analysis should be carried out in order to determine the actions to be taken to remedy the problem. The more widely used approaches to industrial troubleshooting are as follows:

- The 'what changed?' approach
- Conventional approach
- Checklist
- Kepner Tregoe approach
- Morphological approach
- Brainstorming
- Weighted-factor analysis
- Quality circle method (fishbone diagram)
- Relevance tree
- Statistical approach (experiment design)
- Simulation
- Expert systems.

Variations in products will always be present. The causes may be due to design, process, operator or assignable causes. The last two are easy to analyze and remedy. Usually, the 'what changed?' approach will give good results. This is used for products that have been manufactured for quite some time and with good results, but which have changed. If the effect is changed, then a cause must have occurred at the same time. A quick review of what has most recently changed often provides the clue to the underlying problem. Therefore it is recommended to conduct a system event log book at each workstation, as shown in Table 12.2.

Type of Defect	Tally	Total
Crack	𝍸 𝍸	10
Scratch	𝍸 𝍸 𝍸 𝍸 .... 𝍸 //	42
Stain	𝍸 /	6
Strain	𝍸 𝍸 𝍸 𝍸 .... 𝍸 ////	104
Gap	////	4
Pinhole	𝍸 𝍸 𝍸 𝍸	20
Others	𝍸 𝍸 ////	14
Total		200

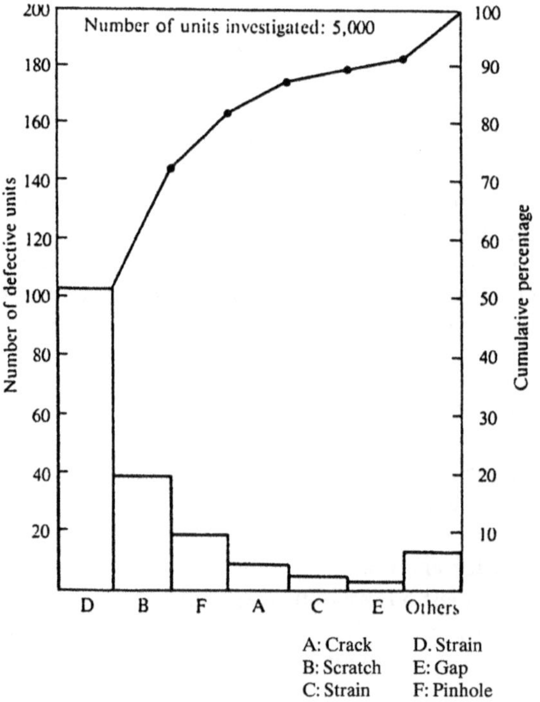

Fig. 12.10 Pareto diagram.

The conventional approach is the one most people use to solve problems. Hypotheses concerning the cause are made, and some potential solutions are developed based on common knowledge. Trials are carried out until the solution is found. This approach is generally not a good approach for new employees or for personnel who are not familiar with the specific operation.

The other approaches are aimed at directing the problem solver to think in a systematic way.

Table 12.2 System event log book

Workstation no. .... Name: ....

Date	*Time/ shift*	*Changes: oper./insp.*	*Changes: tools/material*	*Special notes**

* Special notes on any event such as:
- electricity failure
- visitors interruption
- accidents
- tool breakage
- new measuring instrument
- new set up.

FURTHER READING

Amsden, D., Butler, H. and Amsden, R. (1991) *SPC Simplified for Services*, Chapman & Hall.
DataMyte, (1989) DataMyte Handbook, DataMyte Corp.
Feigebaum, A.V. (1983) *Total Quality Control*, McGraw–Hill.
Hakes, C. (ed.) (1991) *Total Quality Management*, Chapman & Hall.
Lochner, R.H. and Matar, J.E. (1990) *Designing for Quality*, Chapman & Hall.
Lyonnet, P. (1991) *Tools of Total Quality*, Chapman & Hall.
Wetherill, G.B. and Brown, D.W. (eds) (1990) *Statistics Process Control*, Chapman & Hall.

CHAPTER 13

Hole-making procedures

Hole producing seems at first sight to be a simple task. However, there are so many tools and process options available for this operation that care must be taken to select a process which minimizes cost and machining time. The scatter of hole-making time and cost can be within a range of up to 10:1, as shown in Table 6.1.

In order to arrive at the optimum process, the process planner must consider and compute many alternative processes, a time-consuming task; thus it is not a practical approach for small-to-medium production quantities. Therefore, the process planner usually relies on intuition in order to make the process planning time reasonably short.

This chapter describes a method based on technological considerations which can be used to produce an economic process plan in a short period of time. The method can be computerized in order to make the run time even shorter.

13.1 BASIC TECHNOLOGY – CONCEPTS

The process planner should consider all technical engineering specifications of the required holes, such as material hardness, material machinability, solid/cored, open/closed/blind, hole length and diameter, diameter tolerance, surface finish, geometric tolerance (straightness, roundness, concentricity, circular runout), location tolerance, drill marks, core hole details (when applicable) and any special features (slot, thread, chamfer, radius, etc).

There are many tools that can be employed in hole making. Some of these are tool dimension dependent (TDD) such as twist drills, insert drills and reamers, while others are machine dimension dependent (MDD) such as boring and milling tools. Some of these tools can be used to open a through hole in a solid, while others can only enlarge or improve existing holes. Each tool has its capabilities and limitations.

The selection of the right tools for the job is a function of optimization, the machine, the configuration of the center line of the hole, other center lines and the location accuracy of the holes. Therefore, we propose to work with three levels of optimization:

1. *Single hole optimization* At this stage, the best method of producing a hole is computed. Technical boundaries for individual hole operations are defined.
2. *Center line optimization* On each center line, several hole diameters and features may be specified. Each hole diameter will have been computed and optimized as in step 1. In this stage, optimization of all the hole diameters and features on the same center line are made. It is accomplished by selecting a tool type and size that will reduce the number of tools used, and thus, reduce overall hole-manufacturing time (or cost), i.e. reducing machine time and tool change time.
3. *Part optimization* In this stage, the holes on all the center lines are considered. Tool type and tool size may be changed for every center line in order to calculate total manufacturing time (and/or cost).

13.1.1 First-level optimization – single hole

The main concept in this level is that each MDD tool has a maximum depth of cut that corresponds to surface finish and tolerances, and each TDD tool has boundaries of depth of cut, surface finish and tolerance capabilities (a_{max}).

Each tool leaves marks on the machined surface (surface integrity and accuracy). For certain machining requirements, the depth of these marks (a_{min}) must be minimized as described below in order to arrive at the required surface. The maximum depth of cut (a_{max}) of a cutting pass must be larger than the marks left on the part (a_{min}) by the previous operation. If this condition cannot be fulfilled, either an intermediate operation or an alternative operation is required.

First level optimization is as follows. Starting from the finished hole, the sequence of tools to be used is computed for each hole. For each tool (operation), diameter boundaries (DB), as maximum and minimum diameters, are assigned, (Fig. 13.1a). Naturally, these values for the finishing pass are:

$$D_{cmax} = D_{cmin} = D$$

For the pass before the last stage, values are computer as follows:

$$D_{cmax} = D - 2a_{min}$$

$$D_{cmin} = D - 2a_{max}$$

The exact diameter is determined according to the following optimization levels (see section 13.3.4).

The proposed program flow is detailed below.

Stage one

The purpose of this stage is to build two arrays:

1. a list of tools that can produce the finished hole segment (TLSM); and
2. a list of all tools that can be used in the process of making the hole (TLS).

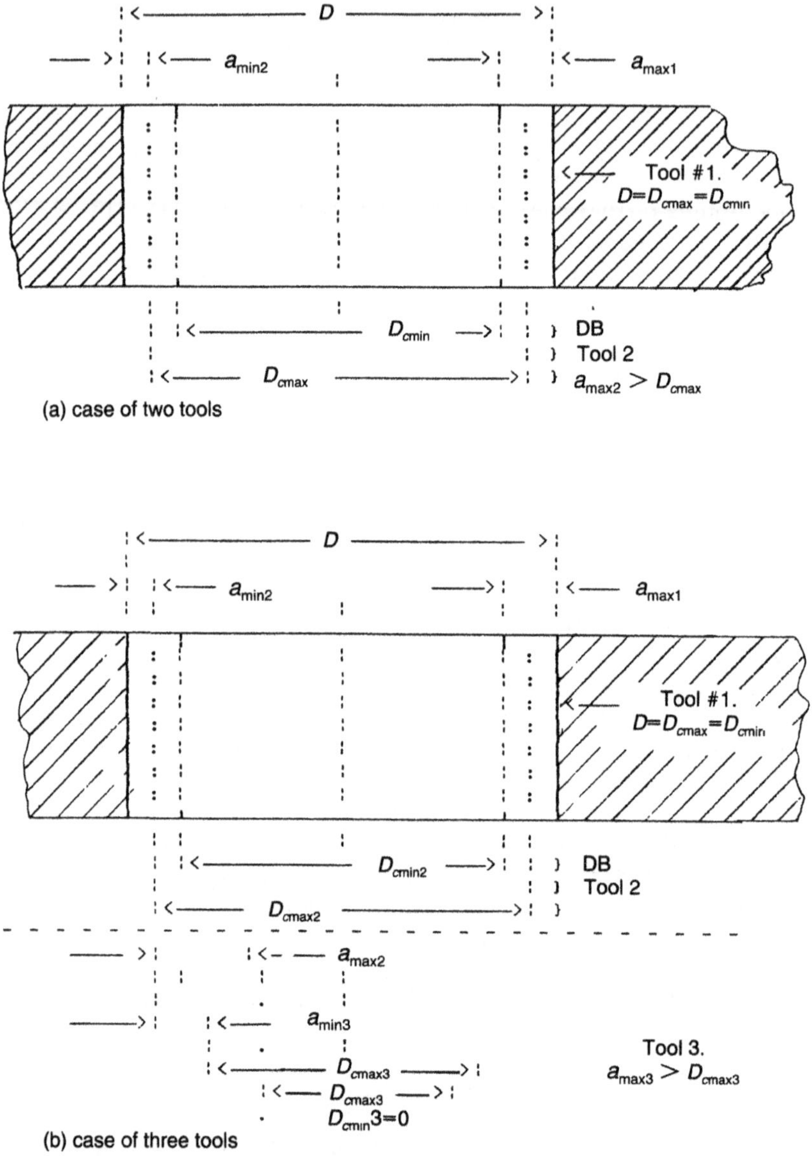

Fig. 13.1 Diametric boundaries.

The arrays are filled by comparing hole specifications to the tool capability table; see section 13.3. For example, if a hole of 20 mm (0.787 in) diameter, 90 mm (3.543 in) deep with 1.5 R_a surface finish is needed, then an insert drill can produce a Ø20 mm hole, but cannot produce a hole of $L/D = 90/20$. Therefore, it will not appear in any array.

A twist drill can make a hole of Ø20 mm, and to a depth of 90 mm, although it cannot produce a hole with 1.5 R_a. Therefore, it will be in array TLS, but not in array TLSM.

A reamer can meet all the above requirements, so it will be in both arrays.

Stage two

The purpose of this stage is to build a working array (TLSW) of the operations needed to produce a hole. The details include tools, sequence of operations, parameters and tolerances for each operation.

The logic of this stage is described as follows:

The first entry to the TLSM array is read. This tool is needed to machine the last operation, therefore it is written in array TLSW. If this tool can also be the first tool, (i.e. it starts a hole from solid, using data from the tool capability table), then it should be used to start and finish the job.

If it cannot be used as the first operation, then its a_{max} value is computed and the a_{min} of the successive operations of the TLS array are computed. If a_{max} is larger than the computed a_{min}, then this tool (from the TLS array) is selected (Fig. 13.1a). If a_{max} is smaller than a_{min}, then the next tool of the TLS array will be examined until a tool is found with a_{min} is smaller than a_{max} (see Fig. 13.1b). When this condition is met, the (tool) operation is written in the TLSW array. Again, if the last written tool (operation) in the TLSW array can be the first operation, the procedure ends. If not, the a_{max} of the last tool is computed, and a search for another tool with a suitable a_{min} is conducted from the TLS array till such an operation is found.

Stage three

The purpose of this stage is to set up cutting conditions, compute cutting time and detail all operations needed to machine a hole.

The input to this pass is array TLSW. It handles one operation at a time, starting from the last entry to the array and ending with the first. The data in this array are, tool type number (corresponding to the tool capabilities table), final diameter, length and surface finish.

The computation of this stage is straightforward. However, due to controlling constraints (machine, chucking, user), there may be a need to modify the operation.

Remarks

1. A cored hole (a hole that was produced by a preliminary process such as casting or forging) is regarded as an operation that should not be carried out, but leaves its marks on the surface, i.e. it has an a_{min} value. This value of a_{min} will be the first value that a_{max} of the recorded operation will examine in order to decide if an intermediate operation is needed.

2. Bottom of the hole. The default is that the tool is allowed to leave drill marks on the bottom of the hole. Thus, it makes no difference if the hole is through or blind and no special operation should be used. If the user specifies other requirements for the bottom, special care should be taken. The concept of the arrays TLSM and TLS is still valid. In this case, stage one will test the tool table for its capability to produce the required hole bottom.
3. Top of hole. If the hole is produced by a TDD tool type, then this operation will be machined as the last operation with a special tool. It is superimposed on the shape and will be machined after the hole operations have been done.

   If the hole is produced by an MDD tool type, the feature will be produced by the motions of the machine. An MDD tool will be better than a TDD tool as it saves extra tools, tool change times and machining time.

13.1.2 Second-level optimization – several holes on one center line

Stage one

Scan the holes on the center line and locate the one with the minimum diameter hole (MDH).

Stage two

Perform passes 1 and 2 of the first-level procedure, i.e. establish the operations needed to machine the MDH. The output will be recorded in TLSW array with the addition of segment number and diameter boundaries (DB).

Stage three

If the MDH is not the last one **and** the hole is closed **or** blind, then the hole can be machined only by an MDD tool. It is assumed that the hole can be machined and that there is an access to it. The algorithm that handles MDD operations will be used to specify the tool, depth of cut and its trajectory.

A 'closed hole' is defined as a hole that must by machined from one side only. A 'blind hole' can also only be machined from one side. An 'open hole' can be machined from both sides. For an open hole, an algorithm will specify how and from which side to machine the hole according to economic considerations. In hole-making operations, the cutting conditions are usually restricted by the strength of the tool, thereby increasing machining time and cost.

The allowable cutting forces are computed for a column or a beam. In both cases, the controlling ratio is that of length (L) to diameter (D) of the tool. For $L/D < 2.5$ the maximum cutting conditions can usually be used. Therefore, it is recommended that if $L/\text{MDH} < 2.5$, the holes should be machined from the side where the top hole diameter is the bigger. Otherwise split the operation to both sides of the through hole in such a way as to minimize the ratio of L/MDH (Fig. 13.2a).

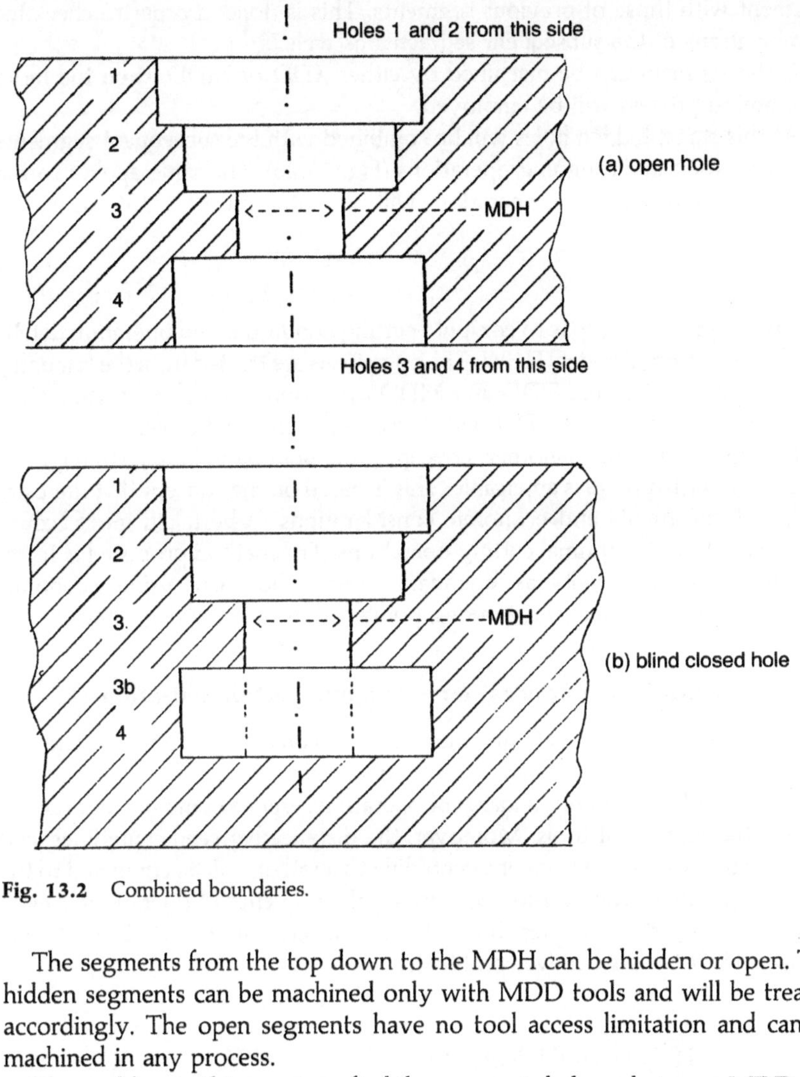

Fig. 13.2 Combined boundaries.

The segments from the top down to the MDH can be hidden or open. The hidden segments can be machined only with MDD tools and will be treated accordingly. The open segments have no tool access limitation and can be machined in any process.

The problem in this case is to find the economic balance between MDD and TDD tools. Using MDD tools reduces the number of tools and reduces tool change time, but may increase machining time. On the other hand, TDD is tool consuming and increases tool change time, but may reduce machining time.

The segments will be treated one at a time. **If** the treated segment can be machined only with TDD tools **and** only one tool is needed, **then** this operation is recorded and the program will proceed to the subsequent segment. However **if** several TDD tools are needed, **then** the last one (finish operation) is recorded, but the diameter of the intermediate tools can be varied within the boundary of minimum and maximum diameter. An attempt is made to combine the boundaries (Fig. 13.2a,b) of the intermediate tools needed for the present segment with those of previous segments. This is done in order to check for combinations of the subsequent segments as well.

If the segment can be machined by either TDD or MDD, **then** the more economical process will be employed.

At this stage, hidden holes will be combined with the subsequent segments and have the same roughing operations (Fig. 13.2b). The same applies to the closed or blind holes. The operations will be written in the TLSW array.

Stage four

The purpose of this step is to compute cutting conditions and operation details for the operations listed in TLSW. The operations are treated from the last entry to the first one. Because TDD and MDD operations are different, they have different treatments. The TDD will be treated as in stage three.

For an MDD tool, another concept (the boundary limits strategy) as discussed in Chapter 8, is applicable. This is based on the concept that there are technical constraints and economic considerations to be taken into account when selecting the optimal cutting operations. The method proposed is to set the technical constraints as boundary limits, and then, using economic considerations, to select the working point within the boundary limits.

13.1.3 Third-level optimization – compute part optimization

In this stage, the holes on all the center lines are considered. Tool type and size may be changed in order to ensure a minimum of total manufacturing time (or cost). The DB (diameter boundary) of operations will be combined in order to reduce the number of tools. Moreover, the distance between centerlines, and the time to move a tool from one centerline to another, will be compared to the tool change time. The decision concerning the sequence of operations will be determined by the minimum time of the two alternatives. MDD tools will replace TDD tools whenever profitable.

13.2 TOOLS FOR HOLE MAKING

There are a huge number of tools that may be used in hole making. To assist the process planner is selecting an economic tool, a short description of the most common tool families is discussed in this section.

The data given in this section are first approximations, and it is up to the process planner to make the final decisions.

13.2.1 Twist drill – TDD from solid

The twist drill (Fig. 13.3a) is tool dimension dependent and can start a hole from a solid. It is made from high speed steel (HSS), or a hard material.

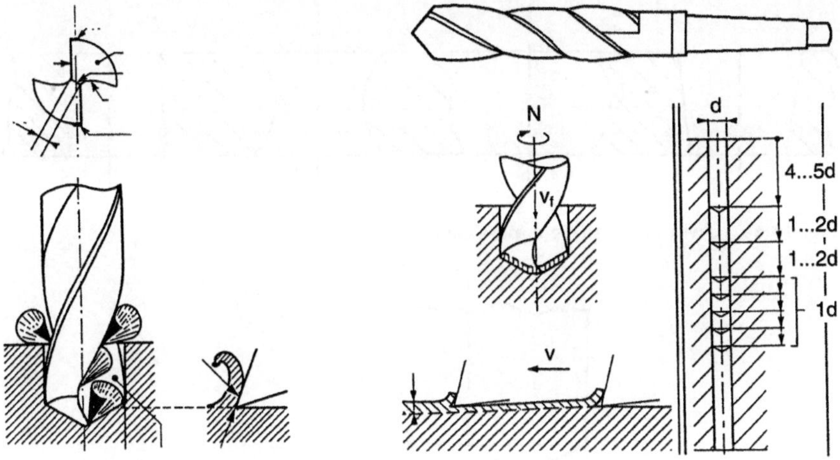

Fig. 13.3a Twist drill.

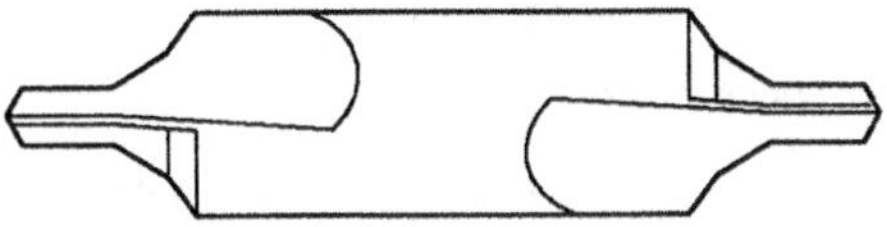

Fig. 13.3b HSS center drills.

A twist drill is a rotary-end cutting tool usually having two cutting lips and two helical flutes for the passage of chips and the admission of cutting fluids.

The cutting edge is the cone end of the drill, usually with an angle of 118°. Due to the assymetry of the two cutting edges, the cutting forces are not distributed equally around the drill centerline and they tend to push the drill off center. Therefore, the hole made is not accurate. The central point, the chisel edge, calls for longitudinal forces in order to feed the tool into the material. Thus, the drill may be considered as a column.

When these forces are large, different shapes of holes may result as shown in Fig. 13.4a. At the starting point of the hole, the chisel is operational so the drill might bend and run out. Therefore, the hole might not keep its location accurately.

To improve location accuracy, especially in making long holes of a small diameter where the drill may deflect, a center drill is used to start the lead of the hole (Fig. 13.3b). It is rugged and short, so it does not deflect. Its size should be such that it will clear the chisel edge of the twist drill that follows.

The shape of the bottom of the hole will be similar to the shape of the twist drill.

In spite of its drawbacks, the twist drill is the most popular tool in the hole-making process, and it is a relatively low cost tool.

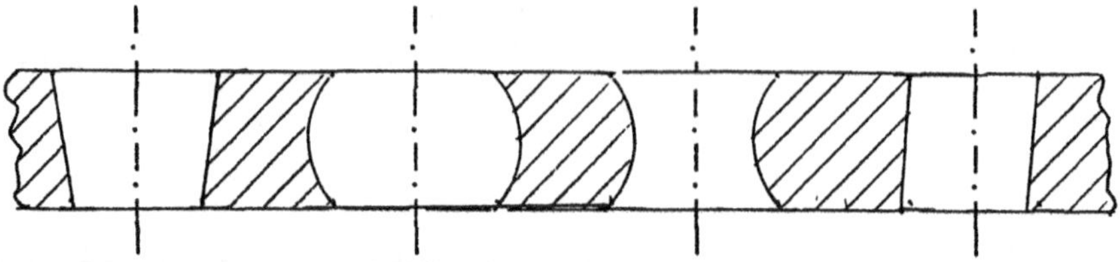

Fig. 13.4a Twist drill – shape of possible holes.

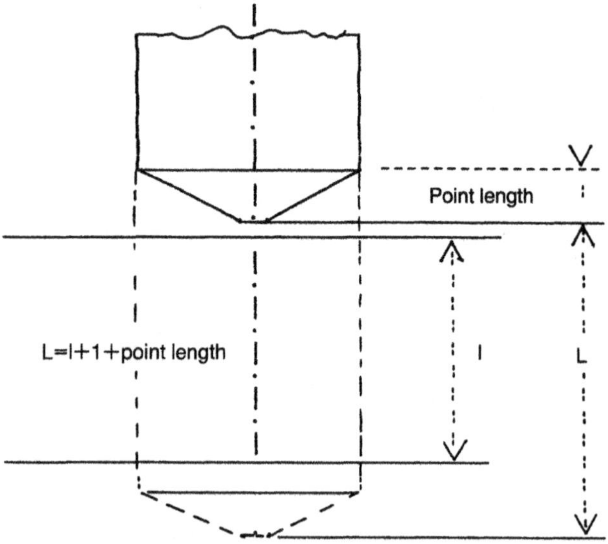

Fig. 13.4b Length of twist drill motion.

The selection of cutting conditions for twist drill operations can be made using the following equations:

$$f = 2.83D^{0.6}R_a^{0.5}\,\frac{\left(\dfrac{1.09-0.04L}{D}\right)}{HBN}$$

For $D<8$

$$f = f\left(\frac{D}{20}\right)^{0.4}$$

$$v_c = \frac{3.38D^{0.4}\left(\dfrac{160}{HBN}\right)^{1.6}}{f^{0.6}}$$

$$T = \frac{0.65D^{1.9}f^{0.7}HBN^{0.75}}{1000}$$

$$Fy = 4.3D\,f^{0.7}HBN^{0.55}$$

For $D_I > 0.1D$ $$F_y = \frac{F_y(D - D_I)^{1.3}}{D}$$

where: f = feed rate (mm/rev)
v_c = cutting speed (m/min)
T = torque moment (Nm)
F_y = feed force (N)
D = hole diameter (mm)
D_I = initial hole diameter (mm)
L = hole length (mm)
HBN = workpiece material hardness in Brinell
R_a = surface roughness of hole (μm)

Remember that the length of the twist drill motion in producing a hole must include, besides free travel, the point length of about $0.3D$ (Fig. 13.4b).

13.2.2 Insert drill – TDD from solid

The insert drill (Fig. 13.5) is tool dimension dependent (TDD) and can start a hole from a solid. It is made of hard metals (carbide).

An insert drill can be used as a rotating drill in a horizontal or vertical position. It is built on a steel shank, with flutes to clear chips. Carbide insert tips are located on either side of the center line. The shape and positioning of these inserts provide balanced cutting forces, which in turn facilitate good drill guidance. The inserts are mounted by screws which clamp the insert to its seat, therefore space must be provided for their mounting as there is a lower limit to the minimum diameter that can be made. Due to the minimum size, the insert drill is more rugged, and therefore its accuracy will depend only on the accuracy of the inserts and their mounting.

Carbide withstands much higher cutting speeds than HSS, so an insert drill will produce a hole much faster than a twist drill and does not need a center drill. The bottom of the hole will be almost flat, depending on the shape of the inserts. Insert drills should be used whenever possible.

The selection of cutting conditions for insert drill operations can be made using the following equations:

$$f = 0.3\left(\frac{D}{56}\right)^{0.424}\left(\frac{R_a}{12.5}\right)^{0.4}$$

$$v_c = \frac{7.32D^{0.6}}{f^{0.3}\left(\frac{HBN}{220}\right)^{0.9}}$$

$$T = \frac{0.4D^2 f^{0.75} HBN^{0.7}}{1000}$$

$$F_y = 0.23D^{1.4} f^{0.8} HBN^{0.75}$$

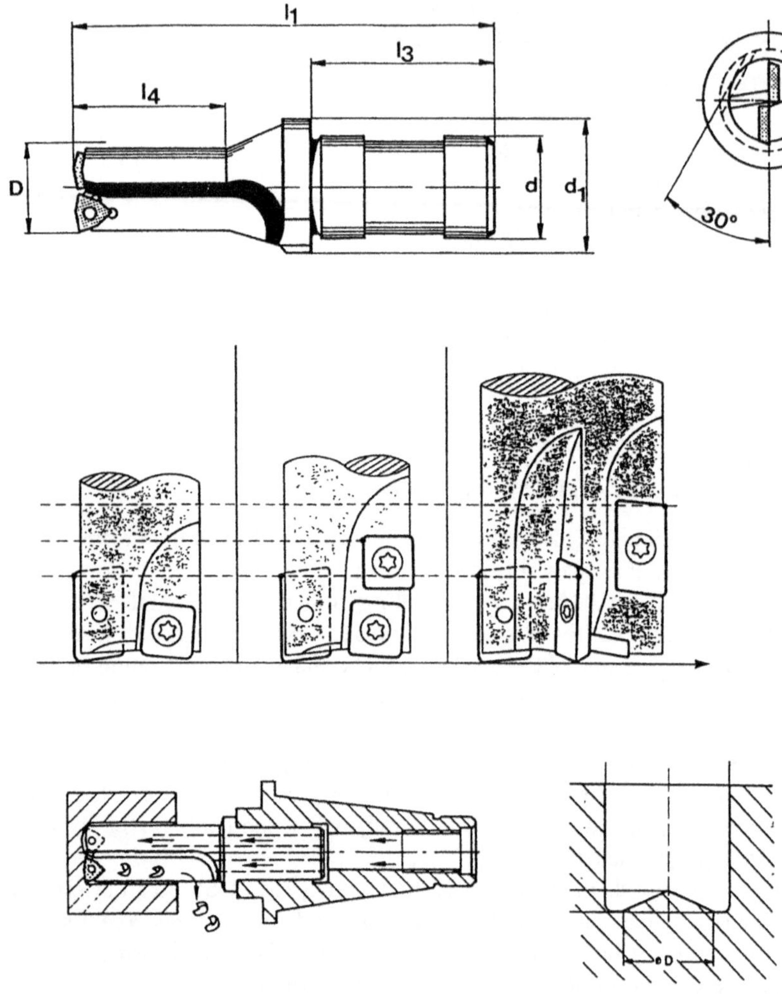

Fig. 13.5 Insert drill.

13.2.3 Solid carbide drill – TDD from solid

A solid carbide drill is tool dimension dependent (TDD) and can start a hole from a solid.

It is built as a twist drill but, as it is made from carbide, it is more expensive. The carbide allows the use of high cutting speeds, as with the insert drill, but as there are no inserts to mount, it can be made to a small drill diameter. Moreover, carbide's modulus of elasticity is much higher than that of steel, so there will be fewer deflections, and thus a higher accuracy than that obtained with an HSS twist drill of the same size.

The selection of cutting conditions for solid carbide drill operations can be made by the following equations:

$$f = 2.83D^{0.6}R_a^{0.5}\frac{\left(1.09 - 0.04\frac{L}{D}\right)}{HBN}$$

For $D < 8$
$$f = f\left(\frac{D}{20}\right)^{0.4}$$

$$v_c = \frac{10.0D^{0.4}\left(\frac{160}{HBN}\right)^{1.6}}{f^{0.6}}$$

$$T = \frac{0.65D^{1.9}f^{0.7}HBN^{0.75}}{1000}$$

$$F_y = 4.3Df^{0.7}HBN^{0.55}$$

For $D_I > 0.1D$
$$F_y = \frac{F_y(D - D_I)^{1.3}}{D}$$

13.2.4 Core drill – TDD improve hole

The core drill (Fig. 13.6) is tool dimension dependent (TDD) and is used to open out existing predrilled holes or to improve hole accuracy.

It is built like a twist drill but without the chisel edge and can come with up to four flutes. The increased number of flutes makes it possible to increase the feed rate of the drilling (keeping the feed rate per tooth constant) and decreases the feed rate forces. These forces are also reduced due to the fact that there is no chisel edge.

The strength of a core drill is no different from that of a twist drill. The hole produced by the twist drill may have shapes as shown in Fig. 13.4a, and does not keep a close location tolerance. The core drill cannot improve the shape and location of a hole because it follows the existing hole pattern; however, due to the reduction of feed forces, diametric accuracy and surface roughness will be improved.

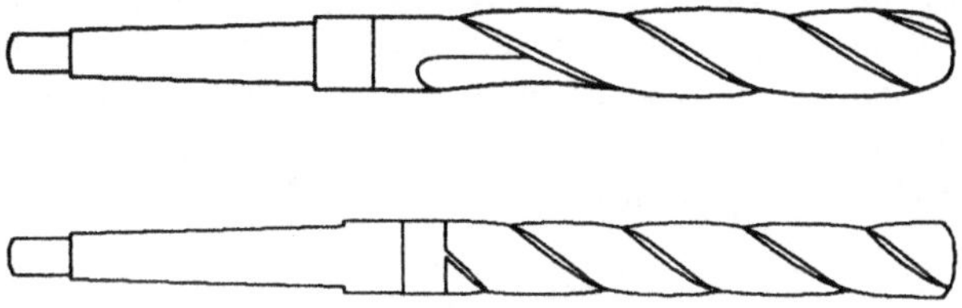

Fig. 13.6 Core drill.

The predrilled hole must be at least the diameter of the straight part of the bottom of the drill.

The selection of cutting conditions for core drill operations can be made using the following equations:

$$f=\frac{2.83D^{0.6}R_{a}^{0.5}\left(1.09-0.04\frac{L}{D}\right)}{HBN\left[\frac{D-D_{I}}{2D}\right]^{0.1}}$$

For $D<8$

$$f=f\left(\frac{D}{20}\right)^{0.4}$$

$$v_{c}=\frac{5D^{0.4}\left(\frac{160}{HBN}\right)^{1.6}}{f^{0.6}\left[\frac{D-D_{I}}{2}\right]^{0.1}}$$

$$T=1.56D\left[\frac{D-D_{I}}{2}\right]^{0.9}f^{0.7}\frac{HBN^{0.75}}{1000}$$

$$F_{y}=4.3Df^{0.7}HBN^{0.55}\left[\frac{D-D_{I}}{2}\right]^{1.3}$$

13.2.5 Reamers – TDD improve hole

Reamers (Fig. 13.7) are tool dimension dependent (TDD) and are used to finish holes to accurate dimensions.

Reamers cannot improve the shape and location of a hole because they follow the existing hole pattern. However, due to the tool's diametric accuracy and the reduction of feed forces, surface roughness will be improved. To obtain the best results when using reamers, it is essential to prepare the hole to the right diameter. As can be seen from Fig. 13.7, the cutting operation of a reamer is usually only on the bevel lead part of the reamer. Material is removed by the cutting edge of the reamer, so the maximum possible depth of cut is limited. Moreover, if too little is left in the hole before reaming, the reamer will scrape and rub and quickly show wear, with consequent loss of diameter.

The selection of cutting conditions for HSS reaming operations can be made using the following equations:

$$f=\frac{0.1}{(D-D_{I})^{0.1}}\left(\frac{220}{HBN}\right)^{1.4}\left(\frac{D}{3}\right)^{0.62}$$

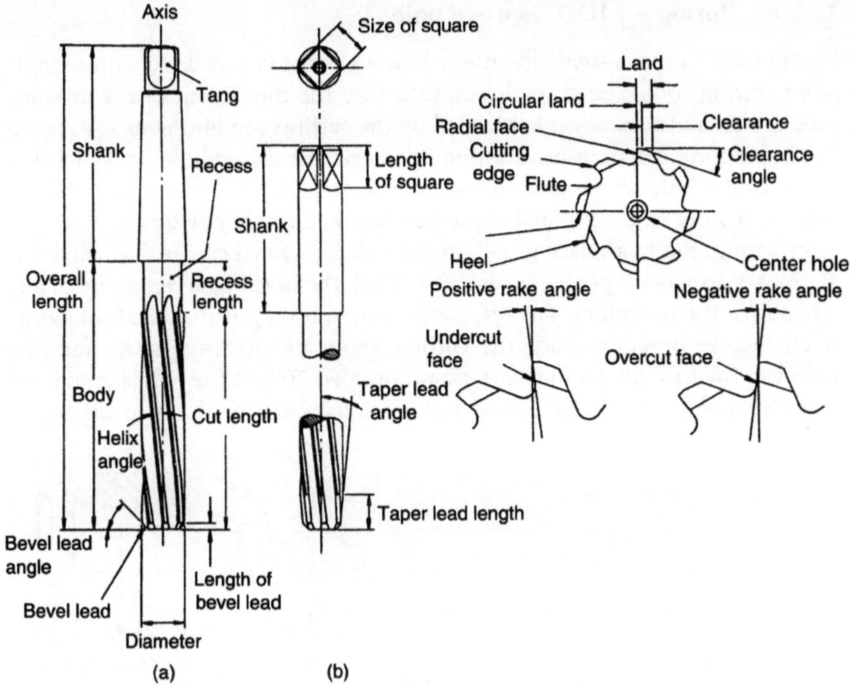

Fig. 13.7 Reamer.

For $R_a > 1.575$ $$f = f\left(\frac{R_a}{3.125}\right)^{0.3}$$

For $R_a < 1.575$ $$f = 0.67f\left(\frac{R_a}{1.575}\right)^{0.15}$$

For $D < 8$ $$f = f\left(\frac{D}{20}\right)^{0.4}$$

$$v_c = 27\left(\frac{220}{HBN}\right)^{0.7}$$

$$T = \frac{1.56D\left[\dfrac{D - D_I}{2}\right]^{0.9} f^{0.7} HBN^{0.75}}{1000}$$

$$F_y = 4.16\left[\frac{D - D_I}{2}\right]^{1.3} f^{0.7} HBN^{0.55}$$

13.2.6 Boring – MDD improve hole

Boring (Fig. 13.8) is essentially internal turning, that is, it is done with a single point cutting tool. The shape is controlled by the movement of the machine tool holder, and accuracy is controlled by the cutting conditions, as in external turning. Boring can improve location tolerance of a drilled hole, as well as the shape of the hole to an accurate cylinder (Fig. 13.4a). It can produce hidden holes and slots (Fig. 13.8) and it can produce a flat hole bottom.

In boring, the workpiece usually rotates along a fixed center line, while the stationary tool is set perpendicular to it. Thus, the hole diameter can be of any size above the minimum. The major problem in boring is that the tool has an overhang. In order to reach the cutting zone, and to leave space for chip removal, the tool holder diameter cannot exceed 70% of the MDH (minimum diameter hole), so slender boring holders must be used. Long overhang in

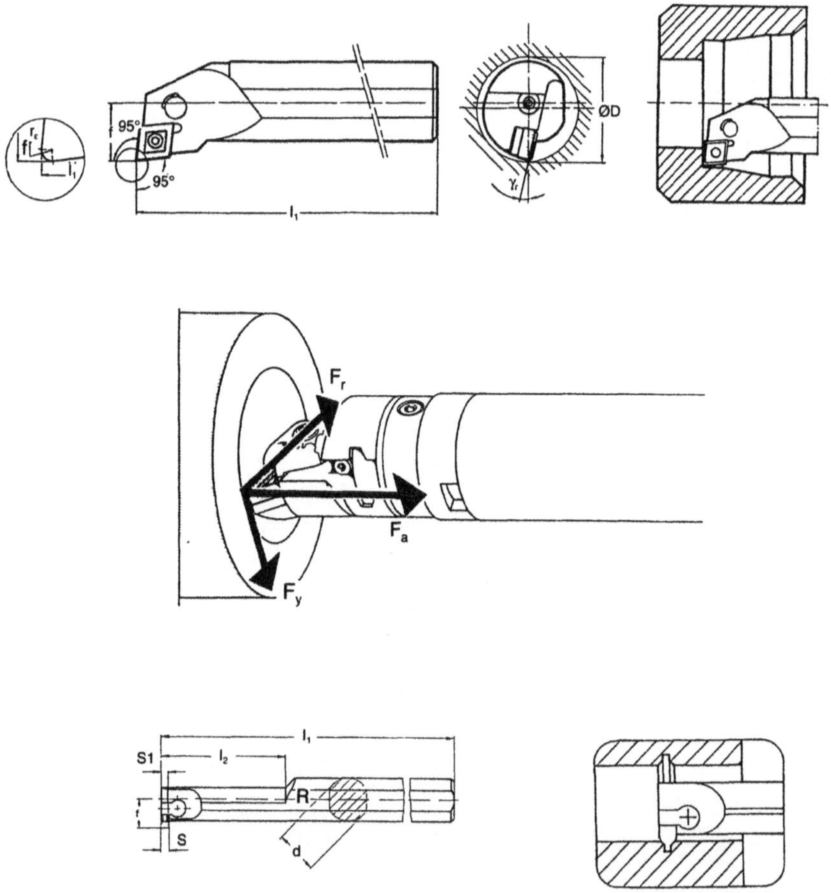

Fig. 13.8 Boring.

machining inevitably leads to vibration, and lack of tool stiffness and deflection. Usually, the deflection of the boring bar acts as a controlling force. Unsatisfactory accuracy, surface roughness and poor tool performance can usually be traced back to some source of instability or lack of rigidity in the machining process.

The selection of cutting conditions for boring operations can be made using the following equations:

$$f = 0.088R_a^{0.5}$$

For slotting and hidden holes,

$$f = 0.66f$$

$$v_c = \frac{90}{a^{0.1}f^{0.25}}\left(3.5 - 2.5\left(\frac{HBN}{150}\right)^{0.18}\right)$$

$$F_y = 33.7af^{0.75}HBN^{0.35}$$

$$T = \frac{F_y}{1000}\left(\frac{D}{2}\right)$$

where a = depth of cut (mm).

Note: remember to check the cutting forces against the allowed forces.

13.2.7 End milling – TDD, from solid and disk milling – TDD, improve hole; MDD, improve hole

In end milling (Fig. 13.9) and disk milling (Fig. 13.10a) the tool usually rotates around its axis, while the workpiece is clamped on a table with x, y and z coordinate movements. Thus, holes can be made anywhere on the workpiece.

An end mill is usually built to operate in a radial direction only, so a hole can be made by internal circular interpolation. In this type of operation, the shape is controlled by the movement of the machine tool holder (MDD – improve) and accuracy is controlled by the cutting conditions. It can improve location tolerance of a drilled hole and the shape of the hole to an accurate cylinder (Fig. 13.4a) and can produce a flat bottom. It can make several holes at different locations on a machining center without removing the workpiece from the fixture. However, it is a time-consuming operation.

Some end mills are constructed so as to be able to operate in both axial and radial feed directions. However, they are limited to shallow operations only. They can be used as TDD tools to open holes from solid, and also to improve and enlarge the hole by internal circular interpolation.

Disk mills are similar to radial operating end mills with the exception that, because of their construction, they can produce slots and hidden holes.

A disk mill can be used to produce special shapes (Fig. 13.10b) such as corner rounding, corner angle, dovetail, etc.

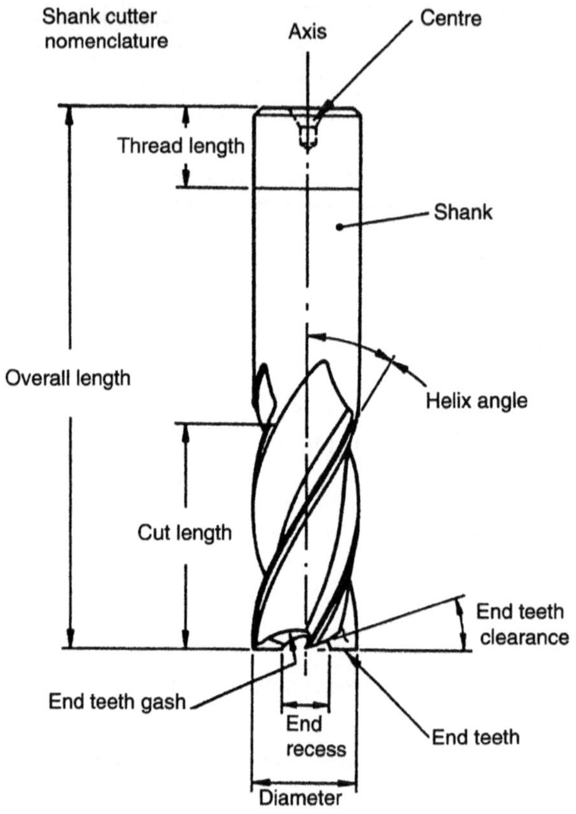

Fig. 13.9 End milling.

The selection of cutting conditions for milling operations can be made using the following equations:

$$a_{max} = (0.11R_a^{1.4} + 0.3)\left(\frac{200}{HBN}\right)^{0.6}$$

$$a_{min} = 0.2$$

$$f = \frac{20.25R_a^{0.4}}{HBNa_r^{0.3}}$$

$$v_c = \frac{3700}{HBN^{0.6}a_r^{0.16}}$$

Solid carbide

$$f_z = 0.7f\left(\frac{D_t}{30}\right)^{1.2}$$

$$z = 4$$

$$v_c = 0.57v$$

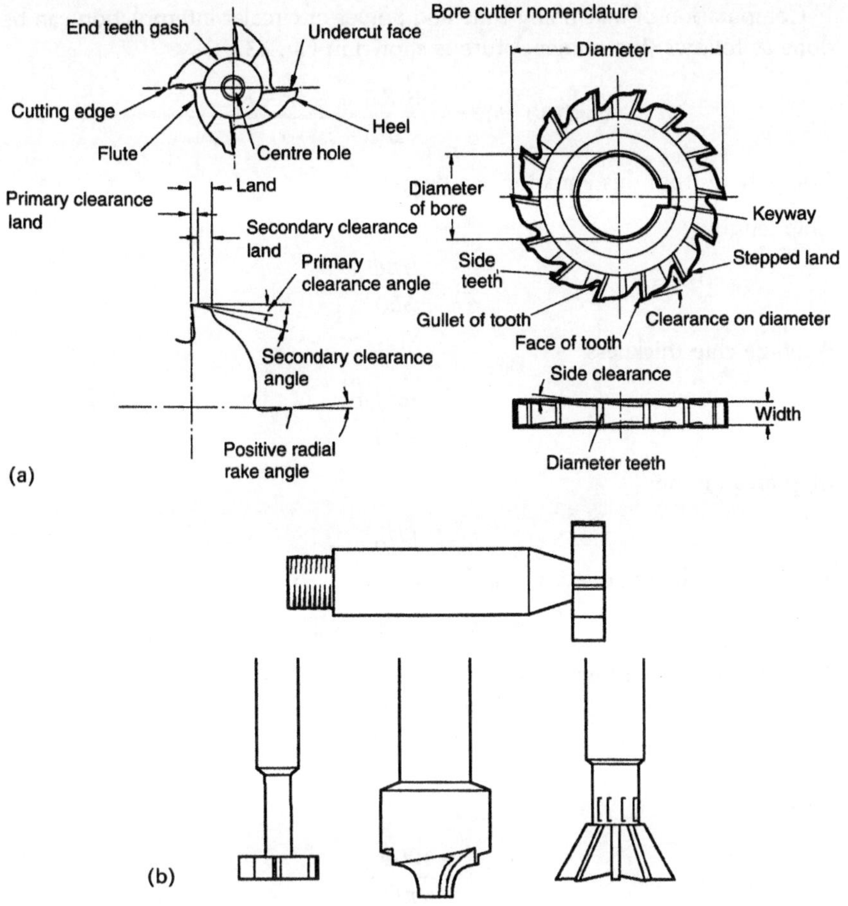

Fig. 13.10 Disk milling.

Disk mill

$$f_z = 0.54f\left(\frac{D_t}{60}\right)^{1.6}$$

$$z = 10$$

$$v_c = 0.3v$$

Carbide tip

$$f_z = 0.8f\left(\frac{D_t}{60}\right)^{0.4}$$

$$z = 2$$

$$v_c = 0.89v$$

where: D_t = tool diameter (mm)
$a_r : a$ = depth of cut = $(D-D_l)/2$

Computation of machining time and power of circular interpolation can be done as follows (the nomenclature is shown in Fig. 13.11).

$$\cos(180-\phi)=\frac{r^2+(R+a_r-r)^2-R^2}{2r(R+a_r-r)}$$

Compute ϕ from this equation, (in degrees):

Chip length

$$L=\frac{2r\pi\phi}{360}$$

Average chip thickness

$$H_m=\frac{360f_z a_r}{\pi\phi D_t}$$

Chip area (in mm^2)

$$a_z=LH_m$$

Total area to be removed (in mm^2)

$$a_t=\pi[(R+a_r)^2-R^2]$$

Time (in min) to remove area a_t

$$t=\frac{a_t}{a_z zn}$$

where

$$n=\frac{1000v_c}{\pi D_t}$$

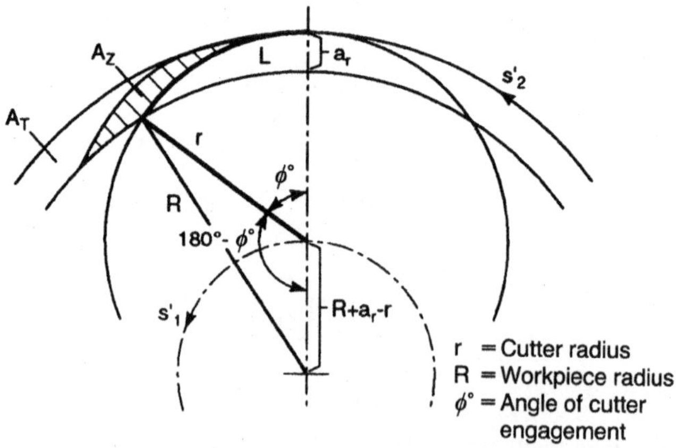

Fig. 13.11 Circular interpolation nomenclature.

Feed per tool center (mm/min)

$$vf_1 = \frac{2\pi(R + a_r - r)}{t}$$

Peripheral feed at contour (mm/min)

$$vf_2 = \frac{f_1(R + a_r)}{R + a_r - r}$$

Specific cutting force (N/mm^2)

$$K_s = (0.216HBN + 298)\left(\frac{0.2}{H_m}\right)^{0.3}$$

Cutting force (N)

$$F_y = LtH_m K_s z\left(\frac{\phi}{360}\right)$$

Cutting power (W)

$$P = \frac{F_y v_c}{60}$$

13.3 DATA FOR COMPUTATION

In section 13.1, the proposed algorithm was described in detail, and the available tool families, with first estimated cutting conditions equations, where given, were described in section 13.2. These equations are based on different sources of data and represent the best compromise between them. The process planner should exercise judgement in making the final decisions in relation to data from different tool manufacturers.

Tool capability data is difficult to collect as it is scattered throughout several sources. Table 13.1 is a collection of tool capability data that seems relevant to the proposed algorithm. Again, it is an estimated average capability of a family of tools, while individual tools might have better or worse capabilities.

13.3.1 Computation of a_{min}

a_{min} is the depth of cut that must be taken in order to remove the inaccuracies of a previous cutting pass. These values are computed by summing the following inaccuracies (marked as I_i) as taken from Table 13.1:

1. Dimension oversize and tolerance (mm) I_1.
2. Surface finish (μm), I_2. The table gives the maximum and minimum values. The resultant surface finish depends on the cutting condition used. As an

Table 13.1 Tool capability table for TDD tools

Twist drill	*Core drill*	*Rough reaming*	*Fine reaming*	*Insert drill*	*Solid carbide drill*
1. Dimension oversize and tolerance (mm)					
$0.05+0.015D^{0.7}$	$0.03+0.012D^{0.5}$	$0.008D^{0.5}$	$0.004D^{0.6}$	0.3	0.05
2. Maximum surface finish (μm)					
12.5	6.25	3.125	1.575	12.5	12.5
3. Minimum surface finish (μm)					
3.125	1.575	1.575	0.4	3.125	2.25
4. Straightness (mm)					
$0.05+0.0005(L/D)^3$	0	0	0	0.03	0.02
5. Roundness (mm)					
0.1	0.05	0.02	0.01	0.07	0.04
6. Parallelism (mm)					
$0.08+0.0006(L/D)^2$	0	0	0	0.08	0.05
7. Concentricity (mm)					
$0.025D^{0.3}$	0	0	0	0.05	0.03
8. Circular runout (mm)					
$0.05D^{0.3}$	0	0	0	0.1	0.06
9. Maximum length to diameter ratio					
10	10	15	15	3.5	2
10. Maximum depth of cut possible (mm)					
$0.5D$	$0.35D$	For $D>9$, $[0.01D+0.2R_a^{0.6}]\left[\frac{200}{HBN}\right]^{0.4*}$		$0.2D$	$0.5D$
		$D<9$, $[0.03D+0.07R_a^{0.6}]\left[\frac{200}{HBN}\right]^{0.4*}$			
		$[0.01D+0.175Rs]\left[\frac{200}{HBN}\right]^{0.4*}$			

11. Surface integrity (mm)					
$D<6 \Rightarrow 0.04$	0.025	0.01	0	—	0.07
$D<10 \Rightarrow 0.05$	0.03	0.015		—	
$D<18 \Rightarrow 0.06$	0.04	0.02		0.06	
$D<30 \Rightarrow 0.08$	0.05	0.025		0.08	
$D>30 \Rightarrow 0.10$	0.06	0.03		0.1	
12. Location tolerance (mm)					
$D<6 \Rightarrow 0.18$	0	0	0	$D<18 \Rightarrow 0.22$	0.05
$D<10 \Rightarrow 0.20$				$D<30 \Rightarrow 0.24$	
$D<18 \Rightarrow 0.22$				$D>30 \Rightarrow 0.28$	
$D<30 \Rightarrow 0.24$					
$D>30 \Rightarrow 0.28$					
13. Deflection of tool per unit length (mm)					
$D<6 \Rightarrow 0.0024$	0.002	0	0	0	0
$D<10 \Rightarrow 0.001$	0.001				
$D<18 \Rightarrow 0.001$	0.001				
$D<30 \Rightarrow 0.001$	0.001				
$D>30 \Rightarrow 0.000$	0.000				
14. Minimum diameter available (mm)					
0	0	0	0	17.5	5
15. Maximum diameter available (mm)					
38	38	38	38	100	20
16. Tool function code (1⇒from solid; 2⇒improve)					
1	2	2	2	1	1
17. Tool bottom shape code (1⇒drill marks; 2⇒flat)					
1	2	2	2	2	2
18. Minimum depth of cut					
0.3	0.25	0.1	0.02	0.3	0.3
19. Tool material (1⇒carbide; 2⇒HSS)					
2	2	2	2	1	1

* For $L/D>3$ divide by $(L/D-2)^{0.33}$

Table 13.1—*continued*

Boring	*Disk mill*	*End mill solid carbide drill and improve*	*End mill carbide tip drill and improve*
1. Dimension oversize and tolerance (mm)			
$0.003D^{0.45}$	$0.035+0.0024D^{0.45}$	$0.03+0.0024D^{0.45}$	$0.05+0.0024D^{0.45}$
2. Maximum surface finish (μm)			
12.5	12.5	12.5	12.5
3. Minimum surface finish (μm)			
0.8	0.8	0.8	3.2
4. Straightness (mm)			
0.01	0.015	0.015	$0.04+0.0004(L/D)^3$
5. Roundness (mm)			
0.01	0.03	0.03	0.07
6. Parallelism (mm)			
0.015	0.03	0.03	$0.06+0.0006(L/D)^2$
7. Concentricity (mm)			
0.01	0.02	0.02	$0.02D^{0.3}$
8. Circular runout (mm)			
0.02	0.04	0.4	$0.04D^{0.3}$
9. Maximum length to diameter ratio – Diameter to cut length ratio			
10	0.5	2	2
10. Maximum depth of cut possible (mm)			

No limits, many cutting passes can be made.

For a single rough cutting pass $a_{max} = \frac{[52D^2]}{[L^{1.5}HBN^{0.5}]}$

For finish pass $a_{max} = 0.11R_a^{1.4} + 0.3$

11. Surface integrity (mm)			
0.03	0.04	0.04	0.08
12. Location tolerance (mm)			
0.01	0.01	0.01	0.28
13. Deflection of tool per unit length (mm)			
0	0	0	$D<6 \Rightarrow 0.0024$ $D<10 \Rightarrow 0.0018$ $D<18 \Rightarrow 0.0012$ $D<30 \Rightarrow 0.0010$ $D>30 \Rightarrow 0.0006$
14. Minimum diameter available (mm)			
10	14	11	17
15. Maximum diameter available (mm)			
no limit	no limit	no limit	no limit

16. Tool function code (1⇒from solid; 2⇒improve)					
2	2	1	2	1	2
17. Tool bottom shape code (1⇒drill marks; 2⇒flat)					
2	2	2	2	2	2
18. Minimum depth of cut					
0.3	0.3	0.3	0.3	0.3	0.3
19. Tool material (1⇒carbide; 2⇒HSS)					
1	1, 2	1	1	1	1

Note – A zero in the table means that there is no effect on the outcome (as in location tolerance) or there is a negligible value.

initial and safe value, it is recommended to use the maximum value. To change the definition of R_a to R_t the following equation can be used:

$$I_2 = 4R_a 10^{-3}$$

3. Surface integrity (mm) I_3.
4. Geometric tolerances (mm) I_4. It is assumed that these tolerances are influencing each other and therefore they should be added on a statistical basis rather than on an arithmetic basis.

   Each of these tolerances: straightness (II_1), roundness (II_2) parallelism (II_3), concentricity (II_4), circular runout (II_5) should be computed separately by the aid of Table 13.1 and computing I_4 by:

$$I_4 = [(II_1)^2 + (II_2)^2 + (II_3)^2 + (II_4)^2 + (II_5)^2]^{0.5}$$

5. Location tolerance and tool deflection (mm) I_5. These two can also be regarded as influencing each other and should be summed in a statistic manner. Thus, we can compute location tolerance (II_6) and tool deflection (II_7) and compute I_5 by:

$$I_5 = [(II_6)^2 + (II_7)^2]^{0.5}$$

The total value of a_{min} is computed by:

$$a_{min} = I_1 + I_2 + I_3 + I_4 + I_5 + I_6$$

where I_6 is minimum depth of cut for the following operation.

13.3.2 Computation of a_{max}

a_{max} for TDD can be computed or retrieved from Table 13.1. For MDD tools, the maximum depth of cut in the table is unlimited. However, this depth of cut must be split into several cutting passes, computed in a method similar to that used for turning. There is a restriction to the depth of cut as a function of surface roughness and in boring due to the deflection of the tool holder in the rough and finishing passes.

These maximum values can be computed thus:
a_{max} for surface roughness (in mm) will be computed by:

$$a_{max} = 0.11\, R_a^{1.4} + 0.3$$

a_{max} for tool deflection

$$a_{max} = \frac{53D_h^2}{L_h^{1.5} HBN^{0.5}} + 0.5$$

where D_h = tool holder diameter (mm)
L_h = tool holder free length (mm).

These values, together with the absolute depth of cut, will be used to compute the number of passes and the depth of cut of each cutting pass.

13.3.3 Location tolerance

The location tolerance is determined by the type of tools used, as given in Table 13.1. Usually the TDD tools cannot keep a close location tolerance, as opposed to the MDD tools, which can.

Thus in the first steps of the algorithm, when there is no way of knowing which tools will be selected, the location tolerance is ignored.

When the selected tools for the job are known, the location tolerance may be checked. If the selected tools do not provide the location tolerance required, the following steps should be taken:

1. *Add center drill* A center drill is usually needed to keep the drill from breaking when there are large deformations, and to improve location tolerance.

   Normally, before using a twist drill, a center drill operation is added in the following cases:
   (a) tool diameter is less than 6 mm (0.236 in);
   (b) the ratio of length to diameter is greater than $2\sqrt{(D/8)}$;
   (c) hole is not perpendicular to drill center; and/or
   (d) location tolerance for twist drill (where a center drill is used) is 50% of the value given in the table.
2. Change tool from a TDD tool to an MDD tool.
3. Add a correction operation with an MDD tool.
4. Change sequence of operations.

13.3.4 Operation diameter – final decision

In the preliminary stages, the diameter to be reached after an operation was given as a boundary. After all the operations have been defined (and before computing cutting conditions), a final decision concerning the diameter and surface roughness of each operation has to be made (the tolerance was taken into account in the calculation of a_{min}). The decision is based upon the diametric boundaries. In a single hole, D_{cmax} and D_{cmin} are as computed. However, in the case of several holes in one center line, an attempt is made to reduce the number of tools (operations) required. This is done by attempting to adjust the boundaries in such a way that they will cover more than one hole segment, as shown in the following example.

Figure 13.12 shows four hole segments in one center line. Each hole segment has its own boundaries, i.e. a_{max} and a_{min}. The final operation has no boundaries, as the specified dimension has to be reached. However, there is freedom in the roughing operations. The figure shows that $D_{cmax}(3)$ (of segment 3) falls within the boundaries of segments 2 and 1, which means it can serve as the upper boundary for all three segments. Moreover, $D_{cmin}(1)$ (of segment 1) is the lowest of all three D_{cmin}, so it can serve as the lower limit for all three segments. The combined boundaries for the last rough operation can therefore be set as $D_{cmax}(3)$ and $D_{cmin}(1)$.

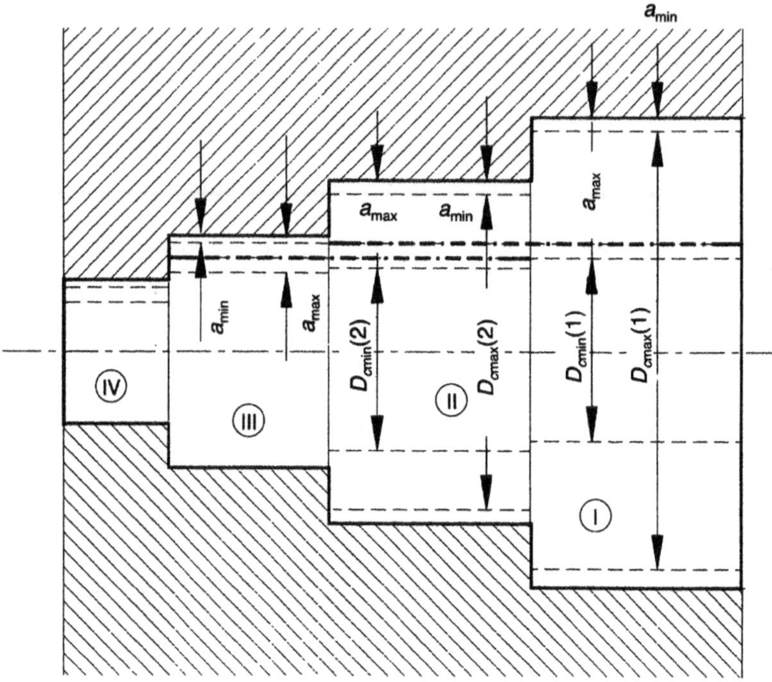

Fig. 13.12 Four holes on one center line.

The boundaries of segment 4 do not overlap any of the previous boundaries, therefore it cannot be combined.

The diameter used in the roughing operation will be decided following these recommendations:

- for TDD solid carbide drill, choose D_{cmax};
- for twist drill followed by TDD tools, choose D_{cmin}; and
- for twist drill followed by MDD tool, choose D_{cmax}.

The surface roughness of the rough cut will be the upper limit of the following operation.

13.4 EXAMPLES

In this section, examples of how to use the hole-making procedure are given. Please note the large differences in machining time when producing a hole with different processes.

In the examples, only the most common tools will be used. These are:

- twist drill
- core drill
- rough reaming

- fine reaming
- insert drill
- solid carbide drill
- boring
- disk milling
- end mill solid carbide, drill and improve
- end mill carbide tips, drill and improve.

These are tool families, and within each family there are many tool variations and standard sizes. The process planner has to use a combination of knowledge, experience and literature sources to select the right tool type.

13.4.1 A single hole from a solid rough hole

A hole as shown in Fig. 13.13 has to be machined. Examination of the tool capability Table 13.1 reveals that any one of the tools can produce the hole as specified. However, only three tools can start the hole from solid.

For the purpose of comparison, the time taken to produce the hole by each one of these tools will be computed. The computations are based on the equation already given:

Twist drill, section 13.2.1

$$f = \frac{2.83 D^{0.6} R_a^{0.5} \left(1.09 - 0.04 \dfrac{L}{D}\right)}{HBN}$$

$$= \frac{2.83 \times 20^{0.6} \times 12.5^{0.5} \times \left(1.09 - 0.04 \times \left(\dfrac{30}{20}\right)\right)}{200}$$

$$= 0.31$$

$$v_c = \frac{\left(3.38 D^{0.4} \left(\dfrac{160}{HBN}\right)^{1.6}\right)}{f^{0.6}}$$

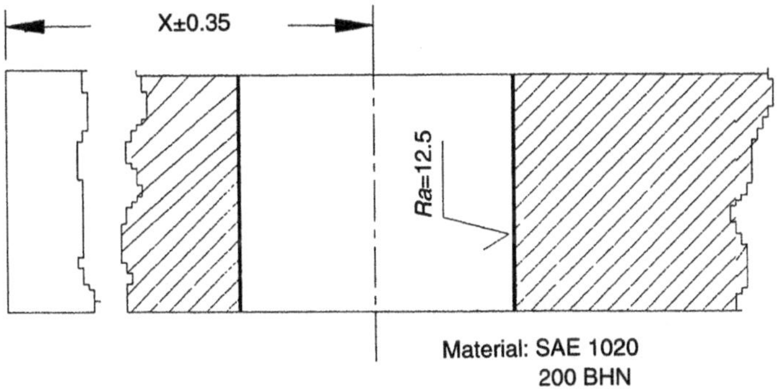

Fig. 13.13 Example of a single hole on a center line.

$$= \frac{\left(3.38 \times 20^{0.4} \times \left(\frac{160}{200}\right)^{1.6}\right)}{0.31^{0.6}}$$

$$= 15.8$$

$$L = 30 + 20 \times 0.3 + 1$$

$$= 37$$

$$t = \frac{\pi DL}{v_c f}$$

$$= \frac{\pi \times 20 \times 37}{15.8 \times 1000 \times 0.31}$$

$$= 0.47$$

Insert drill, section 13.2.2

$$f = 0.3\left(\frac{D}{56}\right)^{0.424}\left(\frac{R_a}{12.5}\right)^{0.4}$$

$$= 0.3\left(\frac{20}{56}\right)^{0.424} \times \left(\frac{12.5}{12.5}\right)^{0.4}$$

$$= 0.19$$

$$v_c = \frac{7.32D^{0.6}}{f^{0.3}\left(\frac{HBN}{220}\right)^{0.9}}$$

$$= \frac{7.32 \times 20^{0.6}}{0.19^{0.3} \times \left(\frac{200}{220}\right)^{0.9}}$$

$$= 79$$

$$L = 30 + 1$$

$$= 31$$

$$t = \frac{\pi Dl}{v_c f}$$

$$= \frac{\pi \times 20 \times 31}{79 \times 1000 \times 0.19}$$

$$= 0.13$$

Solid carbide drill, section 13.2.3

$$f=\frac{2.83D^{0.6}R_a^{0.5}\left(\frac{1.09-0.04L}{D}\right)}{HBN}$$

$$=\frac{2.83\times 20^{0.6}\times 12.5^{0.5}\times\left(1.09-0.04\times\left(\frac{30}{20}\right)\right)}{200}$$

$$=0.31$$

$$v_c=\frac{\left(10.0D^{0.4}\left(\frac{160}{HBN}\right)^{1.6}\right)}{f^{0.6}}$$

$$=\frac{\left(10.0\times 20^{0.4}\left(\frac{160}{200}\right)^{1.6}\right)}{0.31^{0.6}}$$

$$=46.8$$

$$L=30+20\times 0.3+1$$

$$=37$$

$$t=\frac{\pi DL}{v_c f}$$

$$=\frac{\pi\times 20\times 37}{46.8\times 1000\times 0.31}$$

$$=0.16$$

Summary:

Tool	*Cutting speed (m/min)*	*Feed rate (mm/rev)*	*Machining time (min)*
Twist drill	15.8	0.31	0.47
Insert drill	79.0	0.19	0.13
Solid carbide drill	46.8	0.31	0.16

13.4.2 A single hole from a solid rough hole with location tolerance

The same hole as in section 13.4.1 is specified, but with a location tolerance of 0.1 mm. The procedure is as before with the addition of a tolerance location check, as detailed in section 13.3.3.

The location tolerance which can be achieved with a Ø20 twist drill is 0.24 mm (Table 13.1, item 12). This means that a twist drill cannot produce the part as specified. Adding a center drill will improve the location tolerance by 50%, i.e. to $(0.5 \times 0.24 =)0.12$ mm, although it is still not enough, and an improving operation such as boring or milling must be added.

The location of tolerance that can be achieved with a Ø20 insert drill is 0.24 mm (Table 13.1, item 12). This means that an insert drill cannot produce the part as specified and an improving operation must be added, such as boring or milling. Several insert drills may be used on a boring bar as well, which means that no extra tool has to be added.

The location tolerance that can be achieved with a Ø20 solid carbide drill is 0.05 mm (Table 13.1, item 12). This means that a solid carbide drill can produce the part as specified.

The time taken to produce the hole with different processes will be given here for comparison.

The amount of metal to be removed by boring or milling will be the a_{min} of an insert drill, plus the minimum depth of cut for boring (0.3 mm). The computation is based on data from Table 13.1.

$$I_1 = 0.3$$
$$I_2 = 4 \times 12.5 \times 10^{-0.3} = 0.05$$
$$I_3 = 0.08$$
$$II_1 = 0.03$$
$$II_2 = 0.07$$
$$II_3 = 0.08$$
$$II_4 = 0.05$$
$$II_5 = 0.1$$
$$I_4 = \sqrt{(0.03^2 + 0.07^2 + 0.08^2 + 0.05^2 + 0.1^2)} = \sqrt{0.0247} = 0.157$$
$$II_6 = 0.24$$
$$II_7 = 0$$
$$I_5 = 0.24$$

The total value of a_{min} (in mm) is computed by:

$$a_{min} = 0.3 + 0.05 + 0.08 + 0.157 + 0.24 + 0.3 = 1.127$$

Please note the magnitude of the values of the different items and apply judgement with regard to short cut computations. Drilling to diameter 17.80 mm (approx. 0.700 in) is recommended followed by boring or milling to 20.00 mm (0.787 in). The insert drill cutting conditions and machining time is computed as before.

The cutting conditions for boring are computed with the equations given in section 13.3.6:

$$f = 0.088 R_a^{0.5}$$
$$= 0.088 \times 12.5^{0.5} = 0.31$$

$$v_c = \frac{90\left(3.5 - 2.5\left(\frac{HBN}{150}\right)^{0.18}\right)}{(a^{0.1}f^{0.25})}$$

$$= \frac{90\left(3.5 - 2.5 \times \left(\frac{200}{150}\right)^{0.18}\right)}{(1.1^{0.1} \times 0.31^{0.25})} = 103$$

$$t = \frac{\pi \times 31 \times 20}{103 \times 1000 \times 0.31} = 0.06$$

The milling cutting conditions are computed with equations given in section 13.3.7:

$$f = \frac{20.25R_a^{0.4}}{HBNa_r^{0.3}}$$

$$= \frac{20.25 \times 12.5^{0.4}}{(200 \times 1.1^{0.3})}$$

$$= 0.27$$

$$v_c = \frac{3700}{(HBN^{0.6}a_r^{0.16})}$$

$$= \frac{3700}{(200^{0.6} \times 1.1^{0.16})}$$

$$= 151.$$

Choose a solid carbide end mill of 70% of the hole diameter (for chip removal) and $D_t = 0.7 \times 20 = Ø14$ mm.

As the standard sizes of D_t are Ø12 or Ø16, the selected tool diameter will be Ø12 mm. The number of teeth will be $z = 4$ (from catalog). The cutting length of this tool is 18 mm, therefore two passes will be needed to machine the hole of 30 mm long.

$$f_z = 0.7f\left(\frac{D_t}{30}\right)^{1.2}$$

$$= 0.7 \times 0.27 \times \left(\frac{12}{30}\right)^{1.2}$$

$$= 0.063$$

$$v_c = 0.57v$$

$$= 0.57 \times 151$$

$$= 86$$

$$\cos(180-\phi)=\frac{r^2+(R+a_r-r)^2-R^2}{2r(R+a_r-r)}$$

$$=\frac{6^2+(8.9+1.1-6)^2-8.9^2}{(2\times 6\times (8.9+1.1-6)}$$

$$=-0.608$$

$$180-\phi=127.5$$

$$\phi=52.50^\circ$$

Chip length

$$L=\frac{2r\pi\phi}{360}$$

$$=\frac{2\times 6\times \pi\times 52.5}{360}$$

$$=5.5$$

$$H_m=\frac{360f_z a_r}{\pi\phi D_t}=\frac{360\times 0.063\times 1.1}{\pi\times 52.5\times 12}$$

$$=0.0126$$

Chip area (in mm^2)

$$a_z=LH_m=5.5\times 0.0126=0.0693$$

Total area to be removed (mm^2)

$$a_t=\pi[(R+a_r)^2-R^2]$$

$$=\pi\times (8.9+1.1)^2-8.9^2$$

$$=65.31$$

$$n=\frac{1000V}{\pi D_t}=\frac{1000\times 86}{\pi\times 12}=2281$$

Time

$$t=a_t/(a_z Zn)=\frac{65.31}{0.0693\times 4\times 2281}$$

$$t=0.103$$

For 2 passes the time is 0.206 min.

Feed at tool center (mm/min):

$$v_{f_1} = \frac{2\pi(R + a_r - r)]}{t}$$

$$= \frac{2 \times \pi(8.9 + 1.1 - 6)}{0.103} = 244$$

Peripheral feed at contour (mm/min):

$$v_{f_2} = \frac{f_1(R + a_r)}{(R + a_r - r)}$$

$$= \frac{244 \times (8.9 + 1.1)}{8.9 + 1.1 - 6}$$

$$= 610$$

Summary

Tool	*Cutting speed (m/min)*	*Feed rate (mm/rev)*	*Machining time (min)*
Solid carbide drill	46.8	0.31	0.16
Insert drill	74.0	0.185	0.13
Boring	103.0	0.31	0.06
Total time			0.19
Insert drill	74.0	0.185	0.13
End milling	86.0	610*	0.206
Total time			0.336

* mm/min.

13.4.3 Several holes on one center line (Fig. 13.14(a))

The shortest machining time is achieved by making each hole separately. There is no incentive to drill a smaller diameter and then to increase the hole, i.e. drill Ø17 to the length of 60 mm, increase to Ø20 to the length of 40 mm and finish by drilling to Ø25 to the length of 20 mm. This way, the total length of drilling would be twice that needed and the machining time will increase proportionately.

The example in section 13.3.1 has indicated that the insert drill is the most economical tool for making a hole from solid, so this should be the first choice.

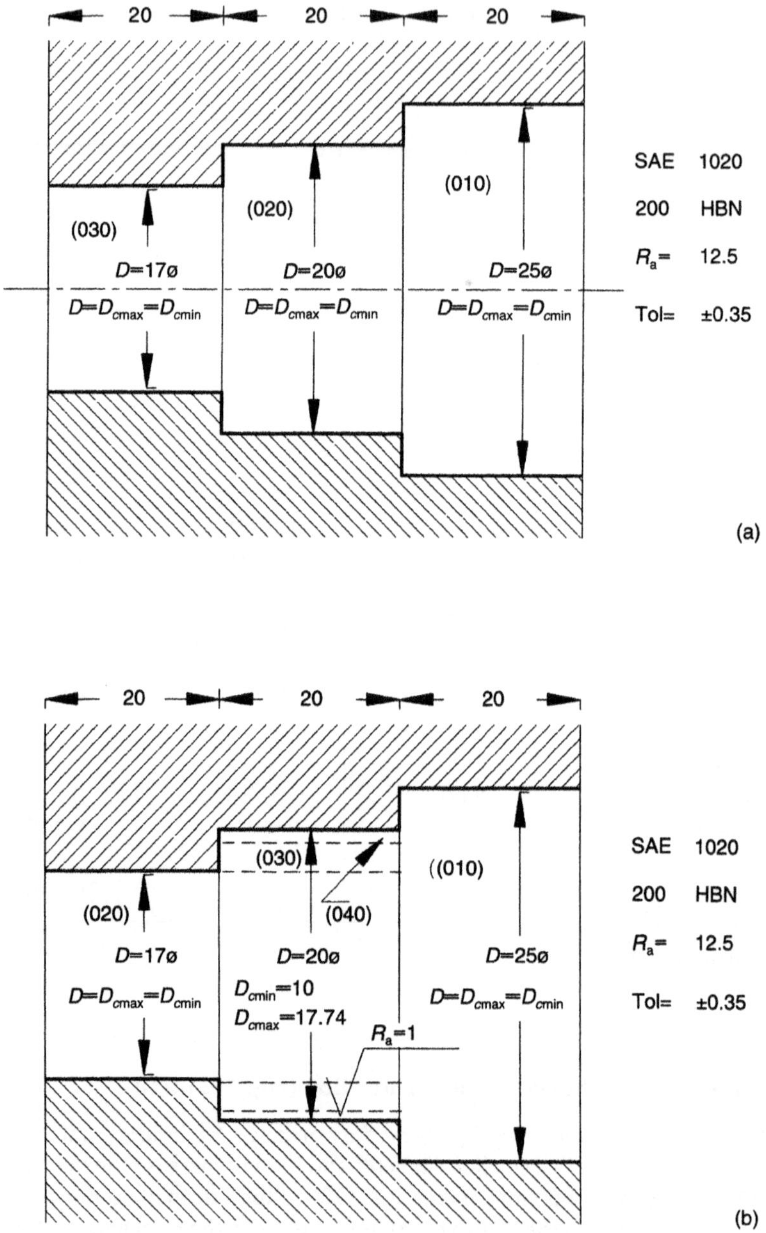

Fig. 13.14 Example of three holes on a center line.

The cutting data for producing these holes is as follows:

Op.	Diam. (mm)	Length (mm)	Speed (m/min)	Feed (mm/rev)	Forces (N)	Torque (mN)	Power (kW)	Time (min)
010	25	20	87	0.21	321	3.20	5.00	0.09
020	20	20	79	0.19	218	1.91	3.36	0.09
030	17	20	73	0.19	182	1.49	2.78	0.09
							Total time	0.26

The power is computed by:

$$P=\left(\frac{F_y v_c}{6}\right)$$

If the required power is greater than that available, then the cutting speed should be reduced. If it has to be reduced by more than 50% (the machining time will be more than doubled), it is advisable to produce it in two cutting passes, one pass to drill the Ø20 mm hole to the length of 40 mm (its time will be doubled) and another to finish it to Ø25 mm.

If the forces or the moment exceeds those available, then the feed rate should be lowered, (reducing the cutting speed does not have any effect on the forces). If it has to be reduced by more than 50%, the hole should be produced in two passes, as in the case above, where there is excess power.

If a surface roughness of $R_a = 1\ \mu m$ is specified for Ø20 (as shown in Fig. 13.14b), then the insert drill cannot produce such a surface and an additional operation is required.

From Table 13.1, it can be seen that this surface finish can be achieved by fine reaming, boring or end milling. The maximum depth of cut for boring and end milling as given in Table 13.1 has no limits, as many cutting passes can be made. However, there is a minimum diameter available, taken as a_{max}. The boundaries for MDD tools are shown in Fig. 13.14b.

For segments 1 and 3, $D_{cmin} = D_{cmax}$ as they are being processed with one machining pass. For segment 2

$$D_{cmax} = D_2 - 2 \times a_{min}$$

(of the insert drill).

$$D_{cmax} = 20 - 1.127 \times 2 = 17.746$$

(see section 13.4.2 for a detailed computation).

$$D_{cmin} = D - 2 \times a_{max} \text{ of the finish operation (MDD tool)}$$
$$D_{cmin} = 20 - 10 = 10$$

As the boundaries of segment 2 overlap those of segment 3, they will be combined as $D_{cmin} = D_{cmax} = Ø17$. Hence the Ø17 will be drilled to the length of 40 mm and segment 2 will then be machined to Ø20. See the results for boring and end milling in the following table.

For TDD tools, fine reaming will be used. Its parameters are:

$$a_{max} = [0.01D + 0.175R_a]\left[\frac{200}{HBN}\right]^{0.4}$$

$$= 0.01 \times 20 + 0.175 = 0.375$$

The a_{min} of the insert drill is 1.127 mm, therefore it is not compatible with fine reaming and an intermediate operation is needed. (The a_{min} of twist drilling is also greater than the a_{max} of fine reaming.)

The next candidate is the core drill, whose a_{min} is calculated as follows.

$I_1 = 0.03 + 0.012D^{0.5} = 0.03 + 0.012 \times \sqrt{20}$	$= 0.084$
$I_2 = 4R_a 10^{-3} = 4 \times 1.575/1000$	$= 0.006$
$I_3 = 0.05$	$= 0.050$
$I_4 = \sqrt{0.05^2} = 0.05$	$= 0.05$
$I_5 = 0.001 \times 40 = 0.04$	$= 0.04$
$I_6 = 0.02$	$= 0.02$
The total value of a_{min} is	0.250 mm

As its a_{min} (0.25) is less than a_{max} (0.375), it can come before the fine reaming. The boundaries are:

$$D_{cmax} = 20 - 2 \times 0.25 = 19.5$$

$$D_{cmin} = 20 - 2 \times 0.375 = 19.25$$

The core drill cannot start a hole from solid, therefore, another operation has to be made.

The a_{max} of the core drill is

$$0.35 \times D = 0.35 \times 19.5 = 6.825$$

This value is larger than the a_{min} of the insert drill, therefore the operations may be:

- fine reaming to a diameter of 20 mm (0.787 in);
- core drilling to a diameter from 19.5 to 19.25 mm; and
- insert drilling to a diameter of $(19.5 - 1.127 \times 2 =)17.246$ mm to $(19.5 - 6.825 \times 2 =)5.85$ mm.

The insert drill boundaries overlap those of segment 3, therefore they are combined.

The results of producing these holes by the three different methods are presented in the following table.

Op.	*Tool*	*Diam. (mm)*	*Length (mm)*	*Speed (m/min)*	*Feed (mm/rev)*	*Time (min)*
1. Boring						
010	Insert drill	25	20	87	0.21	0.09
020	Insert drill	17	40	73	0.19	0.18
030	Rough boring	19.48	20	108	0.30	0.04
040	Finishing boring	20	20	163	0.09	0.09
			Total machining time (min)			0.40
2. End milling						
010	Insert drill	25	20	87	0.21	0.09
020	Insert drill	17	40	73	0.19	0.18
030	Rough end mill	19.48	20	84	526	0.25
040	Finishing end mill	20	20	99	221	0.58
			Total machining time (min)			1.10
3. Fine reaming						
010	Insert drill	25	20	87	0.21	0.09
020	Insert drill	17	40	73	0.19	0.18
030	Core drilling	19.5	20	20	0.40	0.16
040	Fine reaming	20	20	29	0.26	0.18
			Total machining time (min)			0.61

13.5 CONCLUSION

It can be seen, then, that hole producing is not the simple task it first appears to be, and that cost and machining time relate directly to the process selected. The few examples in section 13.4 demonstrate the time spread, and the importance of giving special attention to devising a hole-making process plan. Note, too, that the relative time taken to produce a hole is much higher than the time taken to produce a cylinder by turning, or a flat surface by milling.

CHAPTER 14

Milling operations

The logical approach to process planning, as outlined in the previous chapters, holds true for almost all types of processes. In selecting process details and cutting conditions, there are several topics that are unique to milling such as: tool diameter, tool motion direction, tool motion interpolation, length of cut, etc. These special topics are discussed in this chapter.

14.1 MACHINING TIME

Machining time is computed by the simple equation of length of cut divided by the feed rate of the tool per minute.

$$T=\frac{L}{nf}=\frac{\pi DL}{v_c f} \tag{14.1}$$

where: L = the length of the workpiece (+0.5 or 1 mm for approach)
f = the feed rate per revolution.
n = the number of revolutions per minute $= v_c/(\pi D)$
v_c = cutting speed
D = workpiece or tool diameter.

In turning and boring, there is no problem in using this equation. Equations to compute f and v_c were given in the previous chapters. The diameter D is taken from the drawing. As the tool radius is a fraction of a millimeter, the length L is the length of the segment taken from the drawing.

In milling, the length of the cutting path has to allow for tool entry and tool exit, as shown in Fig. 14.1. As can be seen, the additional path length as a function of the tool diameter, both in short and wide workpieces, can be quite significant. The smaller the tool diameter, the shorter the length will be and thus the shorter the machining time.

Moreover, the tool rpm (revolutions per minute) is also a function of tool diameter. The smaller the diameter, the higher n will be, resulting in a faster feed movement and a shorter machining time.

The feed rate computed is per revolution. However, milling tools have several teeth (z). The feed is therefore $F'=f_z z$ (f_z is the feed per tooth). This equation indicates that the higher the number of inserts, the shorter the

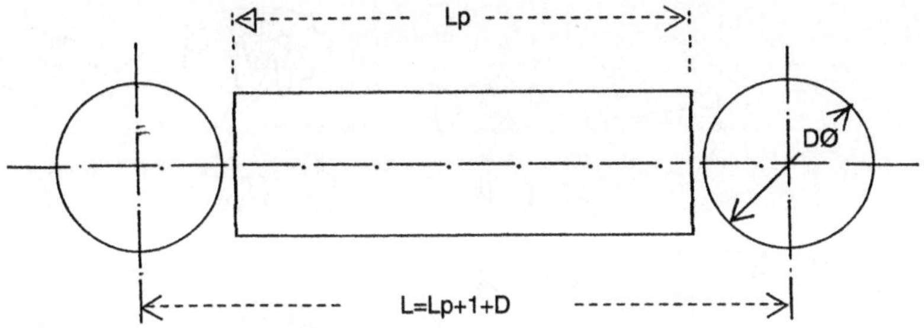

Fig. 14.1 Length of the cutting path.

machining time. The larger the diameter, the larger the number of inserts that can be mounted on the tool.

However, there is a restriction. Chip flow is an important factor in milling. The formed chips cannot leave the chip space until the insert is finishing its cut. Thus, the chip space may restrict the rate of stock removal, i.e. depth of cut or feed speed (Fig. 14.2). The chip space is a function of the cutter diameter and the number of inserts. It is measured by the pitch of the cutter: pitch $=\pi D/z$, hence $z=\pi D/\text{pitch}$.

A coarse pitch gives larger chip space and is best for rough milling. A close pitch gives small chip space and is best for finish milling. Other considerations are the chip type, i.e. spiral or short, and the length of insert contact with the workpiece.

The machining time equation (14.1) for face milling takes the form of:

$$T=\frac{L_p+D}{\left(\frac{v_c}{\pi D}\right)zf_z}$$

$$=\frac{(L_p+D)}{\left[\left(\frac{v_c}{\pi D}\right)\left(\frac{\pi D}{\text{pitch}}\right)f_z\right]}$$

$$=\frac{(L_p+D)\text{pitch}}{(v_c f_z)} \tag{14.2}$$

where L_p = part length.

This equation indicates that the preferred tool should be one with the smallest diameter and the closest pitch possible.

The major effect of the tool diameter on the machining time is the extra cutting length. This extra length may be reduced if the tool movement will be ended when the front end of the milling tool leaves the workpiece, as shown in

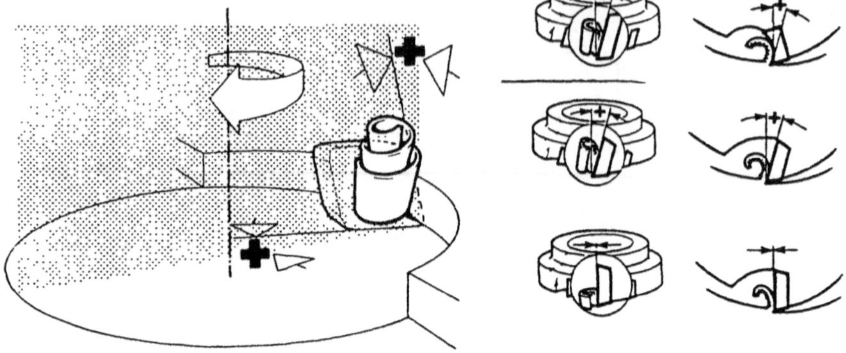

Fig. 14.2 Chip space.

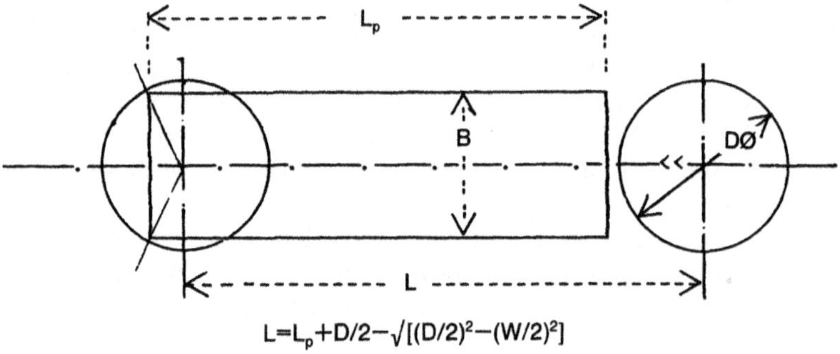

Fig. 14.3 Reduced length of cutting path.

Fig. 14.3. In this case, where the tool center coincides with the workpiece width center, the length will be:

$$L = L_p + \frac{D}{2} - \sqrt{\left[\left(\frac{D}{2}\right)^2 - \left(\frac{B}{2}\right)^2\right]} \tag{14.3}$$

This gives significant time savings. It will, however, leave tool marks on the finished surface. If these marks are tolerable, this method is recommended. It is common practice to use the full length for the finishing passes and the reduced length for roughing passes.

14.1.1 Tool diameter

The importance of selecting an optimal tool diameter was emphasized in the previous section. In order to machine a surface of B mm width and L mm in length, the smallest tool diameter should match the width of the surface, as demonstrated in Fig. 14.4a. In cases where the tool diameter is exactly the width

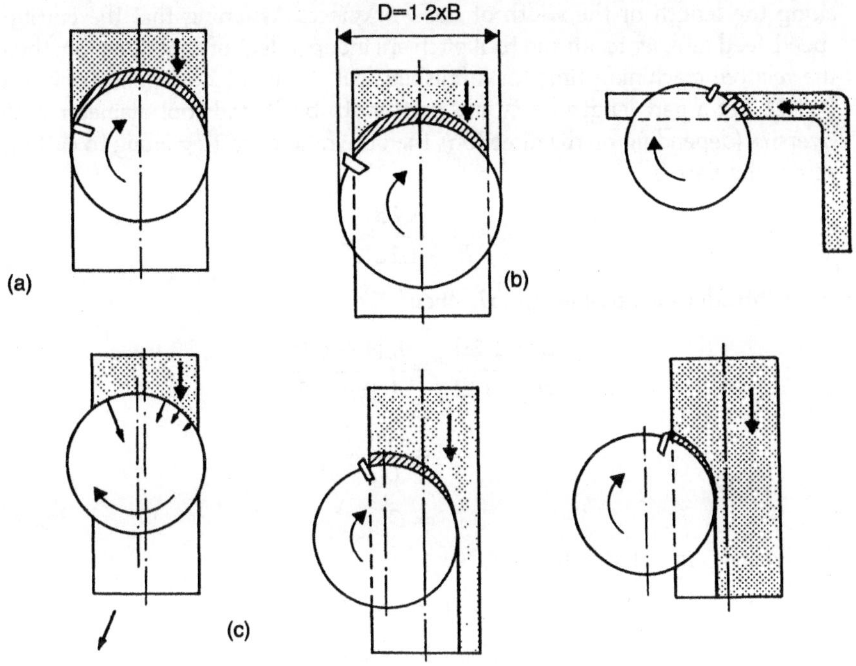

Fig. 14.4 Cutter diameter and positioning.

of the part ($D = B$), the chip increases from zero to its maximum value during the cut. Before the insert starts to cut, it glides over the surface and causes a burnishing of the workpiece. Therefore, the cutter diameter should be selected to be at least 20% larger than the width of the workpiece, as shown in Fig. 14.4b.

Where relatively large tool diameters are used, it can be advantageous to position the cutter off center to put inserts in the cut simultaneously and thus provide smoother running. Milling inserts, contrary to turning or hole-making tools, are subject to an impact load on every revolution when they enter the workpiece, so it is advantageous to position the center of the tool inside the workpiece as shown in Fig. 14.4c. This type of cut is used when the width is too large for one pass, when the cutting forces and the power have to be reduced, or when the part specifications call for it.

The common sizes of face mills are Ø50 to Ø125 mm. Larger sizes are available, but in machining centers, the chain spacing in the tool storage is usually built for these sizes. Larger sizes will occupy more than one slot.

14.1.2 Milling direction (face milling)

If the workpiece is flat and square, the milling direction is not important. If the part specifications call for a shoulder, the tool path should follow the shoulder; however, for flat rectangular shapes, there may be a difference if milling is to be along the length or the width of the workpiece. Assuming that the cutting speed, feed rate per tooth and tool pitch are independent of tool direction, then the relative machining time for a finishing cut, as in (14.2), depends only on $L_p + D$. Let a part length be L_p and part width be B_p and tool diameter 20% oversize (depending on the direction), then the ratio of cutting along to cutting across the part is:

$$\frac{L_p + 1.2B_p}{B_p + 1.2L_p}$$

If we introduce a variable $B_p = xL_p$ then:

$$\begin{aligned}\frac{L_p + 1.2xL_p}{xL_p + 1.2L_p} &= \frac{L_p(1 + 1.2x)}{L_p(x + 1.2)} \\ &= \frac{(1 + 1.2x)}{(x + 1.2)} \\ &= C \qquad (14.4)\end{aligned}$$

The value C in (14.4) will be:

$$C = 1 \text{ for } x = 1$$

$$C < 1 \text{ for } x < 1$$

$$C > 1 \text{ for } x > 1$$

This means that choosing a tool diameter according to the narrow side of the surface and cutting along the length will always, under these assumptions, result in lower machining times.

For a rough cut, using the short cutting length, as in (14.3), and using similar assumptions and computations, the ratio of length to width is:

$$\frac{L_p + \dfrac{1.2B_p}{2} - \sqrt{\left[\left(\dfrac{1.2B_p}{2}\right)^2 - \left(\dfrac{B_p}{2}\right)^2\right]}}{B_p + \dfrac{1.2L_p}{2} - \sqrt{\left(\dfrac{1.2L_p}{2}\right)^2 - \left(\dfrac{L_p}{2}\right)^2}}$$

$$= \frac{L_p + 0.6xL_p - \sqrt{\left[\left(\dfrac{0.6xL_p}{2}\right)^2 - \left(\dfrac{xL_p}{2}\right)^2\right]}}{xL_p + 0.6L_p - \sqrt{\left[(0.6L_p)^2 - \left(\dfrac{L_p}{2}\right)^2\right]}}$$

$$= \frac{L_p + 0.6xL_p - L_p\sqrt{\left[(0.6x)^2 - \left(\dfrac{x}{2}\right)^2\right]}}{xL_p + 0.6L_p - L_p\sqrt{\left[(0.6)^2 - \left(\dfrac{1}{2}\right)^2\right]}}$$

$$= \frac{1+0.6x-x\sqrt{[0.6^2-0.5^2]}}{x+0.6-\sqrt{[0.6^2-0.5^2]}}$$

$$= \frac{1+x(0.6-\sqrt{[0.6^2-0.5^2]})}{(0.6-\sqrt{[0.6^2-0.5^2]})+x} = \frac{1+0.268x}{0.268+x} \tag{14.5}$$

This means that choosing a tool diameter according to the longer side of the surface and cutting along the width will always, under these assumptions, result in lower machining times.

Thus the economic direction of cutting depends on the type of cut and on the surface roughness specified.

14.2 CUTTING FORCES AND POWER

In turning and boring, the cutting edge is in contact with the workpiece throughout the cutting operation and the feed rate and depth of cut are constants. In milling, there are several cutting edges on the periphery of the tool. Each cutting edge enters the workpiece (the impact) and leaves it while the tool rotates around its axis, and the workpiece (or the tool) advances in the direction of the cutting operation. The feed rate per tooth changes as a function of the tool angle θ (Fig. 14.5a). The forces and power vary continually from the point of engagement of a tooth until its exit. Moreover, the direction of the forces vary at any given instant and is a function of the number of teeth engaged and their relative position. Usually, the forces are determined by measurement methods. There follows a method of computing the forces. Power is computed by the standard equation of force multiplied by the cutting speed.

The cutting force per tooth is computed by:

$$F_z = bhK_s \tag{14.6}$$

where: F_z = cutting force per tooth (N)
b = width of chip (mm)
h = thickness of chip (mm) (feed rate)
K_s = specific force (N/mm^2)
κ = tool cutting edge angle
a = depth of cut (mm)

Figure 14.5b shows the chip cross-section. It can be seen that:

$$a = b \sin \kappa \tag{14.7}$$

$$h = f_s \sin \kappa \tag{14.8}$$

Figure 14.5c shows the relationship between f_s and f_z. The exact equation is:

$$f_s = f_z \sin \theta + \left(\frac{f_z^2}{D}\right) \cos \theta \approx f_z \sin \theta \tag{14.9}$$

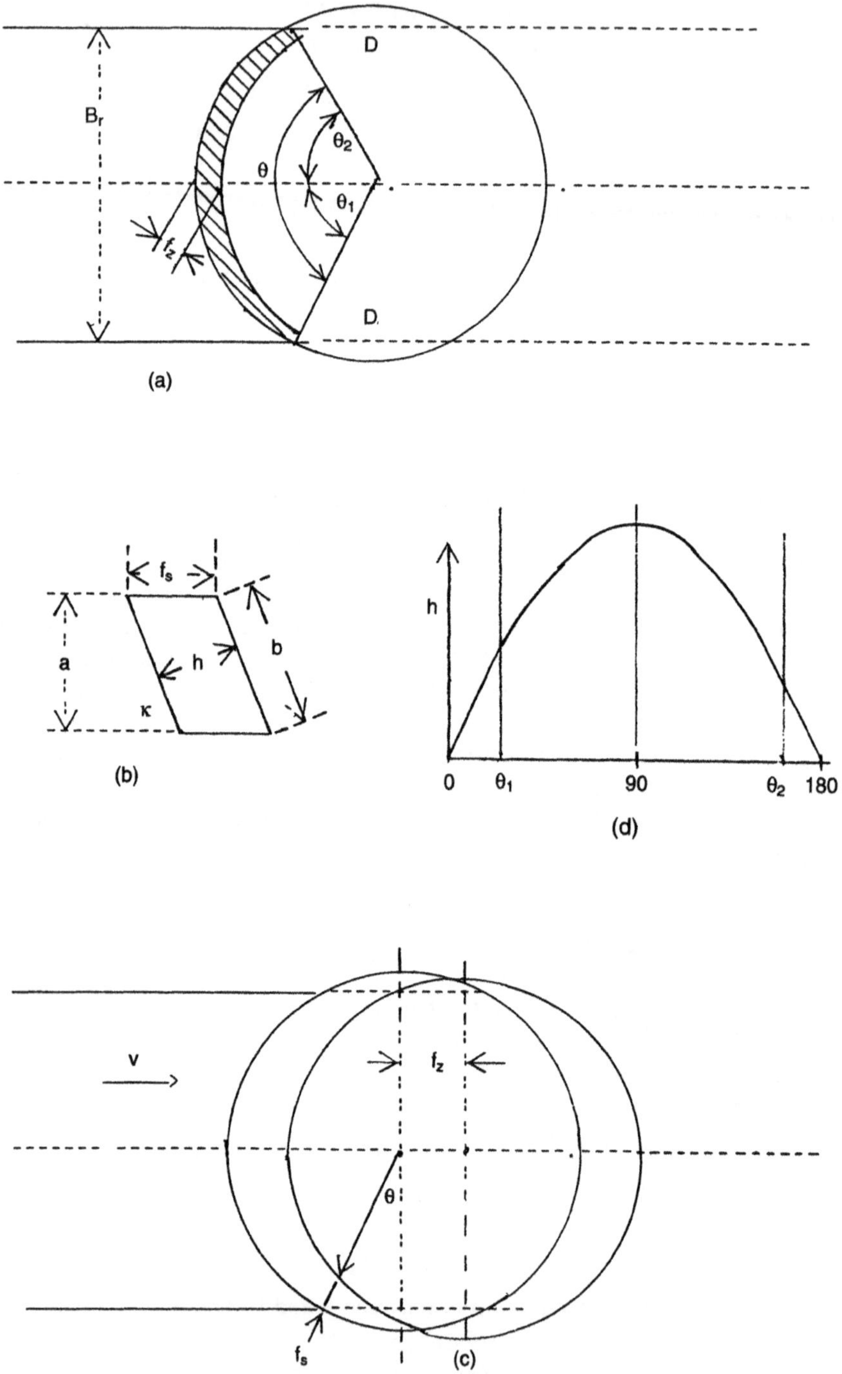

Fig. 14.5 Feed and depth of cut as function of angles.

However the term f_z^2/D is negligible and will be ignored.

Substitution of (14.9) into (14.8) gives:

$$h = f_z \sin\theta \sin\kappa \qquad (14.10)$$

The value of h varies during the tool rotation, as a function of θ, as shown in Fig. 14.5d. The average chip thickness is thus:

$$h_m = \frac{1}{\theta}\int_{\theta_1}^{\theta_2} h \, d\theta = \frac{1}{\theta}\int_{\theta_1}^{\theta_2} f_z \sin\kappa \sin\theta \, d\theta \qquad (14.11)$$

As f_z and κ do not vary with θ, (14.11) can be rewritten as:

$$h_m = \frac{1}{\theta} f_z \sin\kappa \int_{\theta_1}^{\theta_2} \sin\theta \, d\theta \qquad (14.12)$$

Solving the integral results in:

$$h_m = \frac{1}{\theta} f_z \sin\kappa(\cos\theta_1 - \cos\theta_2) \qquad (14.13)$$

For a tool in a central position, as in Fig. 14.5a:

$$\cos\theta_1 = \frac{\left(\frac{B_r}{2}\right)}{\left(\frac{D}{2}\right)} = \frac{B_r}{D}$$

Similarly

$$\cos\theta_2 = -\frac{B_r}{D}$$

and

$$\cos\theta_1 - \cos\theta_2 = 2\frac{B_r}{D} \qquad (14.14)$$

Combining (14.14) with (14.13) and changing the angles from radians to degrees results as:

$$h_m = \left(\frac{360}{2\pi}\right)\left(\frac{1}{\theta}\right) f_z \sin\kappa \left(\frac{2B_r}{D}\right)$$

$$h_m = \left(\frac{360}{\pi}\right)\left(\frac{1}{\theta}\right) f_z \sin\kappa \left(\frac{B_r}{D}\right) \qquad (14.15)$$

The specific cutting force K_s is a function of the workpiece material (K_{sm}), average chip thickness (h_m) and tool rake angle (γ).

The basic value of K_{s_m} is usually given in tables, but may be computed by the equation:

$$K_{s_m} = 0.214(HBN) + 298$$

where *HBN* is the material Brinell hardness.

This K_{s_m} value is for average chip thickness of 0.2 and a rake angle of $-7°$. Therefore a correction factor should be added. For average chip thickness, add $(0.2/h_m)^{0.3}$.

For a rake angle other than $-7°$, a correction factor of $(\gamma + 7)/66.7$ should be employed.

Thus, the specific cutting force is computed by:

$$K_s = [0.214(HBN) + 298]\left(\frac{0.2}{h_m}\right)^{0.3}\left[\frac{\gamma + 7}{66.7}\right] \tag{14.16}$$

Combining (14.7), (14.15) and (14.16) with (14.6) gives the tangential force per tooth as:

$$F_z = bhK_s$$

$$= \left[\frac{a}{\sin \kappa}\right]\left[\left(\frac{360}{\pi}\right)\left(\frac{1}{\theta}\right) f_z \sin \kappa \left(\frac{B_r}{D}\right)\right]$$

$$[0.214(HBN) + 298]\left(\frac{0.2}{h_m}\right)^{0.3}\left[\frac{\gamma + 7}{66.7}\right] \tag{14.17}$$

At any instant, there are $z(\theta/360)$ teeth in engagement, therefore the total tangential force (in newtons) is:

$$F_z' = F_z z\left(\frac{\theta}{360}\right) \tag{14.18}$$

and the power (in kW) is

$$P = \left(\frac{F_z v_c}{60}\right) z \left(\frac{\theta}{360}\right) \tag{14.19}$$

14.2.1 Forces constraints

When the applied forces are larger than the allowed forces, the following steps are recommended to reduce the cutting forces.

1. Change the tool rake angle (γ). This may reduce the forces to about 80% of their initial value, without affecting machining time.
2. Reduce the feed rate by no more than a factor of 0.7. This will reduce cutting forces by up to 22% and increase machining time by 43%.

3. The total force reduction by the above two steps will result in a decrease in the initial cutting forces by 63%. If the cutting forces have to be reduced further, split the depth of cut to two or more passes. As can be seen from (14.17), splitting the depth of cut (a), or the width of cut (B_r) (which means a smaller tool diameter) has the same effect on the forces and the machining time.

14.2.2 Power constraints

When the required power is higher than that available, take the following steps in order to reduce the cutting forces.

1. Change the tool rake angle (γ). This may reduce the forces to about 80% of their initial value without affecting machining time.
2. Reduce the cutting speed to 80% of its initial value.
3. If the power has to be reduced further, split the depth of cut to two or more passes. As can be seen from (14.17), splitting the depth of cut (a), or the width of cut (B_r) (a smaller tool diameter) has the same effect on the forces and the machining time.

14.2.3 Power and forces constraints

If power and forces are constraints, first reduce the forces, then compute the required power again.

14.3 MILLING POCKETS AND SEMI-POCKETS

A pocket is a closed feature consisting of side faces and corner radii. The pocket usually has a bottom face, whose height can vary, although it can be bottomless. The side faces may be of any shape – rectangular or round, perhaps, or a combination of both. Furthermore, a pocket may have inner walls as well. Figure 14.6 shows several pocket shapes.

A pocket allows a tool to approach only from the top of the feature. If the feature allows tool access in horizontal direction, it will be called a semi-pocket, indicating that the feature is not of a closed shape.

The factors which affect machining time in pocket manufacture are tool diameter, tool path and tool approach. These are discussed below.

The tool path is usually supplied automatically by the controller of the machine. This supplies minimum data using NC programs, unless the data is taken directly from a CAD system. This topic is not discussed in this book. The user, the process planner, has to specify the machine for the job and supply technical data such as safety clearance, milling depth, downfeed rate, feed rate and so on. The selection of these data items, and machine selection, were discussed in earlier chapters.

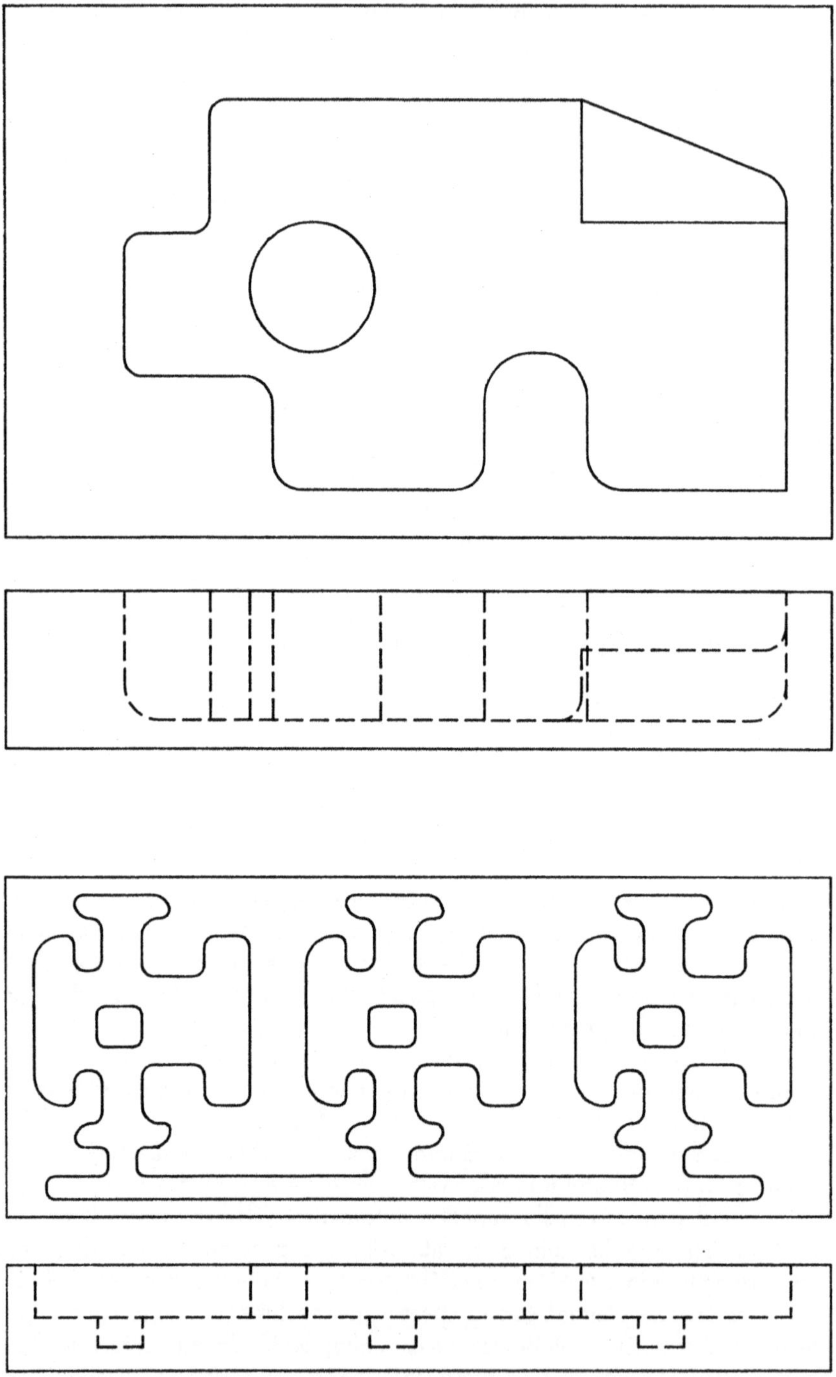

Fig. 14.6 Several shapes of pockets.

14.3.1 Selecting tool diameters

Pockets and semi-pockets are manufactured by end mills. Selecting the tool diameter is determined, in a way, by the pocket specifications. To produce the corner radius, a tool diameter with the same radius or a smaller one **must** be selected. As commercial end mills are not available in all sizes, it might happen that the pocket corner radius is not of a standard dimension. In such cases, select the closest tool radius to that required and the corner will be manufactured by circular interpolation (as discussed in section 13.2.7). In fact, many users prefer to use tool diameters smaller than the corner radius because the resulting radius can be modified using software (tool offset). Several corner radii may be produced with the same tool.

When producing a pocket, the number of cutting passes depends on the tool diameter (Fig. 14.7). Using one end mill whose diameter is determined by the smallest corner radius of the pocket may result in an excessive total path length, low feed rate and depth of cut (as a result of tool strength constraints) and increase the number of vertical cutting passes and thus the machining time. The equations for computing cutting conditions for an end mill are given in section 13.2.7.

The feed rate for a solid carbide end mill is computed by:

$$f_z = 0.7f\left(\frac{D_t}{30}\right)^{1.2} = C_1\left(\frac{D_t}{30}\right)^{1.2}$$

where f = feed rate for a face mill and D_t = tool diameter. The machining time is computed using (14.2), with the correction of cutting length for a pocket being $L_p - D$ (Fig. 14.8):

$$T = \frac{(L_p - D)_{\text{pitch}}}{(v_c f_z)} = \frac{C_2}{f_z}$$

For pockets with L_p much greater than the corner radius and assuming tools with the same pitch, it will be economic to change tools when the machining

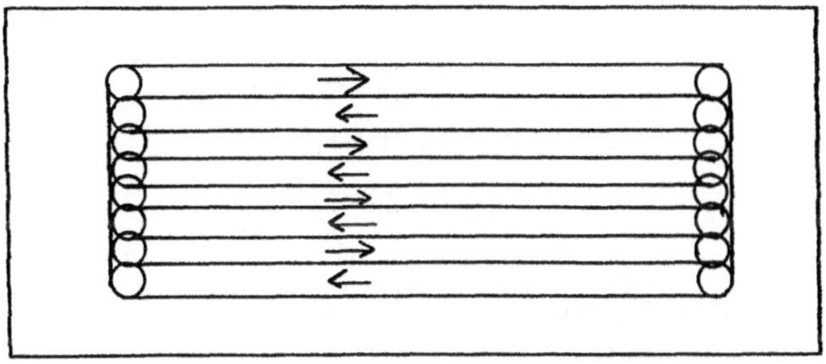

Fig. 14.7 One tool for a pocket.

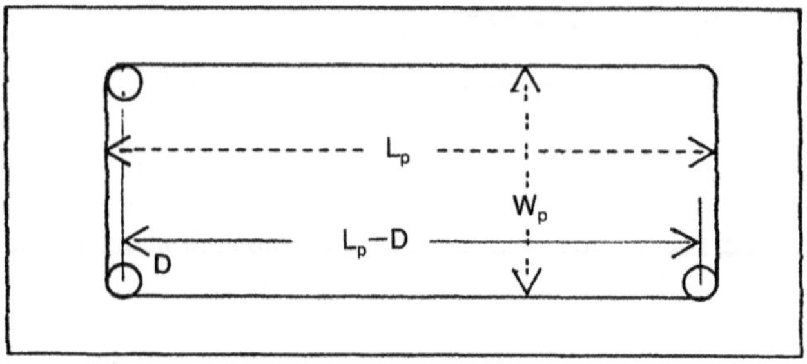

Fig. 14.8 Tool path in pocketing.

time becomes equal to twice the machining time with the tool diameter of the corner radius $d_1 = 2r$ or:

$$\frac{C_2}{f_{z2}} = 2\left(\frac{C_2}{f_{z1}}\right)$$

$$\frac{f_{z2}}{f_{z1}} = 2$$

$$\left(\frac{d_2}{d_1}\right)^{1.2} = 2$$

i.e.

$$\frac{d_2}{d_1} = 2^{0.833} = 1.78$$

The conclusion is that if, under the above assumptions, the width of the pocket is greater than 1.78 the corner radius, it may be economical to start the pocket with a larger diameter tool than the corner radius and change the tool for finishing the pocket.

The maximum initial tool diameter that will leave one straight cutting pass with the tool radius equal to the corner radius is computed as shown in Fig. 14.9:

$$R = R\sqrt{2} - r(\sqrt{2}+1)$$

$$R(\sqrt{2}-1) = r(\sqrt{2}+1)$$

$$R = \frac{r(\sqrt{2}+1)}{(\sqrt{2}-1)} = 5.83r \qquad (14.20)$$

where: R = initial tool radius
r = final tool radius.

It is desirable to have some overlap between the two passes, and its amount has

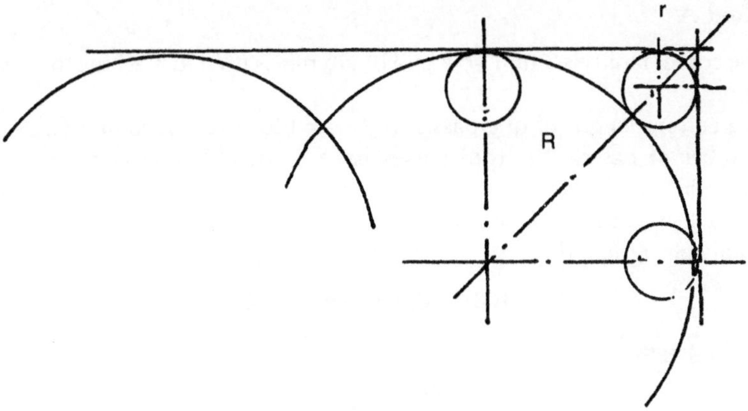

Fig. 14.9 Two tools for a pocket.

to be added to the larger tool diameter. The same ratio can be kept for more than two tools to produce a pocket.

There are many variables to consider when deciding the number of different tools to use for producing a pocket. The process planner will have to use a logic similar to that presented above when making the appropriate decisions.

14.3.2 Example: determining the tool path

When using more than one tool to produce a pocket, the joint of the two tools may leave a small mark on the pocket side faces. This mark may be due to machine inaccuracy or programming error. For a clear side face, it is advisable to have the finishing tool move all along the pocket contour, thus achieving a good surface finish, (in many cases, better than that specified by the drawing). Unfortunately, this can be a very expensive procedure. In the following example, where a pocket as shown in Fig. 14.10 is to be produced, these two methods are demonstrated.

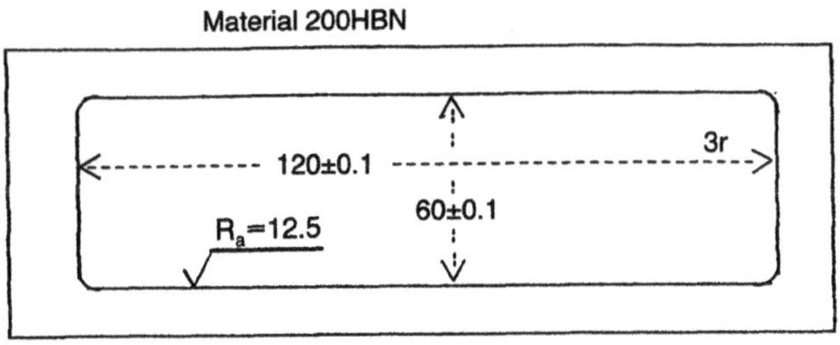

Fig. 14.10 A pocket to be produced.

Method A

- The corner radius is 3 mm, so a tool of Ø6 mm is used to machine the pocket corners.
- The cutting length of one pass is ($L_p - d = 120 - 12 =$)108 mm (4.25 in).
- Number of passes: the tool passes have to overlap one another, so the minimum number of passes must be ($60/6 + 1 =$)11 passes (this means a ($6 - 5.45 =$)0.545 mm overlap).
- Total cutting length:

$$11 \times 108 + 10 \times 5.45 = 1242.5 \text{ mm}$$

- Cutting speed:

$$\frac{0.7 \times 3700}{200^{0.6} \times 3^{0.3}} = 77.5 \text{ m/min}$$

- Basic feed rate:

$$f = \frac{20.25 \times 12.5^{0.4}}{200 \times 3^{0.3}} = 0.2 \text{ mm/rev}$$

- Feed rate:

$$f_z = 0.7 \times 0.2 \times \left(\frac{6}{30}\right)^{1.2} = 0.02 \text{ mm/rev}$$

- Pitch: 9.42 (two teeth)
- Machining time:

$$\frac{1242.5 \times 9.42}{77.5 \times 1000 \times 0.02} = 7.551 \text{ min}$$

Method B1

- The Ø6 mm tool diameter is small compared to the width, so a second tool must be used. It can be up to ($5.83 \times 6 =$)Ø34 mm. However, as the width is 60 mm a tool diameter of Ø32 mm is selected:
- The cutting length of one pass is ($L_p - d = 120 - 32 =$)88 mm
- Number of passes: $60/32 = 2$
- Total cutting length: $2 \times 88 + 1 \times 30 = 206$ mm
- Cutting speed:

$$\frac{0.7 \times 3700}{200^{0.6} \times 3^{0.3}} = 77.5 \text{ m/min}$$

- Basic feed rate:

$$f = \frac{20.25 \times 12.5^{0.4}}{200 \times 3^{0.3}} = 0.2 \text{ mm/rev}$$

- Feed rate:

$$0.7 \times 0.2 \times \left(\frac{32}{30}\right)^{1.2} = 0.151 \text{ mm/rev}$$

- Pitch: 16.75 (six teeth)
- Machining time:

$$\frac{206 \times 16.75}{77.5 \times 1000 \times 0.151} = 0.295 \text{ min}$$

- Clean side faces and corners with tool Ø6 mm
- Cutting length: $2 \times (120-6) + 2 \times (60-6) + \pi \times 6 = 354.8$ mm
- Feed rate and cutting speed as in method A
- Machining time:

$$\frac{354.8 \times 9.42}{77.5 \times 1000 \times 0.02} = 2.156 \text{ min}$$

- Total machining time: $0.295 + 2.156 = 2.451$ min

Method B2

- Use Ø32 mm tool for opening and cleaning the pocket side faces and a Ø6 mm diameter tool to produce the corner radius only. The first tool is as in method B1, giving a machining time of 0.238 min. Cleaning only the corners with Ø6 mm tool, the same cutting conditions as before are used, but the tool cutting length is $(2 \times (32 + (60-6) + 32) =)236$ mm.
- Machining time:

$$\frac{236 \times 9.42}{77.5 \times 1000 \times 0.02} = 1.434 \text{ min}$$

- Total machining time: $0.295 + 1.434 = 1.729$ min.

Summary

Method A 7.551 min.
Method B1 2.451 min.
Method B2 1.729 min.

14.3.3 Starting a pocket

The best tool to use when producing a pocket is an end mill. However, an end mill usually operates in a radial direction, which creates a problem when starting the machining process. This problem does not arise with semi- or open pockets.

Some end mills can operate in both an axial and a radial direction; such end mills may be used to start a pocket. However, since end mills are not designed for drilling, there may be a problem with chip jamming at depths of more than 6–8 mm.

Another method is to drill a hole in the pocket with a diameter larger than the intended end mill diameter and then to use a radial end mill to clear the pocket.

For large, deep pockets, a ramping entrance might be used, as shown in Fig. 14.11. The maximum ramping angle should be about 16°. In this method, an extra pass should be made in order to produce a flat pocket bottom. The feed rate should be reduced to 60% of its original end mill feed rate when the ramp is 16° and may be increased for lower ramp angles.

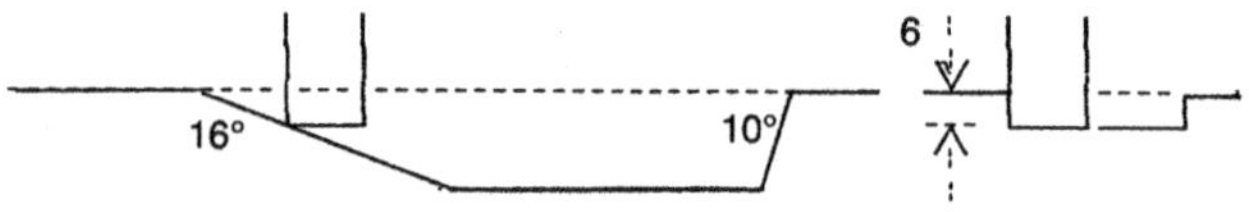

Fig. 14.11 Starting a pocket.

CHAPTER 15

Computer-aided process planning (CAPP)

In spite of the importance of process planning in the manufacturing cycle, it is still predominantly a labor-intensive activity, leaning heavily on experience, skill and intuition. Dependence on practical experience often precludes a thorough analysis and optimization of the process plan and nearly always results in higher-than-necessary production costs, delays, errors and non-standardization of processes.

The desire to increase quality and reduce lead time and cost, or to improve productivity, has led to a widespread interest in computer-aided process planning (CAPP). In the following section, a survey of CAPP developments and approaches is presented and discussed.

15.1 SHORTCOMINGS OF TRADITIONAL PROCESS PLANNING

The traditional approach to process planning, discussed earlier, is based on the process planner studying the part drawing and manufacturing specifications, identifying similar parts or features and recalling past processes. In making decisions and producing drawings, the process planner uses standard data and auxiliary books such as tolerance standards, material code names of different standards, surface roughness standards, machine specifications and capabilities, tool data, cutting condition data, and so on. A huge amount of preparatory work has to be carried out before final decisions about a manufacturing plan can be made.

F. A. Logan has analyzed the process planner's activities (Fig. 15.1) and has found that the planner's time is distributed as follows:

- 15% technical decision making
- 40% data and table look-out and calculations
- 45% text and documents preparation

The main disadvantages of traditional process planning are:

- manufacturing logic is individual – it resides in the process planner's mind;

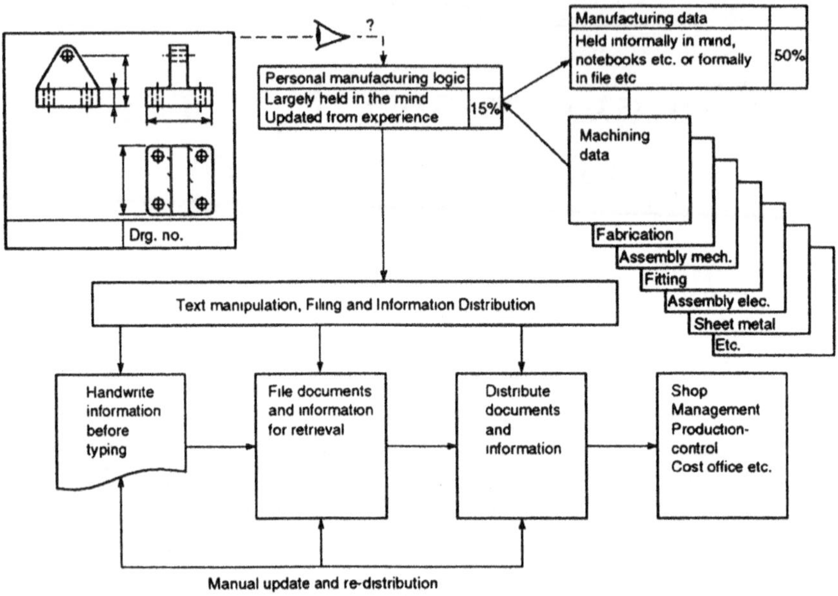

Fig. 15.1 Process planner activities.

- it is a laborious process which must often be shortened or replaced by guess-work or intuition;
- results are often incomplete or inconsistent;
- pre-production lead times are extended; and
- process information is often out of date.

In the following, a gradual development of computer-aided process planning is reviewed.

15.2 CAPP STAGE 1: COMPUTERIZATION OF FILES MANAGEMENT

A computer is a machine built to store and retrieve data at high speed. It is a number cruncher; it performs computations at high speed and with accuracy; it knows how to organize data for printing and how to print; it knows how to sort data. It does not know (yet) how to think.

The current idea is to utilize the power of computers to assist the process planner with clerical work, leaving him/her free for technical work. The idea is to divide the work between the process planner and the computer, letting each perform the tasks they know best. This will reduce the pre-production lead times and increase process planner productivity by 600%, allowing more time to be spent on the planner evaluating alternatives. The result will be complete and consistent processes.

Managing a computerized database is a very straightforward programing task. It relies on the process planner's knowhow and leaves with him/her the responsibility for the content of the stored process plans.

15.3 CAPP STAGE 2: VARIANT (RETRIEVAL) APPROACH

The traditional approach to process planning is to examine a part drawing, identify similar parts produced in the past (usually from memory, or from a filing cabinet), examine process plans for these similar parts and adapt or modify them to suit the specific part on hand.

The variant approach to process planning compares with the traditional manual approach, where a process plan for a new part is created by recalling, identifying and retrieving an existing plan for a similar part from a computerized databank of processes.

The variant approach is derived from group technology (GT) methods where machined parts are classified and coded into families of parts. A family of parts is defined as parts having attributes sufficiently similar to prescribe a common manufacturing method for all of them. For each part family, a standard process plan, which includes all possible operations for the family, is stored in the system; this standard plan is retrieved and edited for the new part.

An incoming part is first coded, and will relate to a certain family of parts. The code is used to retrieve a master plan for that specific family. Initially, the basic process for this family is displayed on the screen. The process planner may edit this process by adding or deleting basic processes which, according to his/her judgement, are required for the specific individual member of the family. The second phase lists the standard operations for the family. Again the process planner may edit by adding, changing and/or deleting any specific operation and adjusting it to the specific individual part dimension, tolerances and any unique requirement, while still keeping it within the family. This is done for all the operations concerned. As a rough example, all round parts with a ratio of length to diameter of not greater than 0.5 belong to the same family. The exact values have to be known when computing the detailed operations.

The process planner may have the edited routine plan printed in the company format.

With this approach, the quality of the process still depends on the background knowledge of the process planner; the computer is merely a tool used to assist the manual process planning activity.

In comparison with manually performed process planning, this approach is highly advantageous in that it:

- uses existing company manufacturing data and company expertise;
- frees the process planner from routine clerical activities;
- applies to all types of manufacturing;
- is capable of updating and reflecting changing manufacturing technologies for new and old parts;

- incorporates company standards; and
- process plans a complete part (and includes all types of processes, such as assembly, painting, storing, marketing, etc.)

The biggest savings can be realized at the preparatory stage of the system. As the system is custom built, a production flow analysis of all the parts that are and were produced in the customer plant is made in order to create the family of parts and the classification system. When this study is done, it is usually found that similar parts are, for no specific reason, being produced in different process plans. The different processes are studied and compared to one another, and the existing process, then a process optimum (according to company policy) is selected as the master process for the group. Applying this master process to all existing parts may result in considerable savings – and this by doing no more than selecting one of the existing process plans as the master process plan and applying it to all similar parts.

15.3.1 Creating families of parts

Before the variant approach can be implemented, families of parts have to be defined. Such families are not difficult to identify in small plants, where the number of components and products are small and a manual visual search of similar parts can be employed. Where the number of parts is high, other tools have to be used.

One of the tools is the 'production flow analysis'. In this system, the process plans are organized into a matrix format, where on one side is a list of all facilities, and on the other, a list of all parts. At a junction point, where a part is using a machine, a mark is made. The two lists are rearranged (by a special technique) to form families.

A more popular method is to use a classification system (Fig. 15.2). Each digit represents a certain feature of the part.

Constructing the classification system is a professional task and must be undertaken by an expert. Industry uses classification systems for inventory, bills of materials, drawing retrieval, etc. Each classification has its own purpose: a classification system constructed to use GT in the design stage, for example, cannot be used for process planning purposes. The difference here lies in the number of parts to be retrieved. If too many part drawings are retrieved for design purposes, they are useless, since no one will scan all these drawings for a specific similarity. For process planning, on the other hand, the more retrieved parts, the better.

A classification number should not be too long (no more than 12 digits) to minimize errors when manipulating the classification and coding. Many methodologies and different classifications are available, and some recommend that each plant should have a custom-made classification system to suit its own needs.

The classification task became the most critical part in implementing variant

Geometrical code

1st digit: Component class

Code	Group	Component class
0	Rotational components	$\frac{L}{D} \leqslant 0.5$
1	Rotational components	$0.5 < \frac{L}{D} < 3$
2	Rotational components	$\frac{L}{D} \geqslant 3$
3		$\frac{L}{D} \leqslant 2$ with deviation
4		$\frac{L}{D} > 2$ with deviation
5		Specific rotational components
6		Flat components $\frac{A}{B} \geqslant 3$, $\frac{A}{C} \geqslant 4$
7	Non-rotational components	Long components $\frac{A}{B} > 3$
8	Non-rotational components	Cubic components $\frac{A}{B} \leqslant 3$, $\frac{A}{C} < 4$
9	Non-rotational components	Specific non-rotational components

2nd digit: External shape, external shape elements

Code	Group	External shape, external shape elements
0		Smooth, no shape elements
1	Stepped to one end or smooth	No shape elements
2	Stepped to one end or smooth	With screwthread
3	Stepped to one end or smooth	With functional groove
4	Stepped to both ends (multiple increases)	No shape elements
5	Stepped to both ends (multiple increases)	With screwthread
6	Stepped to both ends (multiple increases)	With functional groove
7		Functional taper
8		Operating thread
9		Others (> 10 functional diameters)

3rd digit: Internal shape, internal shape elements

Code	Group	Internal shape, internal shape elements
0		Without through bore blind hole
1	Smooth or stepped to one end	No shape elements
2	Smooth or stepped to one end	With screwthread
3	Smooth or stepped to one end	With functional groove
4	Stepped to both ends (multiple increases)	No shape elements
5	Stepped to both ends (multiple increases)	With screwthread
6	Stepped to both ends (multiple increases)	With functional groove
7		Functional taper
8		Operating thread
9		Others (> 10 functional diameters)

4th digit: Plane surface machining

Code	Plane surface machining
0	No surface machining
1	External plane surface and/or surface curved in one direction
2	External plane surfaces related to one another by graduation around a circle
3	External groove and/or slot
4	External spline and/or polygon
5	External plane surface and/or slot and/or groove, spline
6	Internal plane surface and/or groove
7	Internal spline and/or polygon
8	External and internal splines and/or slot and/or groove
9	Others

5th digit: Auxilliary hole(s) and gear teeth

Code	Group	Auxilliary hole(s) and gear teeth
0	no gear teeth	No auxiliary hole(s)
1	no gear teeth	Axial hole(s) not related by a drilling pattern
2	no gear teeth	Axial hole(s) related by a drilling pattern
3	no gear teeth	Radial hole(s) not related by a drilling pattern
4	no gear teeth	Holes axial and/or radial and/or in other directions not related
5	no gear teeth	Holes axial and/or radial and/or in other directions related by drilling pattern
6	with gear teeth	Spur gear teeth
7	with gear teeth	Bevel gear teeth
8	with gear teeth	Other gear teeth
9	with gear teeth	Others

Fig. 15.2 Opitz classification system.

systems, and ultimately, a computerized classification system was developed. This is based on a tablet format, similar to the one in Fig. 15.2, and it works in dialog mode with the user. It is built as a decision tree, where the next question depends on the answer to the previous question. For example:

- Is it a rotary component?
- What is the largest diameter?
- What is the largest length?
- Does the rotary form deviate?
- Is the axis of rotation threaded?

Once the questions have been answered, the computer will assign a coding number to the part and relate it to part name and drawing number. The second step will be to ask the computer to retrieve the parts that belong to this classification number.

Notice that some questions require exact dimensions (instead of asking, as in Fig. 15.2, the ratio of L/D). This data can be used later in editing the master process plan.

Using computerized classification means there is no limitation on the number of digits in the coding system, resulting in a more refined classification. Moreover, it speeds up coding time and eliminates manual coding errors. However, the problems in constructing a classification system have not yet been fully solved.

15.4 CAPP STAGE 3: VARIANT APPROACH – ENHANCEMENT

A classification system, the heart of the variant approach, is a stiff system: once created, it is very difficult to modify, although branches may be added. A classification system is a compromise between the number of features and parameters that may affect the classification number, and the number of digits in the number. However, a process planner must consider all parameters when making decisions and selecting a process.

A computerized classification, based upon a decision tree, may allow flexibility, but when a rearrangement of the families has been done, a complete reorganization of the master process plan files has to be made. Several methods have been proposed to enhance the variant approach and they are discussed below.

15.4.1 Decision tree

The idea here is that instead of using a decision tree to establish a coding number, it is used as a key for retrieving a process, so that the tree leads directly to the process.

A decision tree is a graph with a single root and branches emanating from the root. Each branch communicates with a value. When used in decision making, branches usually carry values or expressions. Each branch represents an 'IF' statement and branches in series represent a logical 'AND'. The process is listed at the junction of each terminal branch. An example of using a decision tree is given in Fig. 15.3.

The process decision tree is structured to duplicate an existing manufacturing facility, its process capabilities, equipment and planning strategies. These items are structured into 'if . . . then' logical situations, with the appropriate decision point set to detect particular parameters. A tree is easy to construct, understand, visualize and debug. It is also easy to update and maintain. Some branches may define type, some define attributes and some perform minor computations. The selected branches may be extended to a considerable depth if necessary, while other branches may be quite short.

A simple computer program, with a lot of 'if . . . then' instructions in it and written in any language, can perform process planning. The content and knowledge regarding which nodes and branches to use, the depth of the branch and the decisions attached to the terminal branch are the user's responsibility.

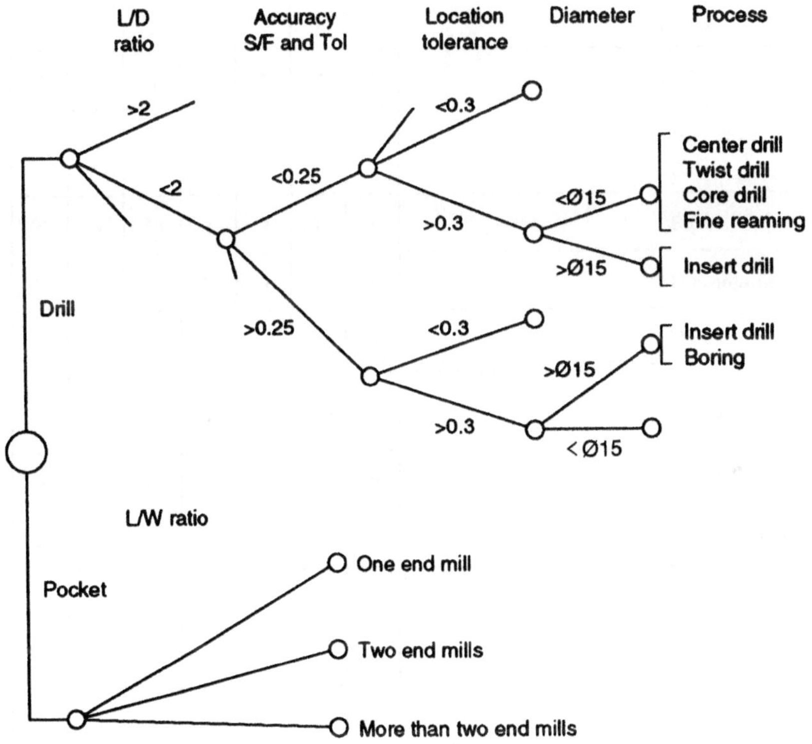

Fig. 15.3 Example of decision tree.

15.4.2 Decision table

A decision table is composed of conditions, data and action, the principle elements of all computer programs. A decision table may be visualized in a table format, as shown in Fig. 15.4. The portion of the left side of the table above the horizontal line specifies the possible conditions, while below, it specifies the possible actions. The right side of the table above the horizontal line specifies condition data, and below are marked the relevant actions.

When constructing a decision table, the factors of completeness, accuracy and consistency must be considered. A decision table which includes all parameters might become too large, which will make it difficult to read and for decisions to be interpreted. In some cases, 'do not care' is inserted as the data of a particular condition. For example, reaming will have 'do not care' entered for location tolerance, as it has no effect on it. In such cases, several processes may

L/D ratio <2	T	T	T	F	F	F	F
>2	F	F	F	T	T	T	T
Accuracy S/F and tolerance <0.25	T	T	F	F	T		F
>0.25	F	F	T	T	F		T
Location tolerance <0.3	F	F	T	F		T	F
>0.3	T	T	F	T		F	T
Diameter <Ø15	T	F	F	F	F	F	F
>Ø15	F	T	T	T	T	T	T
Center drill	1	-	-	1	1	1	-
Twist drill	2	-	-	2	2	2	1
Insert drill	-	1	1	-	-	-	-
Core drill	3	-	-	-	-	-	-
Fine reaming	4	-	-	-	-	-	-
Boring	-	-	2	3	3	3	-

Fig. 15.4 Example of decision table.

be retrieved from the table, all meeting the conditions, and some will be redundant. When using the decision table backwards, from process to conditions, and entering 'do not care' against conditions of the proposed process plan, and recurrent calls to the table are used, an endless loop can result: the process will never terminate from the table.

Decision tables are difficult to maintain and update. The depth of all conditions must be equal, which means using 'do not care' as a value, causing redundancy or loops in the system.

Decision tables and decision trees are reversible; one may be converted to the other by following the condition rules. A computer program may be prepared using any programming language, according to the programmer's preference.

15.4.3 Expert systems

An expert system is a specialized computer program that exhibits the same level of problem-solving skills as an expert for a narrow problem domain. It embodies knowledge and reasoning capabilities that allow it to draw quality conclusions comparable to those drawn by an expert.

The main difference between expert systems computer programs and normal computer application programs is that in most expert systems, the model of problem-solving in the application domain is explicitly in view as a separate entity or 'knowledge base' rather than appearing only implicitly as part of the coding of the program. This knowledge base is manipulated by a separate, clearly identifiable control strategy. Such a system architecture provides a convenient way to construct sophisticated problem-solving tools for many different domains.

Ordinary computer programs organize knowledge on two levels: data and program. Most expert systems, however, organize knowledge on three levels, as follows:

- The first level contains the declarative knowledge about the particular problem being solved, and about the current state of the art in an attempt to solve the problem.
- The second level contains the knowledge base. This is problem-solving knowledge specific to the particular kind of problem which the system is set up to solve and which is used by the system when reasoning the problem.
- The third level contains the control structure. This is a computer program which makes decisions about how to use the problem-solving knowledge found in the knowledge base in order to manipulate the data.

Expert system problems are those for which the precise series of steps necessary to solve the problem may not be known, so it is necessary to search through a space containing many alternate paths, some of which lead to solutions and some of which do not. In such situations, it is useful to encode domain-dependent knowledge in the form of operators or pattern-invoked programs. These are programs which are not called by other programs in the ordinary way, but instead are activated by the control structure whenever certain conditions hold in the data. It is the responsibility of the control structure to invoke the various operators in the knowledge base as they become applicable to the data. Several operators may be applicable at a given time, in which case the control structure must decide which of the applicable operators to employ.

Several different programming languages, such as Prolog and LISP have been developed which allow for pattern-driven invocation of programs as needed for the expert system.

Knowledge consists of symbolic descriptions that characterize the definitional and empirical relationships in a domain, and the procedures for manipulating those descriptions. Relations are expressed as dependencies and associations among objects. There is a set of standard knowledge representation techniques, each having its own strengths and benefits. The three most popular formations today are production rules, frames, and semantic nets.

Production rules utilize the simple 'if . . . then' action format. A production rule is used when the current situation matches the condition of the 'if' part. Successive rules used produce an inference chain. An example of a production rule is shown in Fig. 15.5.

One of the keys to a successful expert system is the acquisition of the knowledge that becomes its very heart and brain. No matter how good the methodology of representing the knowledge, how brilliant the design of the system, or how powerful the shell, **if** the knowledge base content is flawed, it is impossible to have an efficient expert system. Practice has proved that expert process planners usually do not know why they made a certain decision. They also have difficulty expressing themselves in computer terminology. Therefore, the collection of expert knowledge is not as easy as it may first appear.

```
Rule number: 33
Application:  Hole making

IF             minimum diameter        ≥2
    AND        minimum diameter        <3
    AND        surface roughness       ≤3
    AND        bottom type is not flat
    AND        chamfer angle is not present
    AND        thread is not present

THEN        previous operation        =drilling
            surface roughness meeting demand

            CALL DRILL
ENDIF
```

Fig. 15.5 Expert system – example of production rules.

15.4.4 Miscellaneous and remarks

Many CAPP programs have been developed along the lines of the technologies outlined in the previous sections. Some use the top-down approach, while others the bottom-up approach; some start with tool capabilities, some by using type forms, some with constraint bases, and so on.

Basically, two problems arise in applying expert systems:

1. *Redundancy of data* Conflicting rules may cause an endless loop in the decision control structure. Each production rule is true by itself, but a combination of them is not logical. To overcome this problem, each production rule must be tested before it may be entered in the knowledge base files. A knowledge base manager is appointed and is the only person who may update the files.
2. *Knowledge acquisition* The knowledge base includes the knowhow of an expert process planner. However, different expert process planners have different production rules so that the system becomes an expert system of a specific process planner. The problem is to find the right expert process planner. One idea is to have a round table of expert process planners to reach a consensus within the production rules.

Another method under development is to construct a computer program that interrogates the expert process planner, not on production rules, but rather on minor details of occurrences, then for the computer program to formulate the production rules, based upon the collected details.

As with the GT system the above enhancement systems have a problem in handling a part. The rules, whether in tree format, table format or as production rules, refer to a specific feature of the part such as a hole or a pocket. This approach calls for help from two additional research and development fields.

Several systems are therefore being developed which concentrate on combining the separate feature process plans with the part process plan.

In this era of CAD systems, it is strange that the part specification, which is stored in a computer database, has to be manually introduced into the CAPP system. Some advanced CAPP systems call for the process planner to sit near the CAD terminal and to retrieve manually the data needed for CAPP. It is true that CAD systems do not store and possess the supplementary data required for CAPP, but at least the geometry data is avilable. The trend is now towards developing feature recognition programs which retrieve data from a CAD system and transfer it automatically to a CAPP system. Unfortunately, CAD systems are not intelligent enough to recognize features; all they can do is display lines and arcs on a computer monitor, hence the need for a feature recognition system. A process planning part-feature recognition system would distinguish features of a part based on the geometric and topological information stored in the CAD database. Once features and associated manufacturing information are identified, the information can be transferred to the process planning system to generate a process plan.

Most feature recognitions systems are based on expert systems. As an example:

> IF graph is linear, and has exactly one node 'n' with both incident arcs with attribute O THEN feature is SLOT.

Such rules allow recognition of a large number of features such as step, slot, blind step, pockets and holes.

Searching for feature subgraphs is an exhausting procedure, and takes quite a lot of time. Research is going on to find heuristics which will improve the feature recognition system.

15.5 CAPP STAGE 4: GENERATIVE APPROACH

In the generative process planning approach, the computer programs possess metal cutting knowhow and geometric vision of the part. Process plans are generated by means of technology algorithms, decision logics, formulae and geometry base data to perform uniquely the many processing decisions for converting a part from raw material to the finished state.

Manufacturing knowhow and equipment capabilities are stored in a computer system. When using the system, a specific process plan for a specific part can be generated without the involvement of a process planner and without referring to any previous part plan. In the case of changes of manufacturing facilities, or overloaded machines, the generative system will automatically generate alternative process plans. The biggest advantage of the generative process planning approach is that it is fully automatic and does not rely on the specific expertise of a process planner. It is a dynamic system that can be used in CIM systems.

The input, describing part specifications – mandatory for the generative approach – may come from a text input where the user answers a number of questions, or as a graphic input from a stand-alone system or a CAD system. The generative approach is complex and difficult to develop. It requires an in-depth knowledge of manufacturing processes and mechanical engineering. Unfortunately, the field of CAPP development became dominated by computer experts, and artificial intelligence experts; they are developing the frame of the CAPP systems, but neglecting the content of the system. Consequently, not enough real effort is directed toward developing such systems.

15.6 CAPP STAGE 5: SEMI-GENERATIVE APPROACH

The semi-generative approach is an intermediary one, to be used until a generative system is developed. The term semi-generative may be defined as a combination of the generative and the variant, where a pre-process plan is developed and modified before the plan itself is used in a real production environment. It means that the decision logic, formulae and technological algorithms, as well as the geometry and extended specifications, are built into the system. At first sight, the system's working steps are the same as for the generative approach, but the final process plan and intermittent stages of the generative process plan are modified, or supplemented by a process planner. Modifying and process planner intervention is small compared with that needed with the variant approach.

15.7 GENERAL REMARKS ON CAPP DEVELOPMENTS AND TRENDS

The first level of any CAPP system must contain the declarative knowledge about the particular problem being solved, and about the current state of affairs in the attempt to generate a process plan. The natural and obvious source of data should be the CAD system on which the part was designed. Unfortunately, as we have seen, CAD systems were not designed to serve CAPP systems, and unless tedious and exhausting procedures are prepared, they are of no use for CAPP.

CAPP input data should contain the following:

- basic material . . . . .not in CAD;
- form and dimension of blank. . . . .not in CAD;
- geometric model . . . . .in CAD;
- nominal dimensions . . . . .in CAD;
- dimensional tolerances. . . . . .not in CAD;
- positional tolerances. . . . . .not in CAD;
- geometric tolerances . . . . .not in CAD;
- surface roughness. . . . . .not in CAD;
- joining technique . . . . .not in CAD;
- heat treatment . . . . .not in CAD;
- interrelation . . . . .not in CAD;
- order information. . . . . .not in CAD;
- extra information . . . . .not in CAD.

Because organizations install different brands of computers, systems, databases and communications – each with different functions – the capability to exchange and share information is hindered. Data formats differ and so do file and data structure and content. To continue down the path of integration, companies push to develop tools and methods to solve these significant issues. Efforts to improve the data exchange and sharing process between functions found in manufacturing enterprises are gaining momentum.

The Initial Graphics Exchange Specification (IGES) is an engineering data exchange supported by major CAD system vendors. Although it is not an exhaustive description of the data exchange standard, it does show the broad support of data exchange technology mainly in the CAD system of different vendors. Yet, it does not include data related to CAPP systems.

The Product Data Exchange Specification (PDES) is another international standard for the exchange of all data needed to describe fully a product and its manufacturing process.

PDES is an extension to available CAD systems, where the additional and the retrieved data is stored in special physical PDES files. It does not interfere with the CAD system itself.

CAD systems are excellent systems, meeting the specifications set down by the developers. Their most important objectives are to display graphics on a monitor, to allow for kinematics, strength, stress and flow analysis, all of which serve the product designer. However, they do not serve the other functions in a manufacturing organization.

The initial CAD method was the wire-mesh system, which works well and is fast. However, the display is somewhat difficult for a layperson to understand. Thus, CAD systems started to develop solid features, where the display is very clear and understandable, almost like a picture. Unfortunately, the information content was not significantly improved.

The most helpful feature of a CAD system as far as the CAPP system is concerned is the ability of the computer to 'see' geometric shapes, manipulate them, change them, compare them and, if required, to display them. The display is needed for the user; it is of no assistance at all for automatic and generative production management systems.

A good CAD system should also:

- assist in translating a concept into engineering design;
- retrieve existing drawings and designs by concept, key, attributes;
- allow automatic change of design;
- allow automatic design of sub-systems and tooling;
- check and enforce company standards;
- check and recommend design for ease of manufacturing and assembly;
- check and recommend tolerances and surface roughness;
- enable technology transfer;
- transfer data to CAPP systems;
- enable control of FMS systems;
- allow use as input to (robot) vision systems.

All these features can be accomplished; they just have to be recognized and specified as objectives by CAD system designers. Some of the desired features can be added, but only as a patch, and there is a limit to the number of patches that may be added without the collapse of the whole system.

Some trends indicate that CAD systems are moving towards a third revolution in their concept. Unfortunately, the good old faithful drawing standards were not used as the basic concept for CAD systems. This is probably due to the fact that mathematicians, computer experts and management experts are trying to do the job of engineers. Perhaps after several trial-and-error attempts at developing the concept, they will arrive at the right method. The two most popular existing CAD methods are described below.

15.7.1 Feature-based modeling

An engineering feature is defined as 'a physical element of a part that has some specific engineering significance'. It must satisfy the following conditions:

- be a physical constituent of a part;
- be mappable to a generic shape;
- have engineering significance; and
- have predictable properties.

The engineering significance of a feature may involve the function which the feature serves, how the feature can be produced, what actions its presence must initiate, etc. Features can be thought of as engineering primitives relevant to some engineering task.

In feature base design for manufacturing, the part geometry is defined directly in terms of features. It eliminates the need for the complex mathematical and computer science field of research of feature recognition. Feature-based design allows the designer to specify much more about the design than just geometry and topology. The design intent and constraints can be expressed on a different level of abstraction and can be used in downstream applications.

Feature modeling is based on the idea of designing with 'building blocks'. Instead of using analytical shapes such as boxes, cylinders, spheres and cones as primitives, the user creates the product model using higher-level primitives which are more relevant to the specific application. This approach should make a solid modeling system easier to use. However, the fixed set of features offered by the present generation of feature base modelers is much too limited for industrial use. Features must be flexible and represented in a dynamic way, explicitly (evaluated or enumerative) or implicitly. Implicit features are defined only by the description of how to create them and are as such not defined in terms of primitives of a geometric product model. They are merely defined by their location, orientation and parameters.

The most important dilemma of feature base modelers, and the strength of the model, is that many different methods can be used for the synthesizing of parts from features. This implies that the number of possible features is virtually infinite and the system has to know how to handle them.

The need for geometric modeling of features on a higher level of abstraction has initiated the development of solid modelers which are based on non-manifold topology. These modelers enable the geometric processing of 'incomplete' geometry.

15.7.2 Parametric modeling

Parametric modeling allows the user to create product models with variational dimensions. Dimensions can be linked via (conditional) expressions. Bi-directional associativity between the model and the dimensioning scheme allows automatic regeneration of product models after changes in dimensions and automatic updating of the related dimensions. In this way, a flexible product model can be created. Many CAD systems offer limited two-dimensional parametrics while only a few three-dimensional CAD systems have been introduced. Both are based on B-Rep (boundary representation) and enable feature-based modeling with a limited set of primitive features which can be combined to make more complex features.

FURTHER READING

Alting, L. and Zhang, H. (1989) Computer aided process planning: The state-of-the-art survey, *IJPR*, **27**(4), 553–85.

Begg, V. (1987) *Developing Expert CAD Systems*, Chapman & Hall.

Bertoline, G.R. (1988) *Fundamentals of CAD*, Delmar.

Chang, T.C. (1990) Geometric reasoning – the key to integrated process planning. *Proc. 22nd CIRP International seminar on manufacturing systems CAPP*, 1990 University of Twente Enschede, The Netherlands.

Davies, B.J., (1986) Application of expert systems in process planning, *Annals of the CIRP*, **35**(2).

Ham, I. (1988) Computer aided process planning: The present and future, *Annals of the CIRP* **37**(2).

Li, R-K., (1988) A part-feature recognition system for rotational parts, *IJPR* **26**(9), 1451–75.

Warnecke, G., Mertens, P. and Schulz, Ch. (1989) Artificial intelligence in computer integrated manufacturing – A proposal for integration of CAD and CAPP. *Proc. CIRP International workshop on CAPP*, Hannover, FRG, pp. 185–96.

CHAPTER 16

Example of a fully-developed process plan

Having analyzed the different aspects of planning the manufacturing of a mechanical part in the previous chapters, it seems appropriate now to apply these general rules to a specific example. The part used in this example is represented in Fig. 16.1.

This part needs processing by milling and drilling techniques. The preparation of its process planning file will now be carried out according to the procedure described in Chapter 2.

16.1 GLOBAL EXAMINATION OF THE PART

The part is a simple one; it is rigid, without problems of stability. The raw material, produced by sand casting, is not difficult to process, which is why the conventional means of milling and drilling have been selected. The precision required both for dimensional and geometric tolerances is not severe, except for the two holes (6) which have an accuracy of 16H8 and need two machining passes. The surface finish of R_a 3.2 μm is very commonly obtained in production, so it is not necessary to add grinding operations to finish the surfaces. Grinding would increase the cost of manufacturing significantly.

The size of the whole batch is equal to a total of:

$$50 \text{ per month} \times 12 \text{ months} \times \text{three years} = 1800$$

and is an average quantity which justifies special jigs and fixtures or special tooling.

Having described the general framework in which to work, it is now possible to define more precisely the machining conditions necessary to produce the different features in the part.

16.2 DEFINITION OF THE ELEMENTARY MACHINING OPERATIONS

Table 16.1 gives the types of operations and the machining conditions for processing the different features of the part according to Fig. 16.1. The surfaces

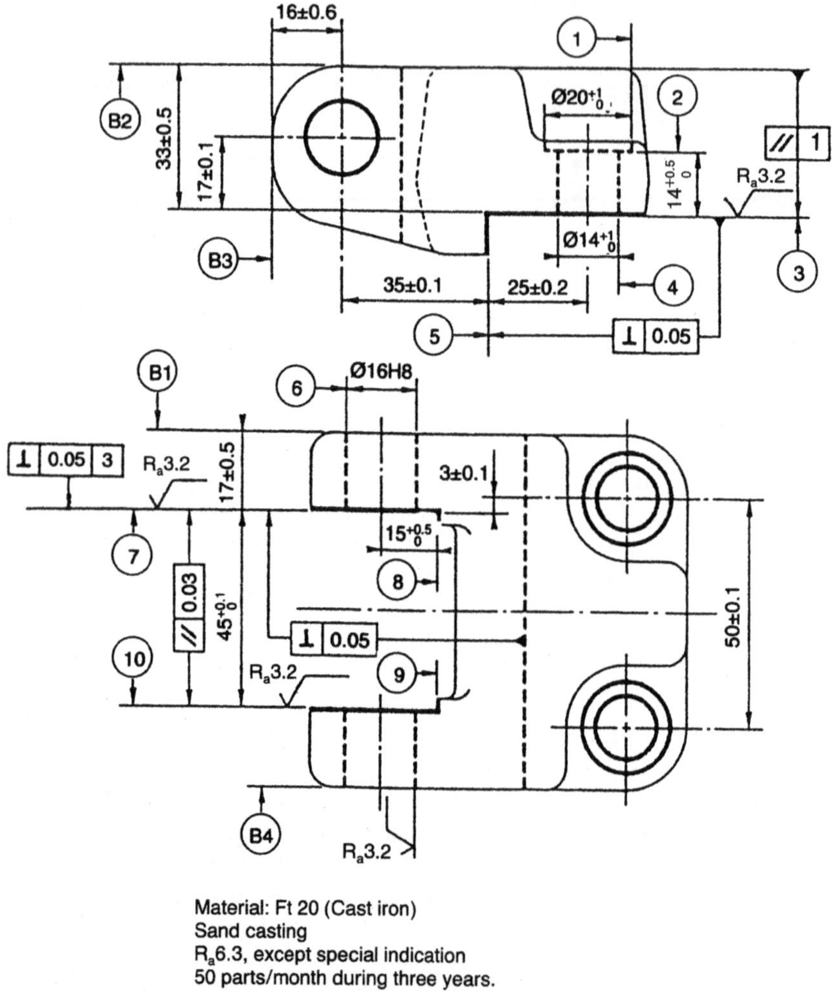

Fig. 16.1 Mechanical part (definition drawing).

to be machined are identified by a number (from 1 to 10). The raw material surfaces are identified by the symbol B and a number (from B_1 to B_4). The selection of the processing conditions is straightforward and requires a single finish pass for every surface, except for surface (6) which needs two passes (one of drilling and one of reaming) because of the accuracy grade 8, i.e. a tolerance of 0.02 in the case of the hole 16H8. Other surfaces have accuracies of 1 to 0.2 mm for the features or the distances between features and so a single pass is sufficient. The geometric tolerances are tighter (e.g. 0.03 for the parallelism between surfaces (7) and (10)); their precision is not given by the machining

Table 16.1 Table of operations

Operation Surface	*Specification*	*Rough R*	*1/2 Finish 1/2 F*	*Finish F*
1a*	R_a 6.3 Conc. 4a** $\varnothing 20_0^1$			Countersinking
1b	R_a 6.3 Conc. 4b** $\varnothing 20_0^1$			Countersinking
2a	R_a 6.3 $14_0^{0.5}$ on (3) Perpend. (1a), (4a)**			Countersinking
2b	R_a 6.3 $14_0^{0.5}$ on (3) Perpend. (1b), (4b)			Countersinking
3	33 ± 0.5 on (B_2) R_a 3.2 // on (B_2)			Milling
4a	$\varnothing 14_0^1$ R_a 6.3 3 ± 0.1 on (7) 25 ± 0.2 on (5)			Drilling with bushing
4b	$\varnothing 14_0^1$ 50 ± 0.1 on (4a) R_a 6.3, 25 ± 0.2 on (5)			Drilling with bushing
5	R_a 6.3 35 ± 0.1 on (6)			Milling
6a	R_a 3.2 $\varnothing$16H8 16 ± 0.6 on (B_3) 17 ± 0.1 on (3)	Drilling with bushing		Reaming
6b	R_a 3.2 $\varnothing$16H8 16 ± 0.6 on (B_3) 17 ± 0.1 on (3)	Drilling with bushing		Reaming
7	R_a 3.2 17 ± 0.5 on (B_1) Perp. (3), (5) // (10)			Milling
8	R_a 6.3 15 ± 0.5 on (6a)			Milling
9	R_a 6.3 15 ± 0.5 on (6b)			Milling
10	R_a 3.2 $45_0^{0.1}$ on (7) // (7)			Milling

* a,b for one of the two holes.
** Added constraint.

process, but, rather, by using the concept of associated surfaces which depend on the tooling chosen or on the machine tool accuracy. The selection of the machining conditions is, of course, made in accordance with the general principles developed in Chapters 13 and 14 on milling and hole making.

Looking through the whole of Table 16.1, it appears that the part can be produced by two processes, milling and drilling. It will be advantageous to group milling operations in one or several jobs and drilling operations in another job.

This is applied in the following analysis, according to more specific constraints.

16.3 DEFINITION OF ASSOCIATED SURFACES AND CHOICE OF JOBS AND MACHINING TOOLS

The analysis of the operations as carried out in section 16.2 also provides some suggestions for the grouping of individual operations in order to respect geometric conditions. However, this analysis does not take into account the problem of sequencing operations; this is dealt with in section 16.4 and thus is not finalized here. In many cases, though, it is a way of thinking of anteriorities and can help to define the precedence of operations found in the next section.

Having selected the operations, therefore, it is logical to arrange the groupings as associated surfaces, thus.

1. The group of surfaces (4 + 2 + 1) is justified by arguments of economy and by technical constraints (coaxiality of 1 with 4, added to the constraints in the drawing, as explained more generally in section 16.4). Also, the grouping of both holes (4) in an associated surface is justified by the accuracy of their mutual distance (50 ± 0.1). Finally, the association of surfaces (1) and (2) is justified by their processing methods, i.e. by using countersinking, piloted by the hole (4).
2. The group (6a and 6b), a and b defining one of holes, is justified both on economic and technical grounds by its coaxiality.
3. The group (3 + 5) is justified technically by the perpendicular tolerance 0.05 and economically by the use of a milling tool which processes both surfaces simultaneously.
4. The group (7 + 10 + 8 + 9) is justified technically by the parallelism (0.03) between (7) and (10), by the alignment of (8) and (9) (although not imposed in the drawing), by the perpendicularity of (8 + 9) relatively to (7) or (10) (an added constraint), and economically by the possibility of using a set of milling tools to machine the whole group simultaneously (7 + 8 + 9 + 10).

Summarizing this analysis, one can already suggest the following jobs:

(a) Job (3 + 5) by milling tools working on two edges.
(b) Job (7 + 8 + 9 + 10) by a milling tool set.
(c) Subjob (1 + 2 + 4) by drilling and countersinking using a special fixture.
(d) Subjob (6) by drilling and reaming using the same fixture as in (c), but in another orientation (Fig. 16.3).

The groupings just defined and the quantity of parts to be produced suggest the types of machine tools to be used. Applying the general rules defined in section 2.4, we come to the following conclusions:

1. Job (3 + 5) can be produced on a vertical milling machine.
2. Job (7 + 8 + 9 + 10) can be produced on a horizontal milling machine.
3. Both subjobs (1 + 2 + 4) and (6) can be produced on a column drilling machine having different tooling improvements to execute the two subjobs in sequence.

The choices of machine tools made earlier can be justified as follows.

1. For machining the group (3 + 5), planing, shaping, broaching or milling machines might come to mind. However, the first two are not productive enough for batch production and broaching needs the preparation of a special tool which is not justified in this case. Of course, the productivity of planing could be enhanced by lining up a number of parts on a table to be produced simultaneously. Nonetheless, the choice of a vertical milling machine is easily justified because of its productivity, which can be enhanced by working in a pendular way: during the machining of one part, another part on the table is removed and replaced by a new part. Alternatively several parts in line could also be machined on one subjob.
2. To machine the group (7 + 8 + 9 + 10), similar evaluations are applicable, i.e. that milling is obviously the most efficient process. In this case, the tool access for planing and shaping would raise difficulties.
3. To machine group (1 + 2 + 4), the choice of a column type drilling machine could be justified by the limited size of the series to be produced, which does not call for an automated drilling unit. Yet a very simple sensitive drilling machine would not be productive enough. Thus a boring machine might be considered instead; its productivity could be improved by using a double headed machine, processing the two holes (1 + 2 + 4) simultaneously. However, the accuracy required does not justify a boring machine. A drilling machine, which is cheaper, is able to respect the required tolerances (not more than ± 0.1 in the most severe case) with the aid of bushings in a special fixture, described in section 16.6. Milling and boring machines would be more precise in positioning, but more expensive, so we can eliminate them. To improve the productivity of the drilling machine, one could use a multitool head such as a revolver head, or a multitool head with the capability of setting the distance between both holes and processing them simultaneously. In one subphase, drilling tools would be used and in the second subphase countersinking tools would be used.

   Another way of improving productivity is to use drilling tools which can be changed very quickly, even during machining, thus saving time. Placing two column drilling machines in line would increase productivity by shifting in sequence the part in its jig to precise locators under the drilling heads. The comparison between these alternatives should be done on the basis of economic evaluations.
4. To machine the group (6), similar evaluations can be made. Changing from a drilling tool to a reaming tool could be carried out by a fast tool changing system and, of course, the use of removable bushings in the positioning jig. Again, one could consider using a milling or a boring machine, which has a higher degree of accuracy, but as mentioned earlier, it is more economical to use a drilling machine with precisely positioned bushings in a fixture which can guarantee the required precision. Also, the use of a supplementary machine would need an additional job which is not justified in this case.

Table 16.2 Table of anteriorities

Const. *Oper.*	*Dimensional*	*Geometrical*	*Technological*	*Economical*
1aF		Concentric 4aF*	Pilot in 4aF	
1bF		Concentric 4bF	Pilot in 4bF	
2aF	3F	Perpendicular 1aF and 4aF*		
2bF	3F	Perpendiculat 1bF and 4bF		
3F	B_2	// B_2		
4aF	5F 7F		8–9–10 associated surf. with 7	
4bF	5F 4aF			
5F	B_3, tolerance transfer	Perpendicular 3F		
6aR + 6bR	3F, 5F, tolerance transfer		4bF by planner decision	
6aF + 6bF	3F 5F		6aR 6bR	Protection of tool (10)**
7F	B_1	Perpendicular 3F 5F		
8F	5F, tolerance transfer	Perpendicular 3F 7F*		
9F	5F, tolerance transfer		8F, aligned with 8F	
10F	7F	// 7F		

* Added constraint.
** A chamfer could be added on (B_1) to protect the reamer.

Coming back to the choice of jobs, it will be noted that the jobs defined have not been placed in any order of precedence because the sequencing constraints have not yet been taken into consideration. This point will now be analyzed, taking into consideration transfers of tolerances, which are the result of decisions based on the anteriorities.

16.4 DETERMINATION OF THE ANTERIORITIES

In Table 16.2, a general review is made of the different constraints which appear in the definition drawing of the part (Fig. 16.1). These constraints will be used to find the proper sequencing of operations to produce a correct part.

The following comments seem appropriate:

1. Concerning surfaces (1a) and (1b), (a,b defining the two different holes), it is logical to add a constraint of coaxiality to surface (4), although no such requirement exists in the drawing. Because no indication of the position of surface (1) is given in the drawing, it is appropriate to add such a constraint, which seems natural for the whole feature (1 + 2 + 4). Also, for surface (2), it is logical to assume that it is perpendicular to surface (1) and to surface (4) which should be its anteriority, although no such constraint is indicated in

the drawing. Later on, a technological solution will be given to these problems by using a piloted countersinking tool, producing simultaneously surface (1) and surface (2), and guided by a pilot introduced in hole (4). From a technological point of view, this solution is ideal. Here, the concept of **simultaneity** between operations is introduced and will be addressed later in this section.

2. Surface (3) is clearly constrained by surface B_2 dimensionally (33 ± 0.5) and geometrically (parallelism of 1), although this latter condition is redundant because it is already contained in the dimensional accuracy.
3. The couple (4a + 4b) is constrained by (5) dimensionally and geometrically by an added perpendicularity to surface (3). One hole, now called 4a, is constrained by surface (7) dimensionally.

   The other hole (4b) is constrained dimensionally by (4a) (50 ± 0.1). The anteriority of surface (8 + 9 + 10) is due to the fact that they will be processed in the same job as (7). Also here, the whole group (7 + 8 + 9 + 10) can be processed simultaneously, as will be seen later, and therefore is an anteriority for the group (1 + 2 + 4).
4. Surface (5) is constrained by hole (6) which in turn, is constrained by the raw surface (B_3). However, applying the rule 'plane before hole', it is decided to transfer 16 ± 0.6 as a resultant dimension of a dimensional chain comprising components C_{m1} and 35 ± 0.14 (Fig. 16.4). C_{m1}, the controlling variable, is easy to determine because the resultant dimension 16 has a large tolerance 1.2. Surface (5) is therefore constrained by surface (B_3). The complete calculation of the transfer is given in section 16.7. Of course, surface (3) is constraining (5) by the perpendicularity 0.05.
5. The surfaces (6a) and (6b) have to be machined in two passes (rough and finish). They are associated surfaces because they have to be aligned. But hole (6a) is first defined in relation to (5) as a consequence of the tolerance transfer seen earlier and is constrained by (3). Technologically, both holes are constrained by the rough operation called (6a*R* and 6b*R*). Also, because planes have priority over holes, the surfaces (8 + 9 + 10) are anteriorities for (6). In the next point, a transfer of dimensions will also be decided on the dimension $15 + 0.5/0$ in application of the rule 'plane before hole'. The controlling dimension will be C_{m4}. Through a supplementary choice made by the process planner, which has no technical justification, it is decided to process the two surfaces (1 + 2 + 4) before surfaces (6).
6. Surface (7) is obviously constrained dimensionally by the raw surface (B_1) and geometrically by the surfaces (3) and (5). Surface (10) is constrained by surface (7) dimensionally and geometrically (parallelism of 0.03). The surfaces (8 + 9) are constrained by (6) according to the definition drawing. However, applying the 'plane before hole' rule, a transfer of dimension is decided and a chain of dimensions as seen in Fig. 16.5 determines the controlling dimension C_{m4}, whereas dimension $15 + 0.5/0$ becomes a resultant dimension. The complete calculation of the resulting dimension C_{m4} is given in section 16.7.

The anteriorities, as defined earlier, are essentially a result of the examination of the definition drawing. However, efficient technological decisions have already been suggested in the comments which influence the order of priorities. There are the transfer of dimensions as explained and, also, the grouping of surfaces which can be processed simultaneously by a common tool set. The surfaces suitable for such a grouping are: (3 + 5), (7 + 8 + 9 + 10) and the couples (1 + 2) and (6) which will be processed simultaneously by a set of tools.

Before proceeding to the determination of the final order of precedence of the operations, it is useful to emphasize the importance of a detailed analysis of the anteriorities so that we can arrive at a correct process plan. This exercise is also excellent for beginners because it demonstrates the many choices available to a process planner and points to the many constraints acting on the definition of an optimal process plan. It also shows the necessity to add constraints which are not necessarily indicated in the drawing and to proceed to tolerance transfers which are imposed by technological rules. Of course, the better the technological background of the process planner, the better the choices he or she will make.

16.5 DETERMINATION OF THE ORDER OF PRECEDENCE OF THE OPERATIONS

Following the recommendations contained in Table 16.2, a matrix of anteriorities can be established (Fig. 16.2). The anteriorities in the matrix strictly follow Table 16.2 to demonstrate the mechanism of finding the order of precedence. However, as mentioned earlier, cases where surfaces are grouped together and executed simultaneously are not directly taken into account in processing the matrix. The association of surfaces becomes obvious however in the bottom line of the matrix defining the order of precedence.

To simplify the graph, the anteriorities relating to raw surfaces (B_1, B_2, B_3, B_4) have not been indicated because they are obvious.

The order of precedence as shown in Fig. 16.2 can be interpreted in a way that takes into account groupings which have already been mentioned. They can be ordered in jobs and subjobs as follows (the numbering of the jobs is compatible with the numbering of the final process plan file in section 16.8):

1. Job 20 (3 + 5) by milling both surfaces with a milling tool working on two edges.
2. Job 30 (7 + 8 + 9 + 10), by using a milling cutter set working on two edges.
3. Subjob 40A by common drilling and countersinking both holes (1 + 2 + 4) simultaneously on a special fixture.
4. Subjob 40B by common drilling of both holes (6) using the same fixture as in point (3), but in a different orientation (Fig. 16.3).

Although these groupings are the same as suggested in section 16.3, they take into account the anteriority constraints.

Execute these operations

Before executing these operations

	1a	1b	2a	2b	3	4a	4b	5	6aR	6aF	6bR	6bF	7	8	9	10
1a						X										
1b							X									
2a	X				X	X										
2b		X			X		X									
3																
4a								X					X	X	X	X
4b						X		X								
5					X											
6aR				X	X			X								
6aF					X			X			X					
6bR									X							
6bF										X						
7					X			X								
8					X			X					X			
9														X		X
10													X			

	1	2	3	4	5	6	7	8	9	10	11	12	13
1a	1	1	1	1	1	1	(1a)						
1b	1	1	1	1	1	1	1	(1b)					
2a	3	2	2	2	2	2	1	(2a)					
2b	3	2	2	2	2	2	2	1	(2b)				
3	(3)												
4a	5	5	4	3	1	(4a)							
4b	2	2	1	1	1	1	(4b)						
5	1	(5)											
6aR	3	2	1	1	1	1	1	1	1	(6aR)			
6aF	3	2	1	1	1	1	1	1	1	1	1	(6aF)	
6bR	1	1	1	1	1	1	1	1	1	1	(6bR)		
6bF	1	1	1	1	1	1	1	1	1	1	1	1	(6bF)
7	2	1	(7)										
8	3	2	1	(8)									
9	2	2	2	2	(9)								
10	1	1	1	(10)									
	3	5	7	8	9	4a	4b	1b	2b	6aR	6bR	6aF	6bF
				10			1a	2a					

Fig. 16.2 Matrix of anteriorities.

16.6 SELECTION OF POSITIONING SURFACES AND CLAMPING POINTS

In agreement with the previously defined order of precedence, a first choice has to be made concerning the positioning of surfaces (3 + 5). As shown in Fig. 16.6 concerning the execution of job 20, the choice of raw surface (B_2) as the starting surface is justified by its size and by its relation in dimension and in parallelism to surface (3). Therefore, three points (1,2,3) are put on surface (B_2) as shown in Fig. 16.6, job 20. Alternatively, surface (5) could have been chosen as the primary surface, but it is smaller and its tolerance in relation to raw surface (B_3) is no better (± 0.5) than the tolerance of surface (3) in relation to (B_2), equal to (± 0.5). Concerning the tolerances of orientation, it appears that the orientation of (3) relative to (B_2) in the larger dimension is equal to 1/72 (dimension measured on the drawing) and that the orientation of (5) towards (B_3) in the larger dimension is equal to it. In the second direction, surface (3) has a tolerance of orientation of 1/35, whereas surface (5) has a tolerance of 1/9. Therefore, the choice of (3) as the primary surface with three points and of (5) as the secondary surface with two points (4,5) is completely justified. The sixth point to respect the isostatism of the part is placed on surface (B_1), but does not play any functional role. Clamping is secured at point S_1 against surface (B_2).

In the next job (job 30), to process the set of surfaces (7 + 8 + 9 + 10) (Fig. 16.6), the choice of surface (3) as the primary surface with three points of contact (1,2,3) is justified by the tolerance of perpendicularity of surface (7) to surface (3). The perpendicularity to (5) is equal to the previous one, but the size of (3) is larger than the size of (5). Alternatively, surface (5) could have been considered as primary surface because of a precise relation to surfaces (8 + 9) (tolerance of $C_{m4} = 0.3$), but, again, its size is smaller than that of (3). Concerning the orientation tolerances, the perpendicularity tolerance between (7) and (3 + 5) which is equal to 0.05 is larger than the orientation tolerances of (8 + 9) to (5) which is equal to (0.3/37 = 0.008). Finally, the previous arguments in favor of surface (3) as the primary surface are retained. The sixth point is placed on the raw surface (B_1) to respect the dimension $C_{m3} = 17 \pm 0.5$ which gives the position of (7). As the tolerance is large (1), there is no reason not to consider surface (B_1) as the tertiary surface. In addition, it can be argued that surface (7) is well positioned already in orientation towards surfaces (3 + 5). Clamping is secured at point S_2 against surface (3).

The last job is executed in two subjobs (40A and 40B) as shown in Fig. 16.6. In the first subjob 40A, executed with a special fixture such as the one in Fig. 16.3, surface (3) is used as the primary surface, surface (5) as the secondary surface and surface (7) as the tertiary surface. The arguments in favor of (3) as primary surface are again its larger size and good access.

From the point of view of precision, the tolerance of surface (2) in relation to (3) is comparable with the tolerance of surface (4) in relation to (5) (0.5 against 0.4). The same can be said for the comparison in orientations: the orientation of (2) towards (3) is equal to 0.5/20, whereas the orientation of (4) towards (5) is

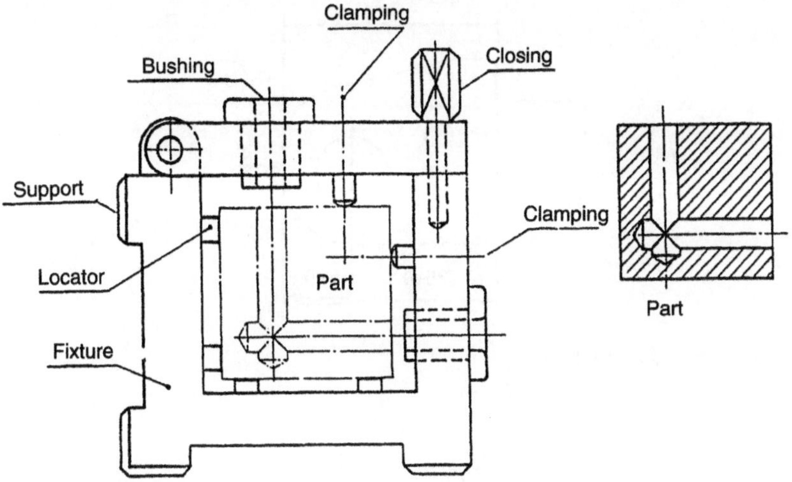

Fig. 16.3 Fixture for two orientations.

equal to 0.4/14. Considering the relation between surface (7) and surface (4), which is more precise (0.2), it would seem justified to choose surface (7) as a primary surface. However, its size and access are not very favorable. Also, it can be said that its orientation towards (3 + 5) is very precise (0.05) and, therefore, a positioning contact in one point (6) is acceptable.

As far as the orientation tolerances are concerned, there is again a better tolerance towards (7) by (4), equal to 0.2/14, whereas the tolerance towards (5) is only 0.4/14. However, the previous arguments in favor of (7) as the tertiary surface are dominant. Of course, in order to improve accuracy, removable bushings are used, as shown in Fig. 16.3.

Concerning surface (6) in subjob (40B), the same system of reference is used because the part is clamped in a special fixture as represented in Fig. 16.3 which enables working in two perpendicular directions. Surfaces (3 + 5) are well suited to serve as primary and secondary surfaces because the relation of surface (6) in position and orientation is the same for both surfaces. This circumstance justifies *a posteriori* the choice of (3 + 5) as primary and secondary surfaces in subjob 40A. Clamping is secured in point I_1 against surface (3).

In the case of this special fixture, it is clear that the location of the features of the part will not only depend on the locators in the fixture as represented in Fig. 16.6, but also on the location of the supports of the fixture in relation to the reference system of the machine tool.

16.7 TRANSFER OF TOLERANCES

The transfer of tolerances which have been decided in jobs 20 and 30, as represented in Figs 16.4 and 16.5, are both justified by the rule 'plane before hole'.

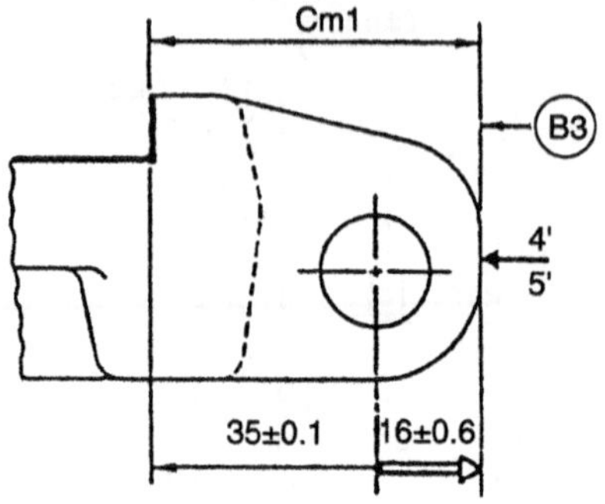

Fig. 16.4 Transfer of tolerances.

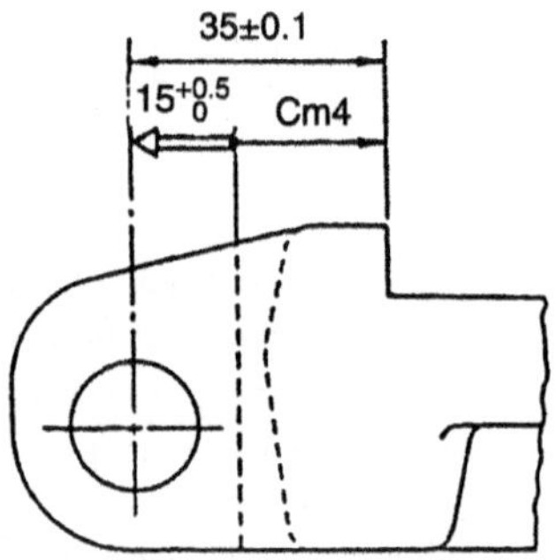

Fig. 16.5 Transfer of tolerances.

In the first chain of dimensions (Fig. 16.4), the dimension 16 ± 0.6 becomes a resultant dimension and the calculation of the controlling dimension C_{m1} is straightforward. For the average dimensions:

$$16 = \text{mean}(C_{m1}) - 35 \quad \text{or} \quad \text{mean}(C_{m1}) = 51$$

For the tolerances (because the dimensions are independent as explained in Chapter 4):

$$\text{IT}(16) = \text{IT}(C_{m1}) + \text{IT}(35) \quad \text{or} \quad \text{IT}(C_{m1}) = 1$$

The final dimension of C_{m1} will be:

$$C_{m1} = 51 \pm 0.5$$

The second chain of dimensions (Fig. 16.5) is processed in the same manner. For the average dimensions:

$$15.25 = 35 - (C_{m4}) \quad \text{or} \quad \text{mean}(C_{m4}) = 19.75$$

For the tolerances:

$$\text{IT}(15) = \text{IT}(35) + \text{IT}(C_{m4}) \quad \text{or} \quad \text{IT}(C_{m4}) = 0.3$$

The final dimension of C_{m4} is therefore:

$$C_{m4} = 19.75 \pm 0.15 \quad \text{or} \quad C_{m4} = 19.6 + 0.3/0$$

All other dimensions in the part are executed as direct dimensions according to the tolerances in the drawing. Therefore, their setting is straightforward and equal to the nominal dimension. The dimensions are machine dimensions such as C_{m1} or tool dimensions such as C_{t45} (job 30) or jig dimensions such as C_{j1} (job 40), as shown in Fig. 16.6. The only exception is the processing of holes (6) which is carried out in two passes. Therefore a calculation has to be made to determine the dimensions of the tools (one in drilling and one in reaming). The principle of the calculation is based on the requirement to have a final pass of a maximum of 0.15 and a minimum of 0.1 chip thickness at the radius. The final dimension is given by the diameter 16H8 or (16+0.027/0). The calculation follows the procedure explained in section 2.7.1.

If C_{m4} is the diameter obtained in subjob 40B, operation b by drilling with a tolerance of 0.1 as usual for drilling and designing the chip thickness by C_p, it becomes:

$$C_{pmin} = 16H8_{min} - C_{m4max} = 16 - C_{m4max} = 0.2$$

at the diameter or:

$$C_{m4max} = 16 - 0.2 = 15.8$$

and for C_{pmax}:

$$C_{pmax} = 16H8_{max} - C_{m4min} = 0.3$$

or:

$$C_{m4min} = 16 - 0.3 = 15.7$$

or:

$$C_{m4} = 15.7\ 0.1/0$$

which is coherent with the tolerance of drilling.

A special case of the transferring of tolerances arises when one of the dimensions to determine relates to the raw blank surface. An example of this is given in section 3.2 where the dimension b has to be limited in its minimum dimension to comply with the requirement of a liaison to the blank of 58 minimum. A similar case with a requirement concerning the raw blank dimension is given in section 4.4 when the chip thickness of the first pass on the raw material has to be of a minimal depth. In the case of the part examined here, these circumstances do not appear because there are no restrictions concerning the liaisons to the raw blank. In the case of drilling and reaming the holes (6), the first pass is done on the full material and there is no problem with disturbed layers at the external surface of the blank. The dimensions of the raw blank in this case are therefore not determined by any constraint. However, because the raw blank has to be inside a suitable envelope in order to be processed without any lack of material in some of its points on the external surface, it is customary

Job	Description of operations	M/T	Positioning clamping	Tooling inspection	Drawing
10	Inspection of raw blank Sizes and specifications in drawing of blank		According to raw blank drawing	3D instrumentation	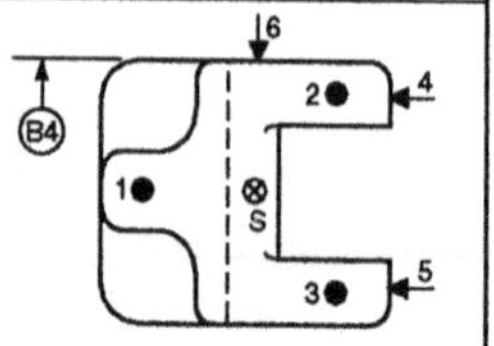
20 a	Milling Simultaneous milling of (3) (5) Finishing sizes: *Cm*1 = 51+/–0.5 *Cm*2 = 33+/–0.5 (3) //1 on (B2) (5) perp. 0.05 on (3) Ra 6.3 on (5) Ra 3.2 on (3)	VMM	Fixture on table Plane support on (B2) (123) Size: 33+/–0.5 Linear support on (B3) (4 5) Size: 51+/–0.5 Punctual sup. on (B4) (6) Clamping on S1	Milling tool 2 edges Checking: Set up for // For *Cm*1 and *Cm*2, DSG	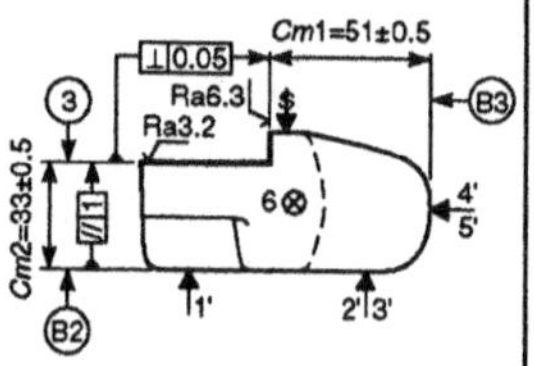
30 a	Milling Simultaneous Finish milling of (7, 8, 9, 10) Sizes: *Cm*4 = 20–0.1/–0.4 *Cm*3 = 17+/–0.5 *Ct* = 45+0.1/0 (10) //0.03(7) (7) perp. 0.05 (3) and (5) Ra 3.2 on (7) and (10) Ra 6.3 on (8) and (9)	HMM	Fixture F30 Plane support on (3) (123) Linear support on (5) (4 5) Punctual supp. on (B1) (6) Clamping on S2	Milling tool 2 edges Checking: Control jig for *Cm*4 and perp. DSG for *Cm*3	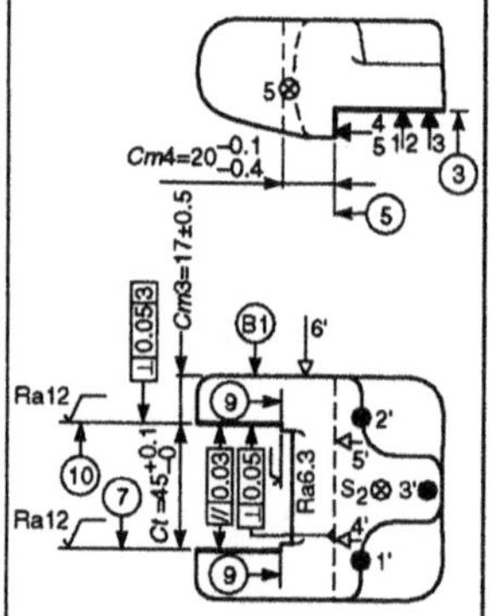
40 A a b B a b c	Drilling Drilling 2 open holes (4) Sizes: 2*Cm*1 = Ø14$^{1}_{0}$ *Cj*1 = 25+/–0.2 *Cj*2 = 3+/–0.1 *Cj*3 = 50+/–0.1 Ra 6.3 Countersink. 2 holes (1) and (2) Sizes: *Cm*2 = 14+0.5/0 2*Cm*3 = 20+1/0 Ra 6.3 Turn over fixture PAQ Drill 2 holes aligned (6) 2*Cm*4 = 15.80/–0.1 *Cj*4 = 17+/–0.1 *Cj*5 = 35+/–0.1 Reaming (6) 2*Cm*4 = Ø16H8 Ra 3.2	CDM	Plane support on (3) (123) Linear support on (5) (4 5) Punctual supp. on (7) (6) Clamping on I1	Fixture PAQ with removeable bushings Drill Ø14 Countersink Ø20 with pilot Ø14 Drill Ø15.8 Reamer Ø16H8 with fast changing tool system	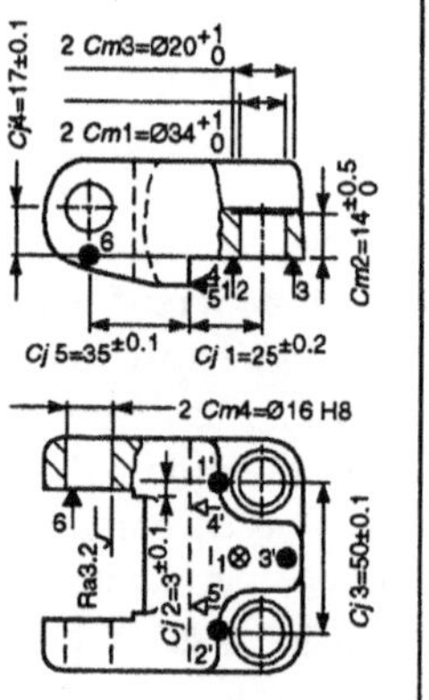
50 A B C	Final checking Dimensional Geometrical Surface roughness			Inspection department	

Jobs are identified by numbers (10, 20, ...). Subjobs by upper case letters (A, E, ...). Operations by lower case letters (a, b, ...).
VMM = vertical milling machine HMM = horizontal milling machine
DSG = double snap gauge F30 = internal code
CDM = column drilling machine PAQ = internal code

Fig. 16.6 Final process sheet.

in this case to add a fixed allowance of material to all the dimensions of the part in all directions. In the general case where some dimensions of the raw blank are toleranced, it is necessary to prepare a special drawing showing the toleranced dimensions to be used for the primary process.

16.8 FINAL EDITING OF THE COMPLETE PROCESS PLAN FILE

The final process plan file is presented in the form of a process sheet as defined in Fig. 2.12. Figure 16.6 shows this process sheet, which gives the following information:

1. In the first column, the identification of the jobs, subjobs and operations is indicated.
2. In the second column, each operation is described in detail, with the addition of the setting dimensions on the machine tool.
3. In the third column, the type of the machine used is indicated, mostly in abbreviated form which is decoded in the bottom of the sheet.
4. In the fourth column, the datum surfaces are identified with their locators and the dimensions which they determine. Clamping of the part is also indicated.
5. In the fifth column, the tooling, the fixtures and the inspection instruments are indicated, also with the special coding explained at the bottom of the sheet.
6. The sixth column is reserved for simplified drawings of the workpiece relating to the job described and showing essentially the surfaces to be processed (in thick lines), the constraints (dimensional and geometric tolerances, surface roughness), the positioning and clamping points and the setting dimensions, ultimately as a result of tolerance transfers.

The information given in the process sheet has to be completed by an operation or an instruction sheet, as explained in Fig. 2.13, giving the machining conditions and an evaluation of the manufacturing times. Detailed information on these subjects is not given here, but can be found in Chapters 6 to 11. The choice of the jigs and fixtures is not detailed either, but as a general rule, they belong to universal clamping fixtures such as machine vises in milling operations (see job 20), or to special fixtures developed for the part to be produced, such as in subjobs 40A and 40B.

Similar remarks can be made for the selection of inspection instruments. Concerning jobs 10 and 50, the instrumentation belongs to the inspection department which is well equipped to check a part from the points of view of dimension, geometry and roughness. Concerning inspection in production, the instruments are mostly universal instruments, such as a double snap gauge to control dimensions C_{m1} and C_{m2} in job 20, or dimension C_{m3} in job 30. In the case of the inspection of dimension C_{m4} in job 30 and the geometric tolerances of parallelism and perpendicularity in jobs 20 and 30, a special jig has to be designed because of the complexity of the control. Other dimensions, given as tool dimensions (C_{t45} in job 30) or jig dimensions (C_{ji} in job 40), are supposed to be precise by construction and do not need inspection in production. Of course, all these characteristics will be checked in the final inspection of the part after completion and will serve as the absolute judge for the acceptance of the part.

16.9 CONCLUSION

The general principles given in Chapter 2, concerning process planning of a mechanical part have been applied to the part shown in Fig. 16.1. It has revealed that, for every stage of a process plan, flexibility should be given to the process planner to make decisions. All the aspects of process planning, such as the selection of the types of operations, the sequence of operations, the choice of the machinery and of tooling, are subject to an analysis which takes into account all the options open for producing the part. The detailed discussion on different options is a very useful exercise when trying to arrive at an optimal process plan. Of course, such evaluations require the process planner to have a good technological background, and there must be good communication channels between process planning and the other departments of the enterprise (design, inspection, production management and so on).

Special attention has to be given to the choice of machine tools and associated tooling, as explained in section 16.3, also taking into account the availability of machines in the company. To help the process planner, a complete file of all the machines existing in the enterprise is necessary and should contain all the characteristics needed: size of the machine, rotational speeds, power, range of displacements of the machine carriages, type of spindle cones, type and location of setting switches, possibilities of fixturing, possibilities of combining several machines in a machining center, and most important of all, accuracy of the machine in its different configurations.

As a general rule, and as explained in section 2.4, the choice of machine tools depends essentially on the size of the series to be produced, on the number of tools which can be mounted on the machine, and, of course, on the accuracy of the machine required. For a more detailed description of machining equipment, the reader is referred to specialist books available in most technical libraries. Because of the fast pace in machinery development, it is also important to pay attention to the technical brochures published by the machinery manufacturers which give an updated view of the most advanced types of machinery.

Another important consideration in the evaluation of different options available when producing a part is the cost of each choice. The calculation has to take into account the hourly costs of machine tools and the operator, the cost of the fixture preparation, of setting the tools, and of loading the part, the cost of the machining time and cleaning, and finally, any other special costs. When all this information is available, it is a straightforward process to make comparative evaluations among different solutions and to choose the most economical one.

Another important factor in a process plan which has not been dealt with in the above analysis is the integration of phases of heat treatments or other mechanical treatments in the plan. These have already been discussed in section 2.9., and the intention here is only to draw attention to the necessity of putting the treatments in the right sequence. Their integration in the process plan is essentially a matter of experience, which is difficult to formalize using a set of instructions.

Finally, because it is almost impossible to predict all the characteristics of a process plan (accuracy of machines and related tooling, relevant anteriorities, influence of positioning, etc), a really optimal process plan can only be defined after observations in the course of real production conditions. Of course, executing pre-series to adjust the process plan is only justified for large series. For unitary or small series, the only way to arrive at the optimum is by trial and error. In certain situations, it is possible to benefit from previous process plans which have been optimized and which are similar to the process plan currently being developed. In this case, the techniques of GT or 'family of workpiece' can be of immense help. The reader is referred to the specialized documentation existing in this field.

In conclusion, it is clear that developing a process plan for a real workpiece, even of a simple configuration such as the part in Fig. 16.1, is a very comprehensive task, based on a huge amount of technological knowledge. A person is able to use his or her personal knowledge in process planning, whereas a computerized process planning system will have problems making the same intelligent decisions, so it is not reasonable to expect CAPP to replace human process planners in the near future.

Nevertheless, as far as special routines in process planning such as computerized tolerancing are concerned, it is obvious that computerized planning modules can be of great help to process planners.

Exercises

CHAPTER 1

1.1 What is the main difference between the forming process by material removal and the other processes?

1.2 In introducing a new product, management wants to know whether additional floor space will be required. Describe qualitatively and with a modest degree of accuracy, how you would set about answering.

1.3 What is the difference, if any, between large and small plant as far as process planning requirements are concerned?

1.4 You, as a process planner, have to define a process for a part that has to be manufactured in a quantity of 30 pieces. Your salary is $10 per hour. After examining the part drawing for about five minutes, you decide to use a machine whose overall hourly rate is $18. The production time will be nine minutes. However, from your experience, you know that when you invest more time in analyzing the part drawing you usually reduce the production time by a factor of three.
(a) Compute the optimum production time.
(b) Compute the economic thinking time.

1.5 How would you:
(a) define a good process planner?
(b) measure the performance of a process planner?

1.6 Why is it important for the product designer to consult with the process planner at the design stage?

1.7 With whom should the product designer confer during the design stage, and why?

1.8 Why is it important to have alternative process plans?

1.9 Assess the chain of activities approach to production management.

CHAPTER 2

2.1 The heavy lines in the drawings below indicate the surfaces to be machined and the tools to be used. Add the setting dimensions to the drawings as a function of the tools used.

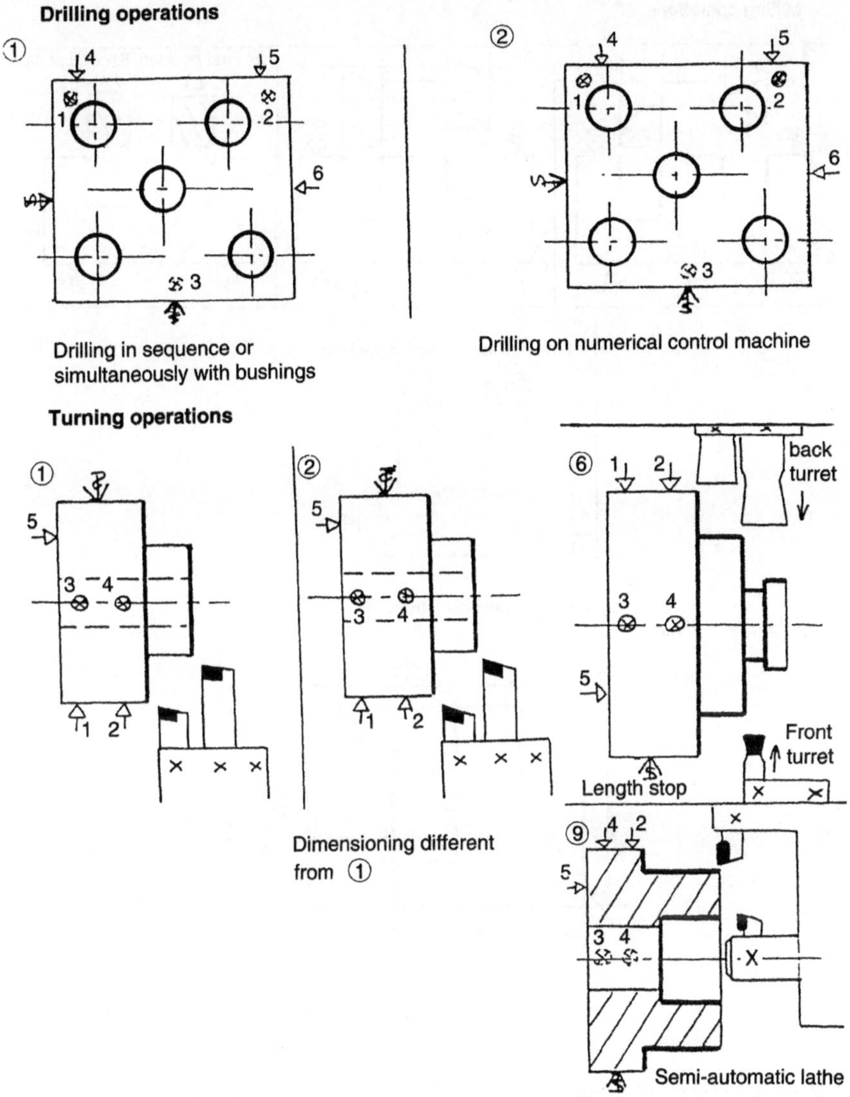

Fig. E2.1

2.2 Assuming that surfaces A and B have been processed before, that adjacent surfaces can be associated and that the surfaces are produced in subjobs with a two-edged milling tool, give:
(a) the appropriate process plan;
(b) the positioning points for each job; and
(c) the optimal setting dimensions from an economical viewpoint (according to the elementary model of tolerancing in Chapter 4).

Milling operations

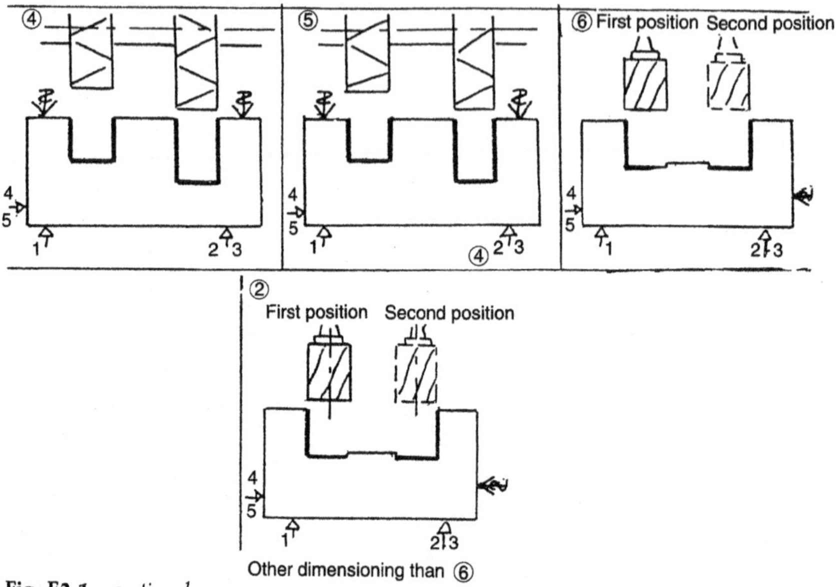

Fig. E2.1—*continued*

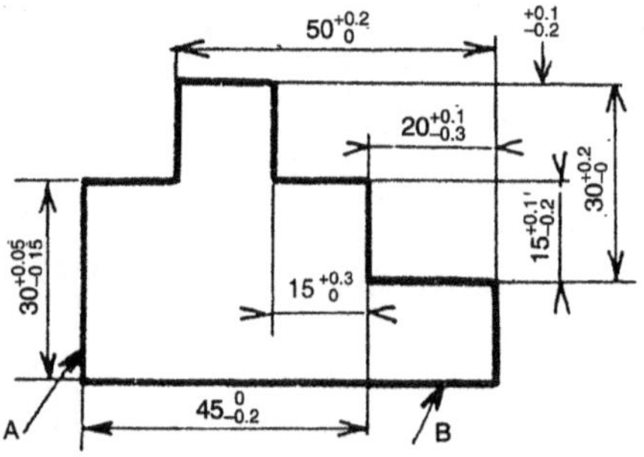

Fig. E2.2

2.3 Determine the machining conditions for every elementary operation of the part represented below (surfaces with numbers have to be processed). According to the choices made and the constraints of anteriorities, prepare a table recording, for every operation, the anteriorities classified as dimensional, geometric, technological and economical. The surfaces identified by a letter are raw, unprocessed surfaces.

2.4 For the part given in the drawing below, the sequence of the jobs is indicated in the adjoining table. Before machining, the parts are produced by casting and have a tolerance of 1 mm up to a size of 25 mm and a tolerance of 1.5 mm for sizes above 25 mm.

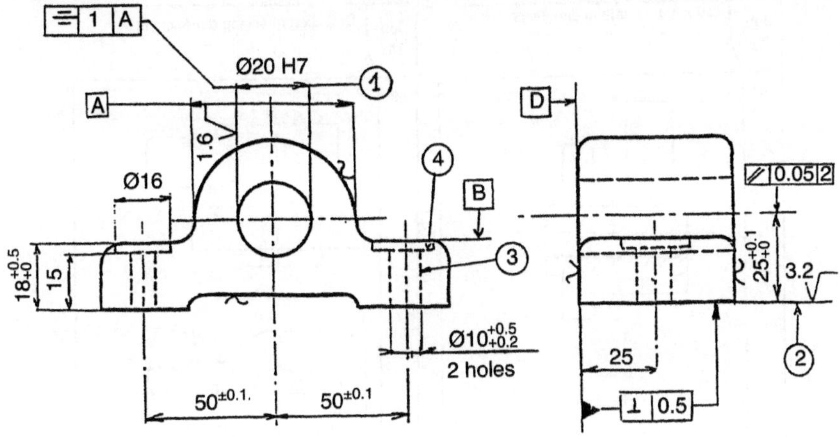

Fig. E2.3

(a) For every job, enter in the table the setting dimensions with their type.
(b) Show the tolerancing chains necessary to calculate the setting dimensions.
(c) Determine the values of the setting dimensions and the tolerances (according to the elementary model).

2.5 For the part in Exercise 2.4, prepare a table giving the choice of machining conditions for the elementary operations.

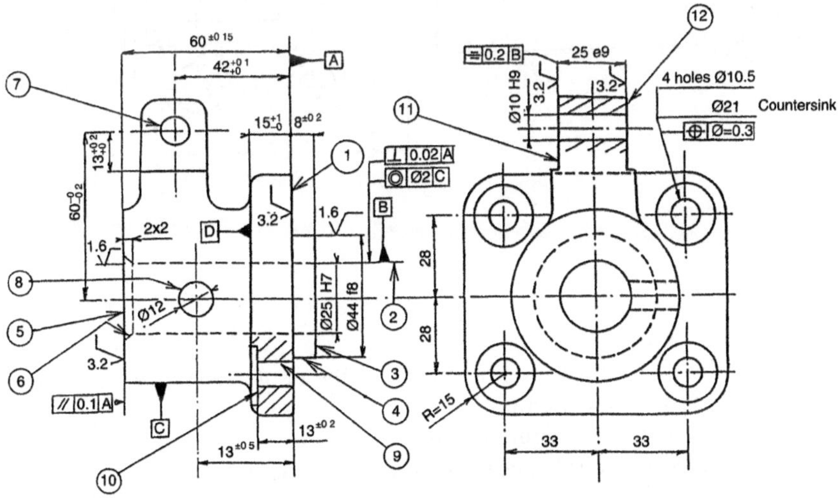

Fig. E2.4 Redrawn from Krulikowski, A. (1990) *Geometric Dimensioning and Tolerancing, Self Study Workbook*, published by Effective Training Inc., 20968, Wayne Road, Westland, MI 48185.

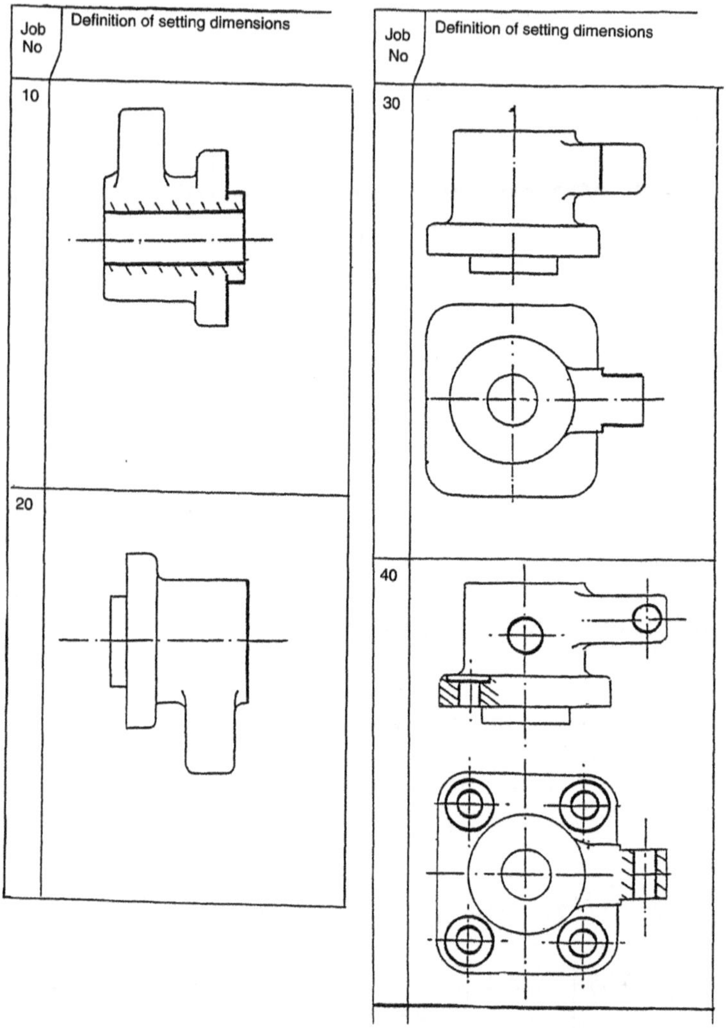

Fig. E2.4—*continued*

CHAPTER 3

3.1 For the part represented below, fill in the bonus tolerance values and give the virtual condition.

3.2 For the part in the figure below:
   (a) Describe the tolerance zone for the circularity call out.
   (b) What is the largest permissible value of circularity in the figure?
   (c) Without a circularity tolerance, what would control the circularity limits?

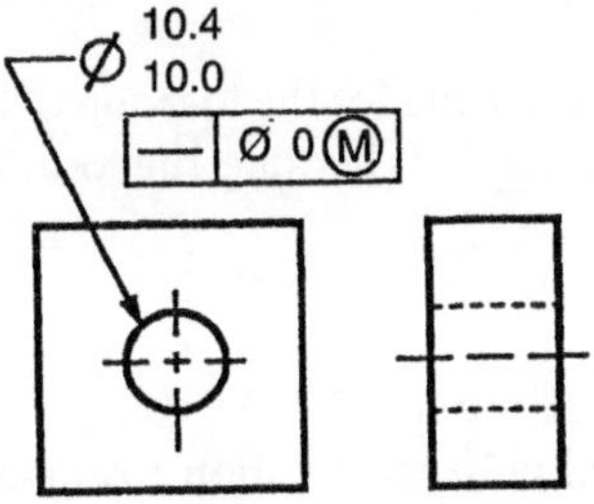

Actual hole Ø	Bonus tolerance is ...
10.4	
10.3	
10.3	
10.1	
10.0	
9.8	

Fig. E3.1 Redrawn from Krulikowski, A. (1990) *Geometric Dimensioning and Tolerancing, Self Study Workbook*, published by Effective Training Inc., 20968, Wayne Road, Westland, MI 48185.

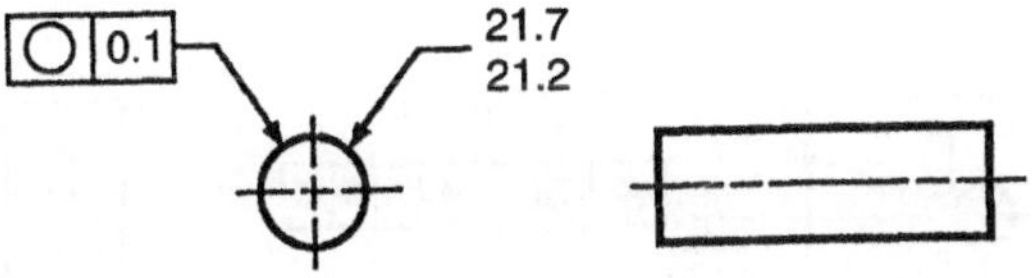

Fig. E3.2 Redrawn from Krulikowski, A. (1990) *Geometric Dimensioning and Tolerancing, Self Study Workbook*, published by Effective Training Inc., 20968, Wayne Road, Westland, MI 48185.

3.3 For the three situations (A, B and C) of the figure below, give the number of contact points for each positioning. Give the maximum and minimum values of size 10.

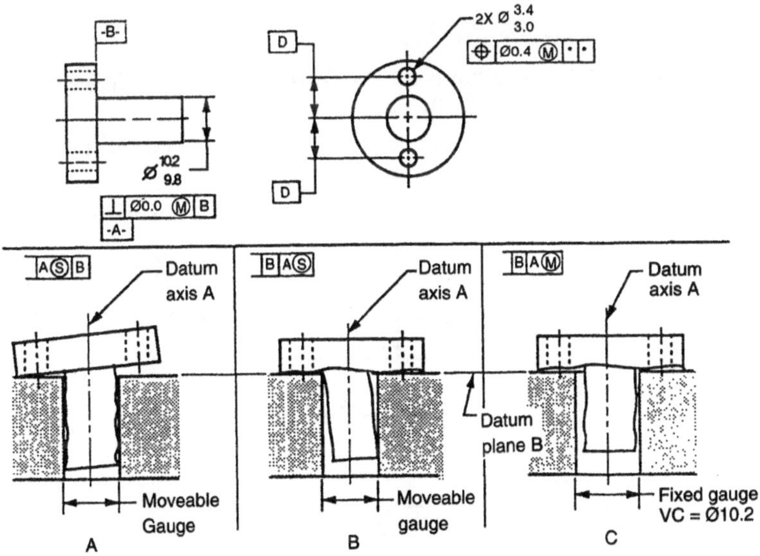

Fig. E3.3 Redrawn from Krulikowski, A. (1990) *Geometric Dimensioning and Tolerancing, Self Study Workbook*, published by Effective Training Inc., 20968, Wayne Road, Westland, MI 48185.

3.4 In the figure below:
(a) Which datum is primary?
(b) What is the shape of the tolerance zone for the location of the hole?
(c) What is the diameter of the gauge for checking the location of the hole?
(d) What is the location of the gauge pin with respect to datum references B and C?
(e) What is the virtual condition of the hole?
(f) How much bonus is associated with the location tolerance?
(g) What is the tolerance zone diameter for the axis of the hole when the hole is at LMC (least material condition)?
(h) What are the minimum and maximum values of distance 'X'?
(i) Draw a sketch to check the hole location.

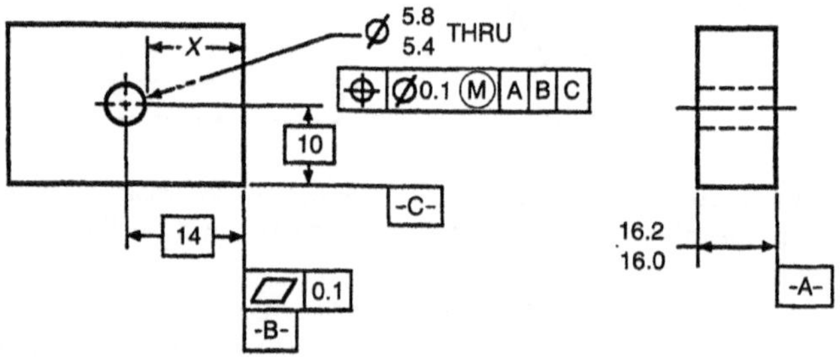

Fig. E3.4 Redrawn from Krulikowski, A. (1990) *Geometric Dimensioning and Tolerancing, Self Study Workbook*, published by Effective Training Inc., 20968, Wayne Road, Westland, MI 48185.

3.5 In the figure below what is the size and shape of the tolerance zone for the location of diameter B?
(a) How much bonus is associated with the tolerance of position?
(b) What is the virtual condition of diameter B?
(c) Draw a sketch for checking the tolerance of position.

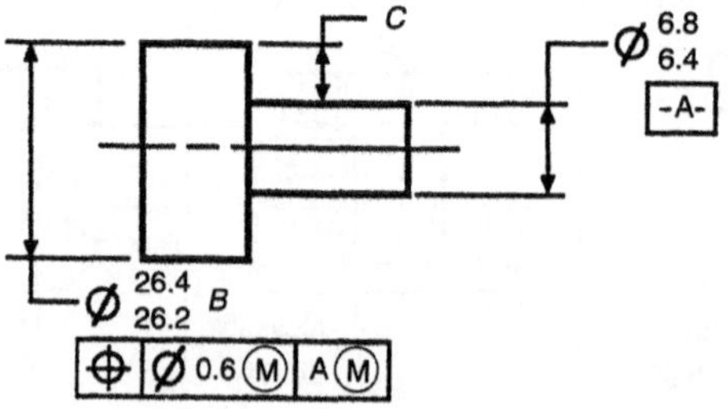

Fig. E3.5 Redrawn from Krulikowski, A. (1990) *Geometric Dimensioning and Tolerancing, Self Study Workbook*, published by Effective Training Inc., 20968, Wayne Road, Westland, MI 48185.

CHAPTER 4

4.1 Draw the setting chains with their tolerances under the drawing, assuming that all settings are made in relation to the locator 4–5.

The surfaces with thick lines are produced with the same setup and the same locator surface 4–5. Give the tolerance of the dimension 15 0/—. (Use the elementary model of tolerancing.)

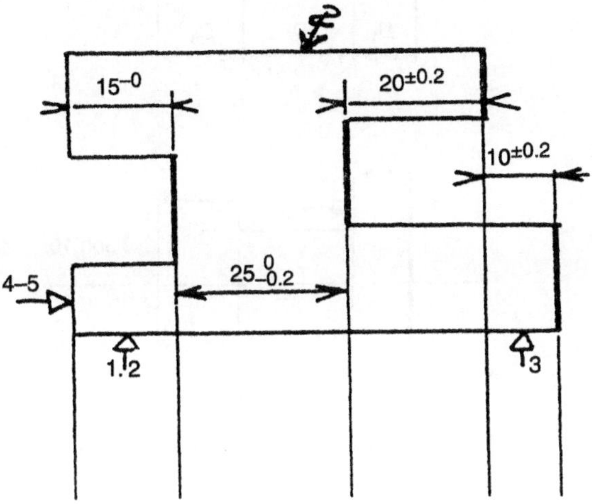

Fig. E4.1

4.2 Surfaces 1, 2 and 3 have to be processed according to the following process plan (see sketch below):
(a) Job 10: positioned on 4, processing of 1.
(b) Job 20: positioned on 1, processing of 3.
(c) Job 30: positioned on 3, processing of 2.
Determine the relations necessary to calculate the setting dimensions.

4.3 The surface 3 has to be processed in relation to the raw surface 2 respecting the dimension $D=30\pm1$. The tolerance of the raw material is IT = 1 mm and the tolerance of positioning is IT(raw/4) = 0.25. Define the setting dimension D. (Use the generalized model of tolerancing.)

4.4 In the figure below, the setting dimensions are:

$$C_{m1}=D_1\pm\frac{t'}{2}$$

$$C_{m2}=D_2\pm\frac{t'}{2}$$

(a) Determine t' in relation to $\varnothing t$.
(b) What seems to be wrong in the symbolization of the figure?

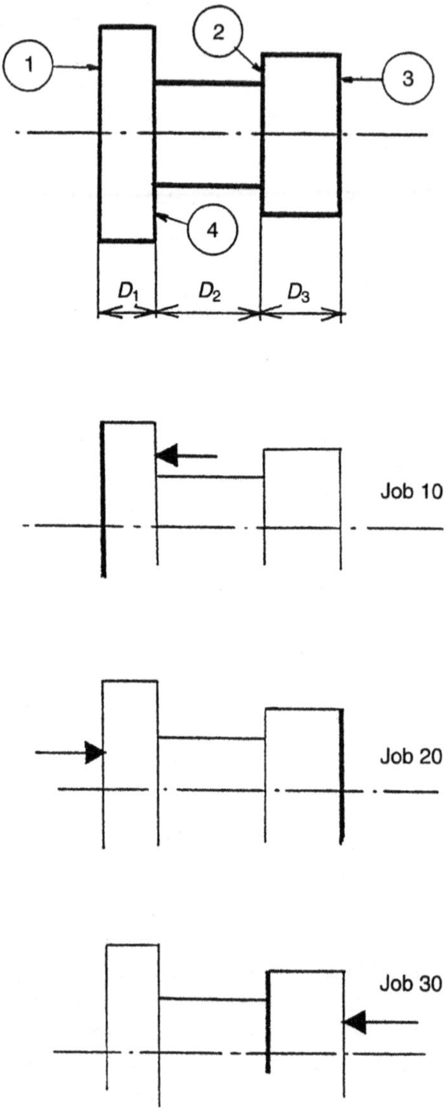

Fig. E4.2

4.5 In the figure below, the part is produced by the following process plan:

Job 40: processing of 1: plane support on reference surface SR1 and linear support on reference surface SR2.

Job 50: processing of 2 by revolving the part, same referencing as in job 40. Which tolerance should be given to the dimensions D_1 and D_2 to guarantee the tolerance of coaxiality?

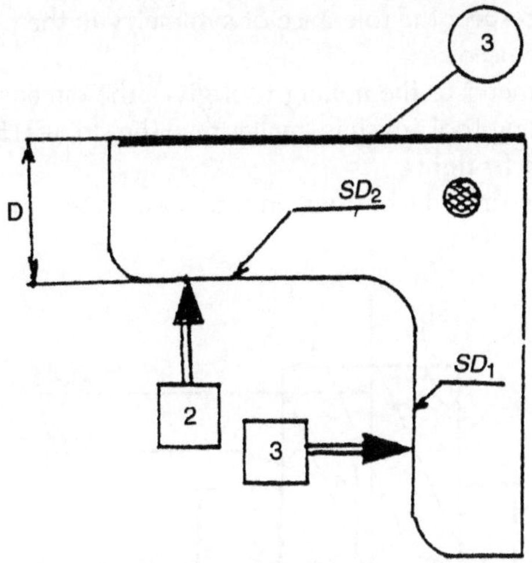

Fig. E4.3

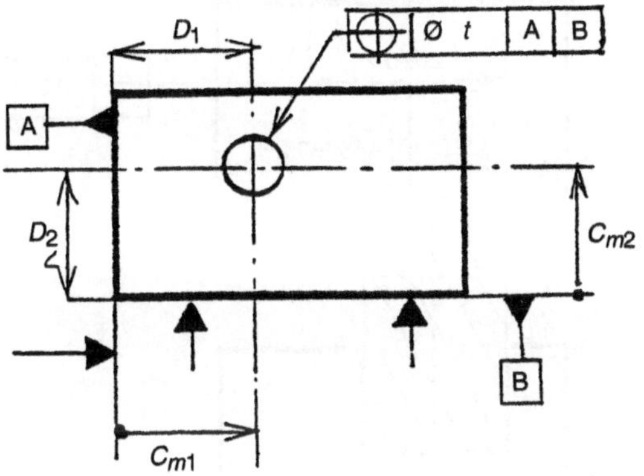

Fig. E4.4

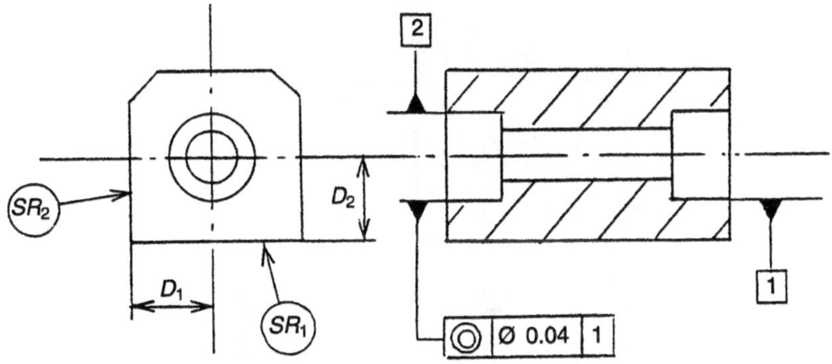

Fig. E4.5

4.6 In order to respect the tolerance of symmetry in the figure below, two methods are used:

(a) The diameter of the milling tool gives the dimension of the slot.

(b) The milling tool which is smaller than the slot (Ø15 ± 0.2) is shifted from left to right.

Give the setting dimensions in the two cases in relation to the positioning surface.

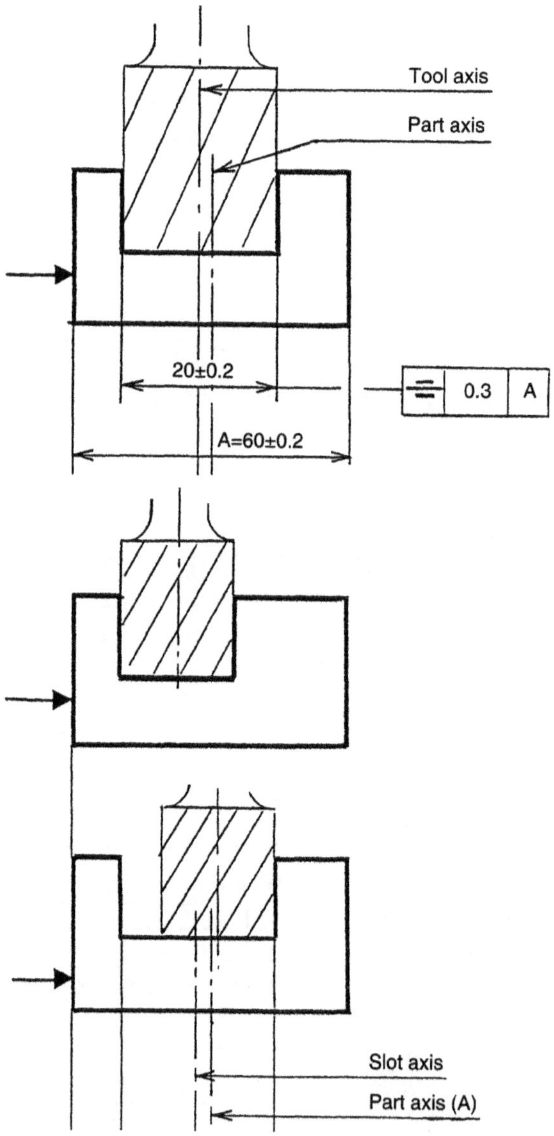

Fig. E4.6

4.7 Complete the charting of the given part by the setting dimensions in accordance with the design and manufacturing requirements, taking into account the given process plan (see Table 4.2 in Chapter 4). Are the dimensions $A, B, \ldots B_2$ correct?

4.8 In the figure below, the dimension C_m is realized on a lathe by a setting dimension L_r and by using slip gauges and a calibrated cylinder in the

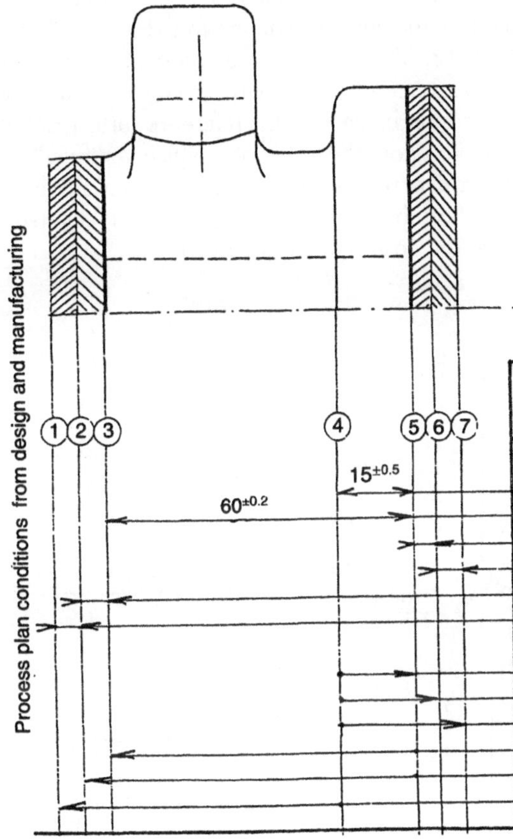

Rep	Conditions Maxi	Mini	IT
1	15.5	14.5	1
2	60.2	59.8	0.4
3		0.2	
4		1.5	
5		0.2	
6		1.5	
A	15.5	14.5	1
B	16.1	15.7	0.4
B_1	18.6	17.6	1
C	60.2	59.8	0.4
D	60.8	60.4	0.4
B_2	48.8	47.8	1

Fig. E4.7

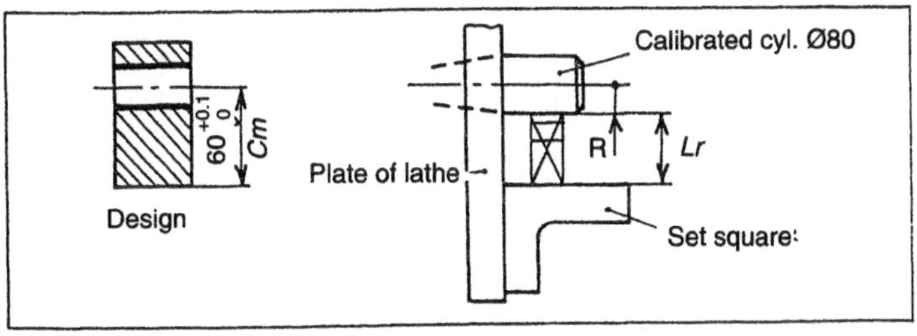

Fig. E4.8 Redrawn from Karr, J. (1979) *Methodes et Analyses de Fabrication Mecanique*, published by Dunod Bordas, Paris.

spindle. Determine the dimensions and tolerances of L_r. The size of the calibrated cylinder is Ø80 + 0/ − 0.01 and the coaxiality error with the axis of the spindle is Ø0.02.

4.9 In the figure below, the part is processed on surface A in relation to the axis of cylinder Ø75 by the dimension 100 ± 0.05. However, the setting of the tool will be executed in relation to the support having a vee shape in the form of a 90° angle, as indicated by the dimension L_r. The auxiliary dimension C_L is introduced to guarantee the dimension 100 and 12 min and has a value of: 126 ± 0.3. This value is the result of a chain of dimensions comprising L_r and the distance of the contact point B of diameter Ø75 with the surface of the vee and the reference surface of the table *SR* (called x). In order to know the value of x which is difficult to measure directly and which depends on the dimension z, an auxiliary measurement is performed with a calibrated cylinder of diameter Ø75.010 ± 0.01 giving the dimension y, and indirectly x.

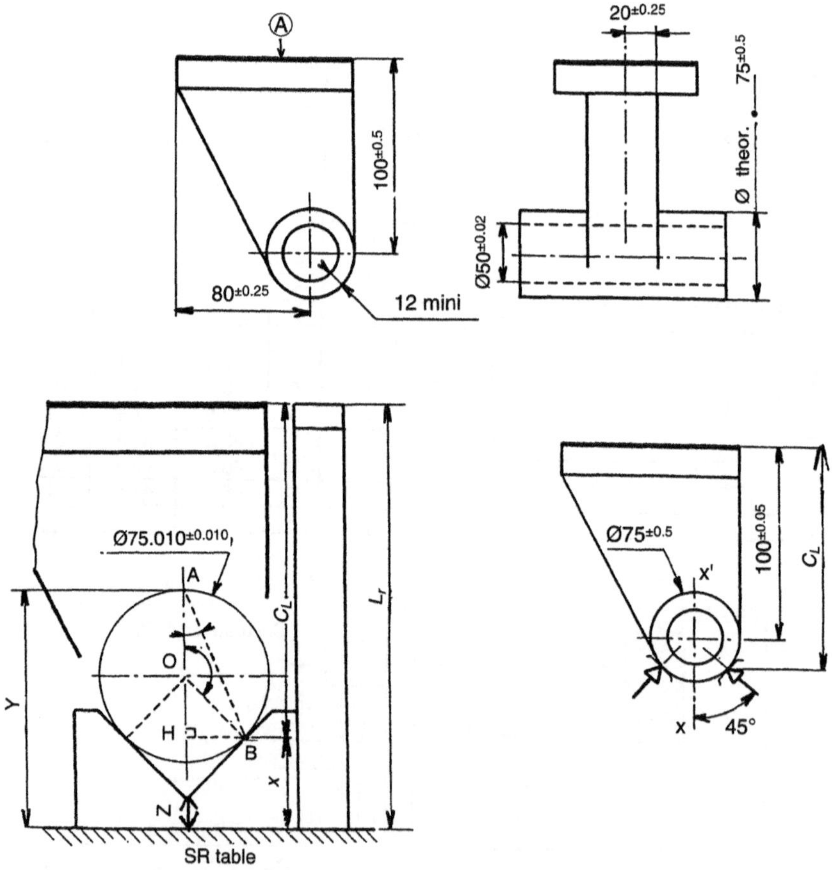

Fig. E4.9

(a) Determine the dimension x as a function of the measurement of y in the calibration phase.
(b) Determine L_r so that the liaison dimension to the raw material C_L will be 126 ± 0.3 for any value of the diameter 75 ± 0.5.

Remark: the dimension z can be used to position the vee in relation to the table.

4.10 Given a positioning system using the principle of locating, determine the width E of the locating pin as a function of its diameter D_4 for the cases (a) and (b). Determine from this result the value of diameter D_4 when $E = D_4/2$.

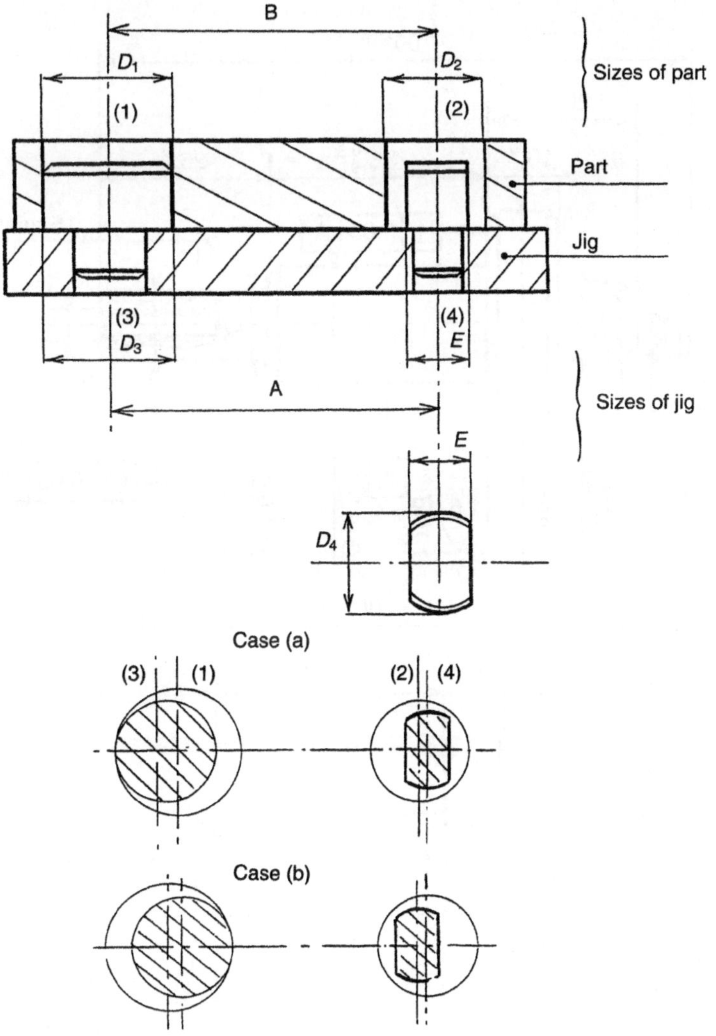

Fig. E4.10

Data:
Diameter $D_1 = D_1 + a', + a$ (part)
Diameter $D_2 = D_2 + b', + b$
Distance between axis: $B = B + c', + c$
Diameter $D_3 : D_3 + d', + d$ (locating)
Distance between axis: $A + e', + e$.

4.11 (According to the generalized model of tolerancing.) In the figure below, a part is represented with its dimensions and tolerances. In the lower part of the figure, the dimensions and tolerances in the direction of the axis of the component are given and are considered exclusively in this exercise.

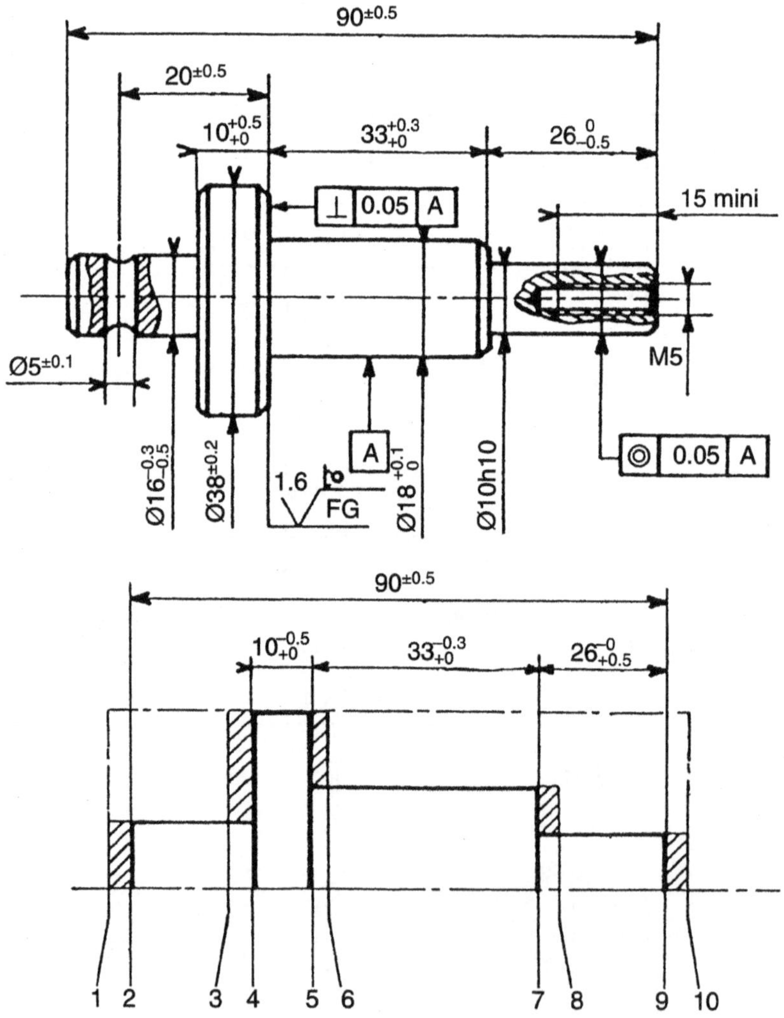

Fig. E4.11

The purpose is to determine the setting dimensions l_i, δl_i and the manufactured dimensions C_i and δC_i for the surfaces 1 to 10.

The process plan is as follows:

Job 00: preparation of the raw stock by sawing from a bar.
Job 10: positioning on surface 10 and rough processing of surfaces 2 and 4.
Job 20: positioning on surface 4 and rough processing of surfaces 6 and 8. Finish processing of surfaces 5, 7 and 9.

Data to be used concerning the tolerances:

Finish processing δl_i: 0.03.
Positioning on fine surface: $\delta l'_i$: 0.05.

Questions:

(a) Draw the setting dimensions l_i, δl_i for the finishing passes only. Prepare the table for the determination of the final tolerances. Determine the optimal values of δl_i and $\delta l'_i$ by giving every phase in the optimization process and add in one line the new tolerances which were found.
(b) In order to process the part correctly, it is advisable to limit the tolerances of the thickness of the final chip, as follows:

$\delta lR_{3-4} = \delta lR_{1-2} = 1$
$\delta lR_{9-10} = 1.2$
$\delta lR_{5-6} = \delta lR_{7-8} = 0.5$

Determine the tolerances δl_3, δl_6 and δl_8 of the surfaces 3, 6 and 8.

(c) Using the tolerances of the chip thickness given before, determine the tolerances of the raw material.
(d) Given the minimal thickness of the chip, it is possible to determine the maximal, minimal and mean thickness of the chip. Use: $\min(C_{p3-4}) = 0.2$; $\min(C_{p5-6}) = \min(C_{p1-2}) = \min(C_{p9-10}) = 0.4$. Give the dimension of the raw material C_{1-10}.
(e) Calculate all the setting dimensions l_i. Calculate all the manufactured dimensions C_i and compare with the original dimensions and tolerances.

CHAPTER 5

5.1 What are the functions of the process planner, who defines his/her objectives and in what form is the task definition received?

5.2 Is the process planner responsible for the function of the product? If yes, in which way? If not, then who is?

5.3 What are the process planner's responsibilities and objectives?

5.4 What effect has the product designer on the process planner's decisions. Who controls product costs? Explain.

5.5 Give an example of how a change in design may reduce production costs without affecting product functionality.

5.6 What is the difference between committed costs and production costs?

5.7 What is the main constraint in designing a part that is intended to be produced by forming from liquid?

5.8 What is the main difference between casting, forging and powder metallurgy? Can casting and forging be interchangeable processes, and if so, in which cases?

5.9 Why is forming from solid by metal removal popular in batch manufacturing of small to medium quantities? How can this basic process be used economically for mass production?

5.10 What is the main difference between welding, soldering and adhesive joints? List the pros and cons of each one.

5.11 What is the main advantage of forming by material increase and how is it implied?

5.12 What process had the designer of the connecting rod (see Fig. E5.1) in mind?

How would you change the design if the selected process is: (a) by welding; (b) by material removal?

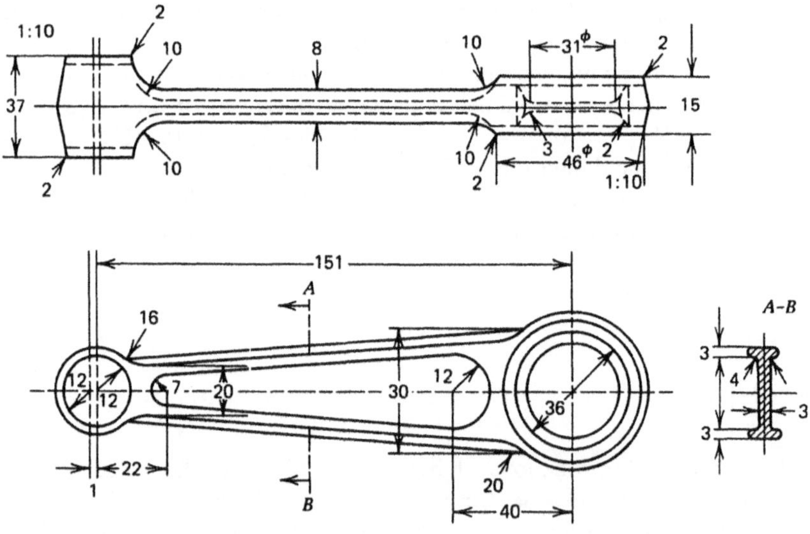

Fig. E5.1

5.13 A part may be produced by the following options:

	Machining time (min)	*Hourly rate ($)*
Metal cutting	72	15
Forging	0.5	30
Casting	3.0	27

What process would you select on an economic basis, if the required quantities are:
(a) 1000 pieces.
(b) 2500 pieces.
(c) 5000 pieces.

5.14 For each of the parts in Fig. E5.2, state the shape complexity.

5.15 The part shown in Fig. E5.2b is made of aluminum 7075-T6. Its circumference is 40 mm, web thickness 3 ± 0.4 mm. The surface roughness is 1.2 μm and the length 300 mm. An order of 1200 pieces was received.

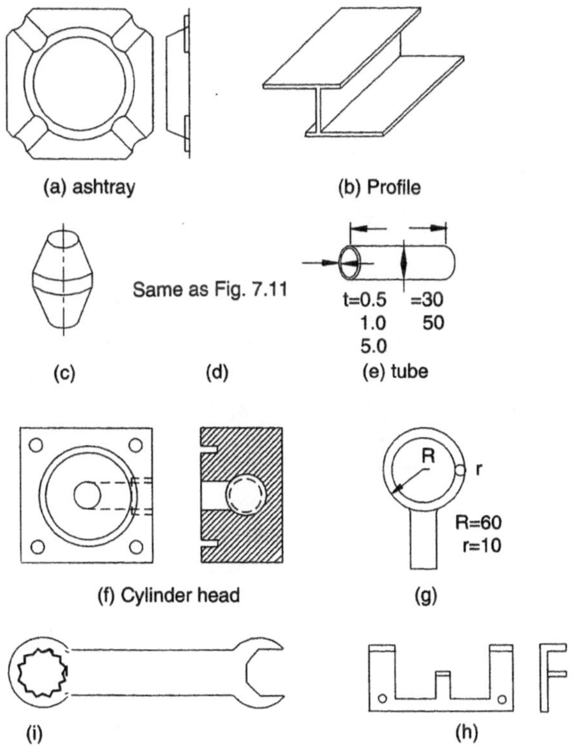

Fig. E5.2

(a) What is the economic process for manufacturing this order?
(b) Are subsequent processes required? If yes, state what they are and what operations should be performed.
(c) What process would you recommend if the order quantity called for only 50 pieces?

5.16 The ashtray in Fig. E5.2a is made of alloy steel 0.4 mm thick. Its circumference is 80 mm. What process would you recommend to produce 10 000 ashtrays?

5.17 A cylinder head as shown in Fig. E5.2f is made of low alloy steel. What process would you recommend to produce:
(a) 10 cylinder heads.
(b) 5000 cylinder heads.
Are subsequent processes required? If so, which?

5.18 A prototype (quantity of one) of a part as shown in Fig. E5.2d is required. Select the method to produce this prototype.

5.19 A part as shown in Fig. E5.2g is ordered in a quantity of 60 pieces.
(a) What process should be selected?
(b) Does a change in design reduce the part cost? If yes, show your recommended design.

5.20 Parts shown in Fig. E5.2e belong to a family of parts the dimension of which are given in the table. What process would you select to produce quantities of:
(a) 50 pieces?
(b) 1000 pieces?
State your reasons in each case.

CHAPTER 6

6.1 List the parameters which have to be considered when selecting a specific material removal process and explain how your choice affects your final decision. Support your explanation with an example.

6.2 Discuss the difference between maximum profit optimization criterion and minimum cost criterion.

6.3 What is the main difference between 'main shape' and 'superimposed shape'? What purpose do they serve?

6.4 The part shown in Fig. E6.1 is to be produced in a quantity of 150 pieces. Based upon economic considerations:
(a) What basic forming process would you recommend?
(b) What process/processes would you recommend using for each one of the line of dimensions as given in the table?
(c) If you recommend using an intermediate process, give a sketch of the part, including dimensions and tolerances, for each stage of the process.

6.5 The part shown in Fig. E6.1 is to be produced in a quantity of 5000 pieces. Based upon economic considerations:
(a) What basic forming process would you recommend?
(b) What process/processes would you recommend using for each one of the line of dimensions as given in the table?
(c) If you recommend using intermediate processes, give a sketch of the part, including dimensions and tolerances, for each stage of the process.

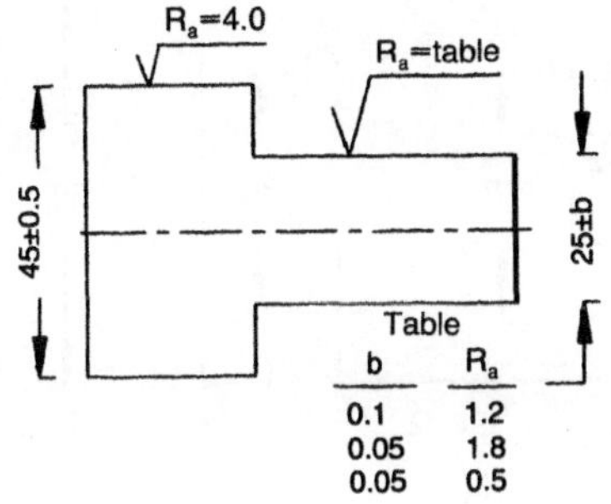

Fig. E6.1

6.6 The part shown in Fig. E6.2 is to be produced in a quantity of 150 pieces. Based upon economic considerations:
(a) What basic forming process would you recommend?
(b) What process/processes would you recommend using for each one of the line of dimensions as given in the table?
(c) If you recommend using intermediate processes, give a sketch of the part, including dimensions and tolerances, for each stage of the process.

6.7 The part shown in Fig. E6.2 is to be produced in a quantity of 5000 pieces. Based upon economic considerations:
(a) What basic forming process would you recommend?
(b) What process/processes would you recommend using for each one of the line of dimensions as given in the table?

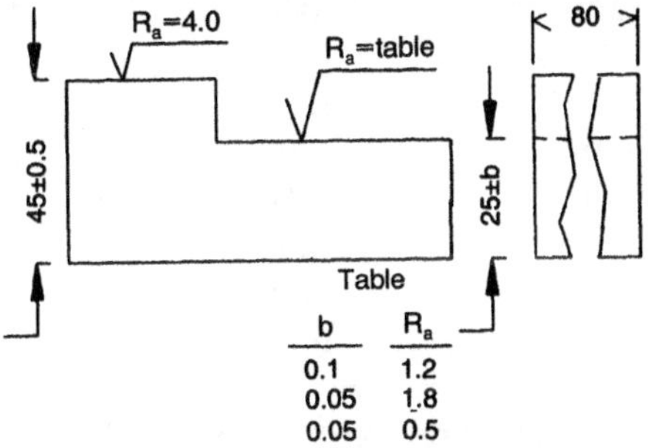

Fig. E6.2

(c) If you recommend using intermediate processes, give a sketch of the part, including dimensions and tolerances, for each stage of the process.

6.8 The part shown in Fig. E6.3 is to be produced in a quantity of 100 pieces. List the processes that you recommend to produce this part, and their sequence.

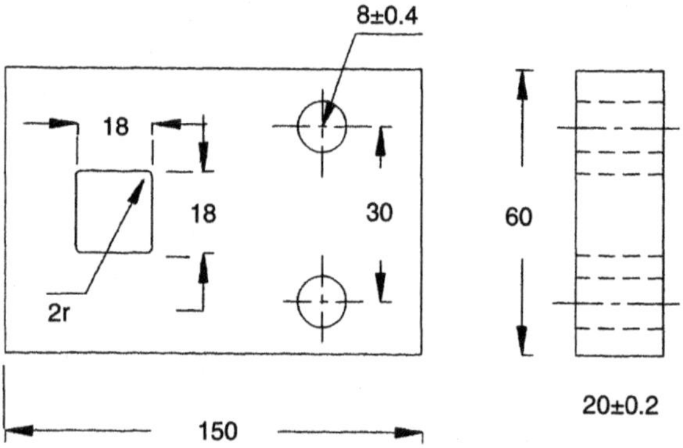

Fig. E6.3

6.9 The part shown in Fig. E6.4 is to be produced in a quantity of 75 pieces. Based upon economic considerations:
(a) What basic forming process would you recommend?
(b) If you recommend using intermediate processes, give a sketch of the part, including dimensions and tolerances for each stage of the process.

6.10 Your company receives the following order:

Item	*Part no.*	*Quantity*	*Due date*
1	A15b	1010	May 15
2	d38	950	June 1
3	A24X	1000	May 8
4	513	1300	April 28
5	CAM5	1000	May 10
6	PP28	900	June 3
7	MP12	1250	May 7
8	GJ578	1000	May 27
9	HAL94	1100	April 26
10	CH152	1000	May 2

The drawings of the parts are shown in Fig. E6.5. Select processes that will result in the minimum cost of producing the order.

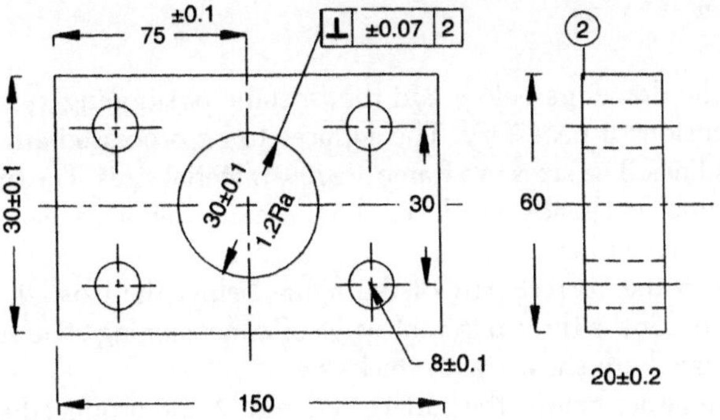

Fig. E6.4

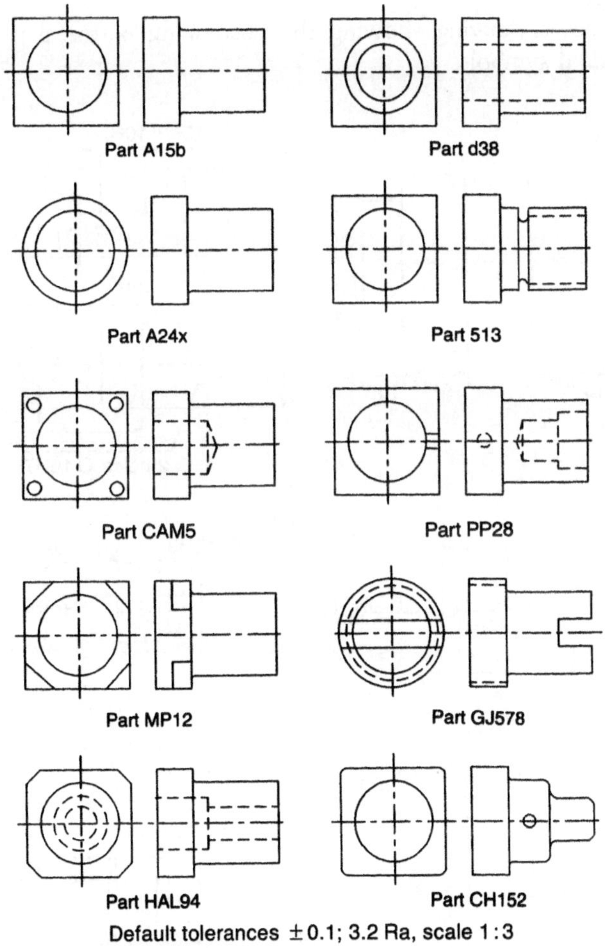

Fig. E6.5

CHAPTER 7

7.1 On the drawings below add the isostatic positioning system, and the dimensions if necessary. The surfaces to be processed are shown with thick lines. The sizes in a frame are raw material sizes. The other sizes are for surfaces already produced. For case 26, the planar faces have been processed earlier.

7.2 Give in the correct part of the figure below the isostatic positioning system for the three different ways of dimensioning. The surfaces to be processed are shown with thick lines.

7.3 In the figure below, the surfaces (1) and (2) are produced as associated surfaces in a common subjob. Justify the positioning decisions by preparing a table of evaluation of tolerances as shown in Table 7.3, Chapter 7.

Produce a drawing showing the positioning of the part using the standard symbols.

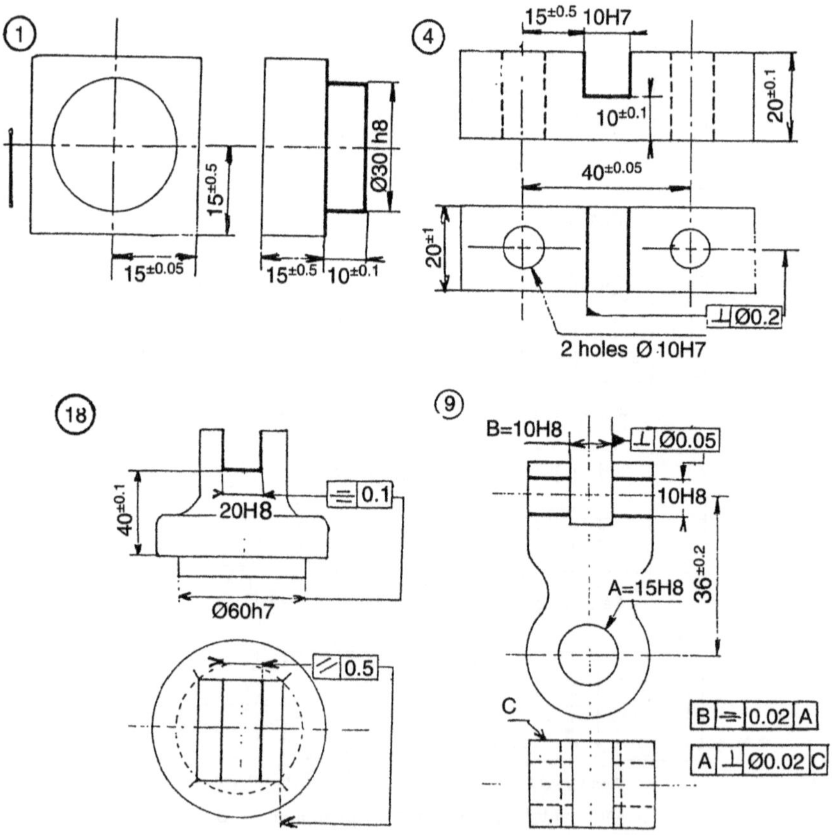

Fig. E7.1

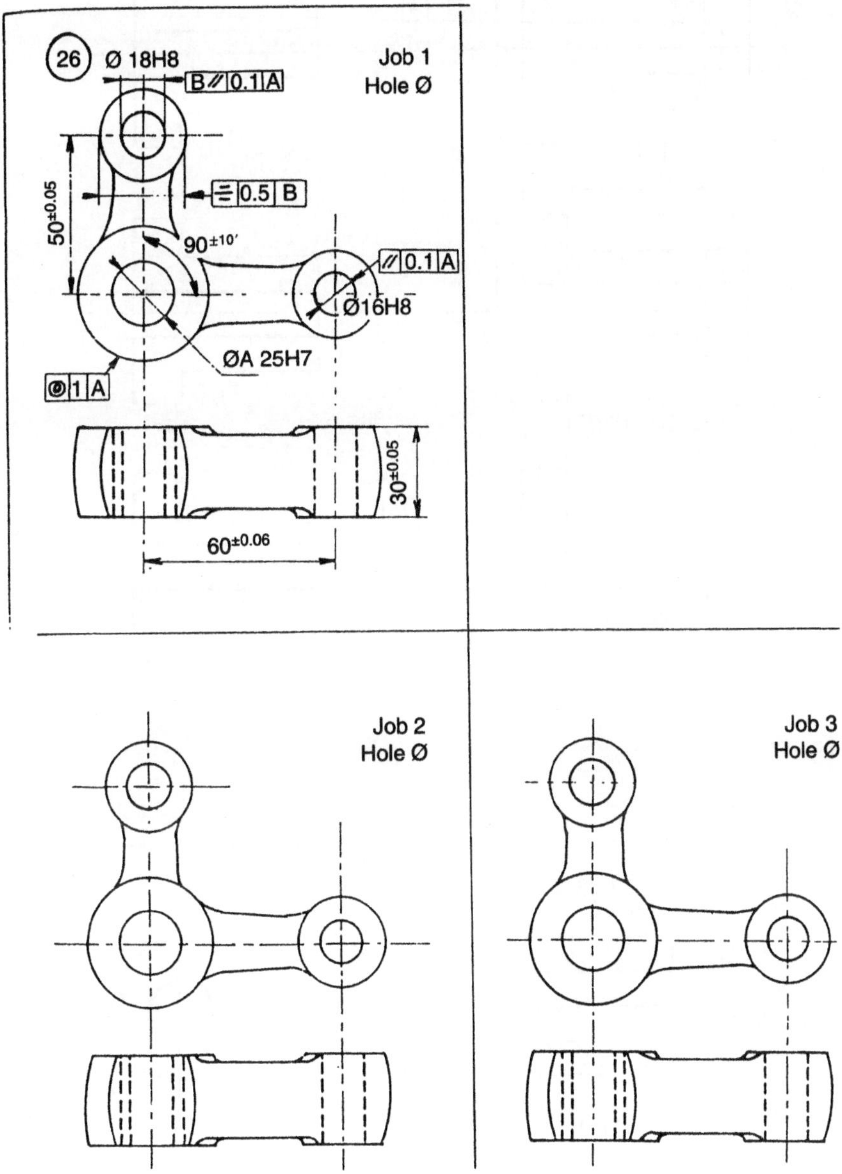

Fig. E7.1—*continued*

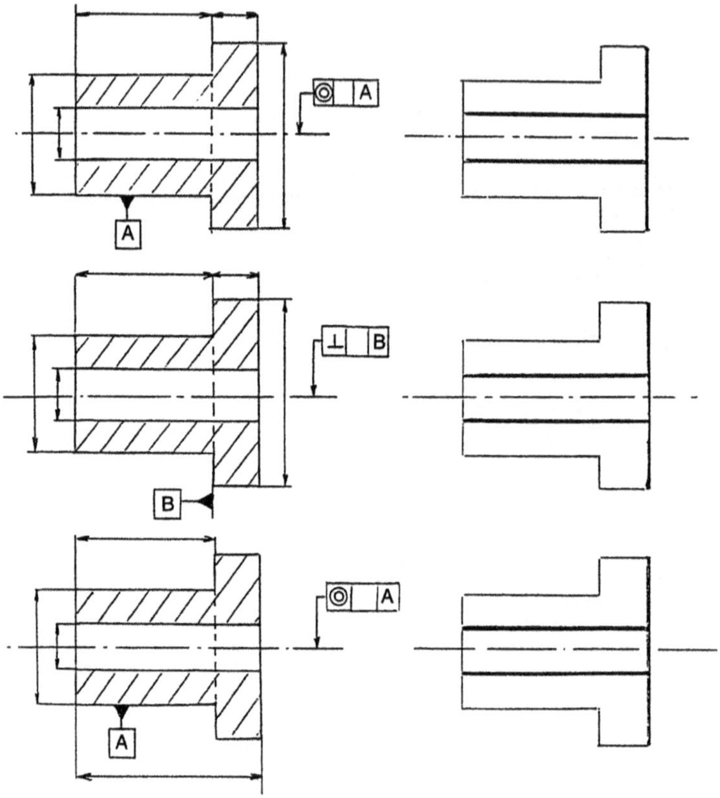

Fig. E7.2

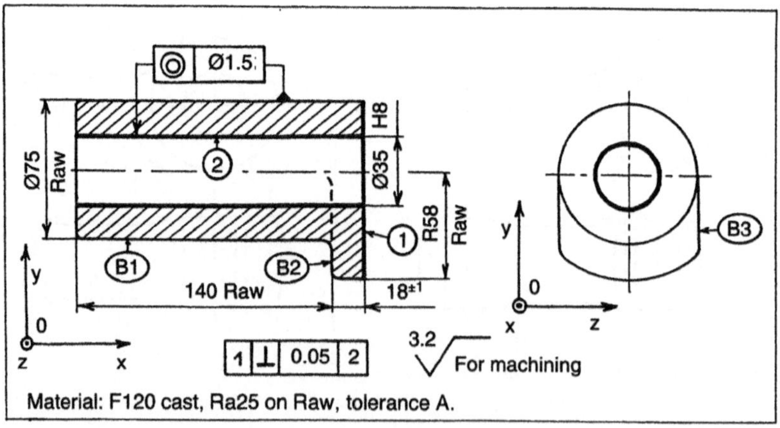

Fig. E7.3 Redrawn from Karr, J., *Methodes et Analyses de Fabrication Mecanique*, published by Dunod, Bordas, Paris, 1979.

7.4 For the part represented in Fig. E2.4 and the different subjobs chosen as shown in Fig. E7.4 below, indicate both the isostatic positioning system and the clamping system using the standard symbols. Put the design requirements on the drawings and justify the choices made for the positioning points.

The surfaces to be processed are shown with thick lines. The raw parts are produced by casting.

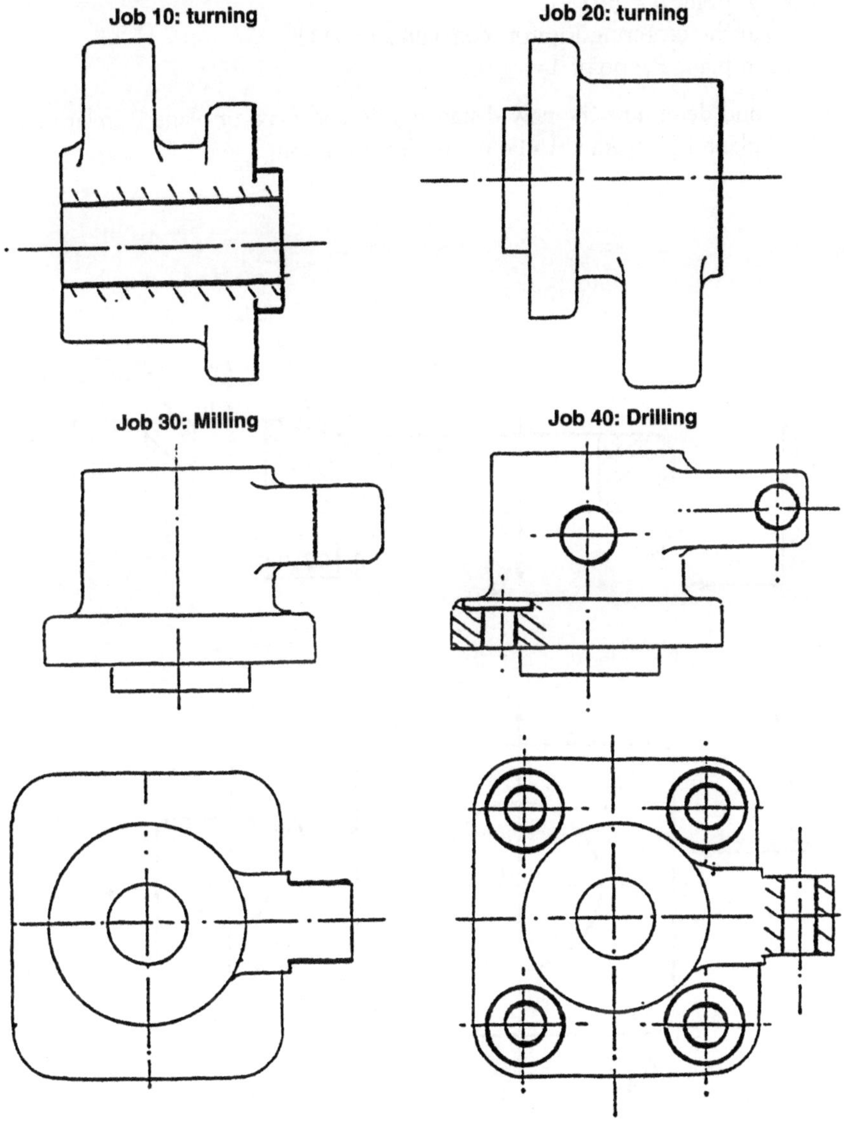

Fig. E7.4

7.5 Given a part as shown in the figure below without dimensional or geometric errors, placed on plane P_2 in three points (1,3,6) on a jig having a centering locator in O_2 and a punctual contact in plane P_1 in point 1 and in direction z, determine the displacement vector of point O_1 in three directions (x,y,z) as a function of the errors in the locator positions as follows (relative to an ideal position):

in plane P_2: in points 1 and $3=0$
in point $6=\pm 0.1$
in the centering locator: coaxiality of Ø0.5
in plane P_1: point $1=\pm 0.1$

and determine the new distance (min and max) of plane P_1 relative to plane P_2, assumed to be in an ideal situation.

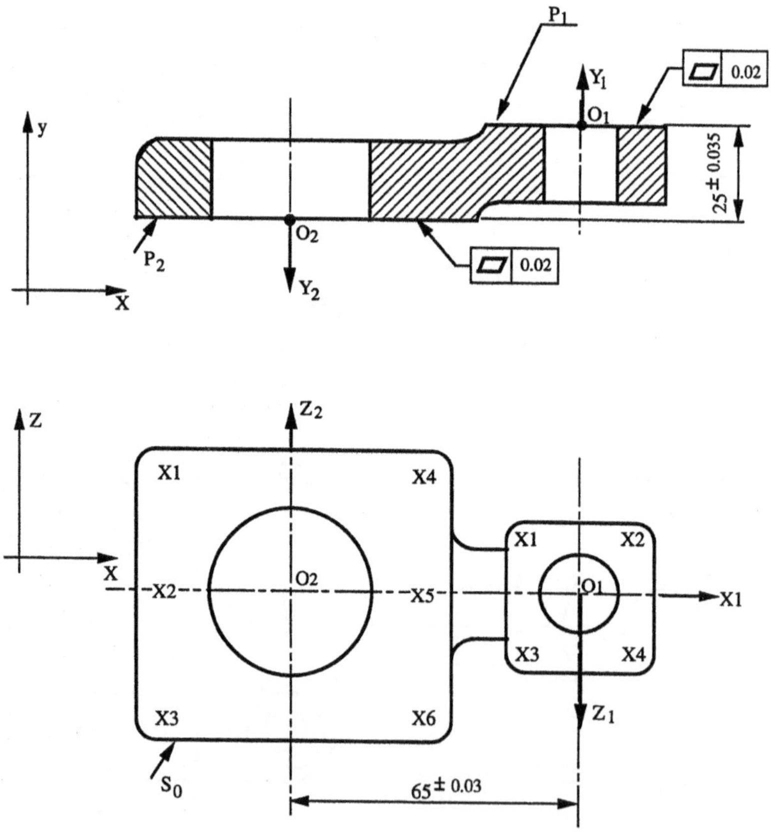

Fig. E7.5

CHAPTER 8

8.1 Explain boundary limits and give examples.

8.2 Explain why BTU (built up edge) may introduce vibrations.

8.3 Discuss the pros and cons of vibrations during the process of material removal.

8.4 What is the most influential means of determining surface roughness, and why?

8.5 Recommend the maximum feed rate for turning operations to produce surface roughness in R_a (μm) scale of the following values:

12; 9; 6; 4; 3.2; 2; 1; 0.8.

8.6 Recommend the maximum feed rate for a face milling operation to produce surface roughness of 32 RMS μinch.

8.7 Recommend the maximum feed rate for face milling operation to produce N7 surface quality.

8.8 Recommend the maximum feed rate for peripheral milling operations to produce surface roughness of R_a (μm) scale of the following values:

4; 1.6; 0.8

when using:

(a) tool diameter of 60 mm with four teeth; and

(b) tool diameter of 100 mm with eight teeth.

8.9 (a) What are the pros and cons of up milling and down milling?

(b) Which one will produce a better surface quality?

(c) Will the two sides of a slot and its bottom, produced by an end mill in one pass, have the same surface quality? Discuss.

8.10 Explain why the depth of cut should be restricted in order to produce a quality surface.

8.11 What is the maximum depth of cut that can be used in order to produce a surface roughness of $R_a = 1.2$ μm when machining alloy steel of 200 Brinell hardness?

8.12 A medium carbon steel ($C_p = 220$) bar of 40Ø mm is turned to Ø36 with a feed rate of 0.7 mm/rev. What is the cutting force?

8.13 A medium carbon steel bar of 40Ø mm is chucked on a lathe held by a three-jaw chuck at one end and free at the other. In turning the bar to a diameter of 36Ø mm, what is the maximum length of cut possible, if the maximum allowed deflection is 0.1 mm?

8.14 In a turning operation the maximum allowable cutting force is 175 N. The maximum feed rate allowed is 0.4 mm/rev and the minimum is 0.1 mm/rev. What is the maximum and computed minimum depth of cut (a_{cmax} and a_{cmin})?

8.15 To produce a surface of R_a 6 μm, 1.5 mm has to be removed. The part is made of 200 Brinell hardness steel, the maximum allowable cutting forces are 400 N, the minimum depth of cut is 0.2 mm and the maximum feed rate is 0.8 mm/rev. Determine the number of passes to use and their cutting conditions (depth of cut, and feed rate).

8.16 To produce a surface of R_a 5 μm, 2.5 mm has to be removed. The part is made of 200 Brinell hardness steel, the maximum allowable cutting forces are 700 N, the minimum depth of cut is 0.2 mm and the maximum feed rate is 0.8 mm/rev. Determine the number of passes to use and their cutting conditions (depth of cut, and feed rate).

8.17 To produce a surface of R_a 3.2 μm, 5 mm has to be removed. The part is made of 250 Brinell hardness steel, the maximum allowable cutting forces are 318 N, the minimum depth of cut is 0.2 mm and the maximum feed rate is 0.8 mm/rev. Determine the number of passes to use and their cutting conditions (depth of cut, and feed rate).

8.18 To produce a surface of R_a 0.8 μm, 13 mm has to be removed. The part is made of 250 Brinell hardness steel, the minimum depth of cut is 0.2 mm and the maximum feed rate 0.8 mm/rev. Due to constraints, the maximum and minimum computed depth of cut are 5.3 and 0.2 respectively. Determine the number of passes to use and their cutting conditions (depth of cut, and feed rate).

CHAPTER 9

9.1 'A change in cutting speed by a factor of X results in a change of machining time by the same factor.'
   (a) Is this a correct statement?
   (b) Does this statement hold true if the term 'machining time' is replaced by 'machining cost'?

9.2 Explain:
   (a) what is meant by machinability rating;
   (b) to which cutting conditions it refers;
   (c) whether it can have a value greater than 100%.

9.3 Can the recommended values specified in different sources of machinability handbooks be used as is? Justify your answer.

9.4 What is the difference between Taylor's equation and the extended Taylor's equation? Explain.

9.5 What are the meaning and units of C_v in Taylor's equation?

9.6 Discuss the term 'tool life'.

9.7 Discuss the validity of the cutting speed optimization technique.

9.8 When calculating optimum tool life, can the same result be derived from either Taylor's equation or the extended equation? Prove and explain.

9.9 How would you expect the optimum cutting speed to change in a case where the machine operator was given a raise in wages but all the other parameters remained unchanged? Answer by intuition and justify your answer.

9.10 Develop an equation for the optimum cutting speed for the maximum production criterion.

9.11 The equation for machining time (or converting cutting speed to RPM) includes the term 'diameter'. Which diameter would you insert:
(a) for a milling operation?
(b) for turning?

9.12 A part is made of alloy steel – medium carbon of 200 Brinell hardness. It calls for removing 3.5 mm (depth of cut) with a feed rate of 0.3 mm/rev for a length of 250 mm and diameter of 60 mm.
Assume that:
Cutting tool edge costs $3
Direct operating cost is $32/hour
Tool entering angle is .... 0.45°
Tool change time is .... 10 seconds
Allowable flank wear is .... 0.6 mm
(a) Compute the optimum cutting speed for the minimum cost criterion.
(b) Compute the total cost and time to machine this operation.
(c) Repeat (a) and (b) once by using Taylor's equation and then the extended equation and compare the results.
(d) Repeat (a), (b) and (c) for a depth of cut of 0.7 mm, a feed rate of 0.08 mm/rev and flank wear of 0.3 mm.

9.13 A part is made of alloy steel – medium carbon of 200 Brinell hardness. It calls for removing 3.5 mm (depth of cut) with a feed rate of 0.3 mm/rev for a length of 250 mm with diameter of 60 mm.
Assume that:
Cutting tool edge costs $3
Direct operating cost is $32/hour
Tool entering angle is .... .45°
Tool change time is .... 10 seconds
Allowable flank wear is .... 0.6 mm
(a) Compute the optimum cutting speed for maximum production criterion.
(b) Compute the total cost and time to machine this operation.
(c) Repeat (a) and (b) once by using Taylor's equation and then the extended equation, and compare the results.
(d) Repeat (a), (b) and (c) for a depth of cut of 0.7 mm, a feed rate of 0.08 mm/rev and flank wear of 0.3 mm.

9.14 Compare and discuss the results of Exercises 9.12 and 9.13.

9.15 How many parts can be produced with one cutting edge when using the cutting speed recommended in Exercises 9.12 and 9.13?

9.16 A surface of 300 × 80 mm is to be machined by removing 2 mm at a feed rate of 0.3 mm/rev. The part is made of nickel alloy cast of 125 BHN.
(a) What is the recommended cutting speed?
(b) What is the machining time?
Note: if some data are missing, then assume reasonable values for the missing data.

9.17 A part is made of Ø50 mm diameter stainless steel-ferritic of 185 BHN.

The plant receives an order to produce parts of Ø46 mm, 300 mm long. It was decided to use 0.2 mm/rev feed rate. Direct operating cost is $40/hour, cutting tool edge costs $2 and tool change time is 15 seconds.

(a) Compute the optimum tool life, machining time and part costs for the minimum cost criterion of optimization.
(b) Compute the optimum tool life, machining time and part costs for the maximum production criterion of optimization. Discuss the results.
(c) How many parts can be machined per tool edge?
(d) If the order was for 12 pieces, can the minimum cost be reduced? If so, prove.
(e) If the order was for eight pieces, can the minimum cost be reduced? If so, prove.

9.18 An order for 1000 pieces of part A was received. To produce this order one turret lathe with four different tools is needed. To change a tool it takes two minutes for setup time (moving the turret to the change location, preparing for tool change and returning it to its working position) and 45 seconds change time. The direct operating cost is $45/hour and the cutting tool edge costs $1.50. The tool life exponent of the material is 0.25.

The data of the four operations are as follows:

Operation	*Tool life (min)*	*Machining time (min)*
010	7.25	0.13
020	10.40	0.80
030	7.25	0.36
040	21.28	0.19

(a) Compute the number of parts that can be produced by each tool and total the machining time of the order (including tool change time).
(b) It was decided to stop the production after every 15 parts and change all required tools.
   (i) How will this decision affect the machining time of each operation and the entire order? (Cutting speed may be reduced but not increased.)
   (ii) Compute batch machining time and cost.
(c) (Optional) Is the decision to change tools at integer multiplier of 15 pieces the most economic decision? If not, what would you recommend in order to achieve maximum production time? (Cutting speed may be reduced, but not increased.)
   (i) Compute batch machining time and cost.

CHAPTER 10

10.1 What does 'artificial constraint' mean? Give an example.

10.2 Does a machine with higher power, speed and accuracy than required have an advantage over a machine with the exact requirements? Discuss.

10.3 Explain the term 'Priority code'.

10.4 Figure E10.1 shows 17 operations that are required to produce a part.
(a) Assign a priority code to each operation.
(b) Give at least three sets of possible sequences of operations.

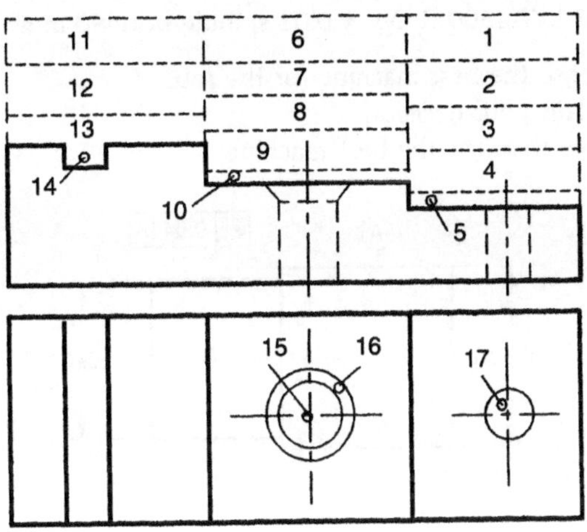

Fig. E10.1

10.5 What measures can be taken to reduce cutting forces while maintaining minimum machining time? Which one is preferred and why? Prove with an example and equations.

10.6 What measures can be taken to reduce cutting power while maintaining minimum machining time? Which one is preferred and why? Prove with an example and equations.

10.7 An operation calls for 18 kW and takes 0.42 minutes. If the operation is performed on a 15 kW machine:
(a) What will be the machining time?
(b) What measures should be taken to adjust the operation to the machine?

10.8 The optimum operations for turning a Ø38 bar of medium steel ($Cp = 220$) to a diameter of Ø30 and 300 mm long and a feed rate of 0.5 mm/rev and a cutting speed of 454 m/min.
(a) Compute the optimum machining time.
(b) What will be the machining time if a machine of 35 kW, 80% efficiency and maximum 2500 RPM were selected for the job?
(c) What will be the machining time if a machine is 15 kW, 80% efficiency and maximum 4500 RPM were selected for the job?

(d) What will be the machining time if a machine of 20 kW, 80% efficiency, maximum 4500 RPM and maximum allowed cutting forces of 3500 N were selected for the job?

10.9 Can a machine with a runout of 0.12 mm produce a part with a tolerance of 0.03mm?

10.10 A part as shown in Fig. E10.2 can be produced by one of the three available machines, having the following specifications:

M1 – 20 kW 3600 RPM ±0.02 spindle bore 90 mm
M2 – 20 kW 3600 RPM ±0.02 spindle bore 30 mm
M3 – 25 kW 4500 RPM ±0.01 spindle bore 90 mm

(a) Choose the best machine for the job.
(b) Explain your decision.
(c) Prove that it is the best machine.

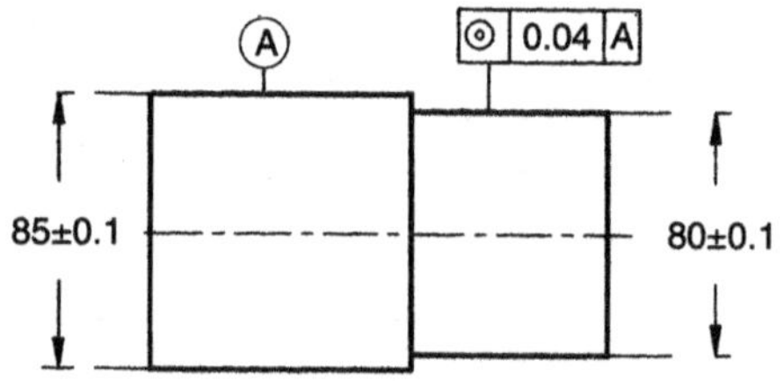

Fig. E10.2

10.11 A part is made of material where $Cv=333$; $Cp=220$; $n=0.25$ and five theoretical operations (ATO) as follows are required to produce the part:

Oper.	*Diameter*	*Length*	*Depth*	*Feed*	*Accuracy*	*Priority*
010	40	250	12	0.8	0.3	0
020	40	250	0.4	0.07	0.07	1
030	30	300	5	0.8	0.3	1
040	30	300	0.8	0.15	0.1	3
050	30	100	1.2	0.5	0.1	3

Four machines are considered for the job and their specifications are as follows:

Mach.	*Power kW*	*RPM*	*Force daN*	*Accuracy*	*Cost $/h*
M1	25	3600	600	0.07	33
M2	15	6000	400	0.02	30
M3	30	2500	800	0.2	15
M4	18	4500	315	0.06	27

Assume that the setup per batch is 30 minutes and the hourly rate of the setup operator is \$30/h. (For other missing data, if any, assume reasonable values). Neglect handling time.

(a) Build an operation–machine matrix.
(b) Fill in the theoretical machining time and compute the total theoretical machining time.
(c) Fill the matrix (adjust operation time on each machine).
(d) Compute the total machining time of each machine and compare them to the theoretical one.
(e) Recommend an economic sequence of operations and machine for each operation if the batch size is 10 pieces and the criterion is maximum production.
(f) Recommend an economic sequence of operations and machine for each operation if the batch size is 300 pieces and the criterion is maximum production.
(g) Recommend an economic sequence of operations and machine for each operation if the batch size is 10 pieces and the criterion is minimum cost.

CHAPTER 11

11.1 List all the desired characteristics of a cutting tool.
(a) Can all characteristics be incorporated in one tool? If so, why?
(b) What characteristics are required for a finish cut?
(c) What characteristics are required for a rough cut, and why?

11.2 What are the determining parameters in selecting an insert grade? What is the ISO grade designation?

11.3 Which ISO insert grade is tougher, P50 or P01? Explain the meaning of tougher.

11.4 Select the ISO insert grade for the following cutting operations performed on carbon steel:
(a) Depth of cut = 0.7 mm; feed rate = 0.1 mm/rev.
(b) Depth of cut = 2.0 mm; feed rate = 0.1 mm/rev.
(c) Depth of cut = 4.0 mm; feed rate = 0.3 mm/rev.

11.5 Select the ISO insert grade for the following cutting operations performed on aluminum alloys:
(a) Depth of cut = 0.7 mm; feed rate = 0.1 mm/rev.
(b) Depth of cut = 2.0 mm; feed rate = 0.1 mm/rev.
(c) Depth of cut = 4.0 mm; feed rate = 0.3 mm/rev.

11.6 Select the ISO insert grade for the following cutting operations performed on alloy steel:
(a) Depth of cut = 0.7 mm; feed rate = 0.1 mm/rev.
(b) Depth of cut = 2.0 mm; feed rate = 0.1 mm/rev.
(c) Depth of cut = 4.0 mm; feed rate = 0.3 mm/rev.

11.7 What are the benefits of using a tool that covers a wide range of ISO insert grades?
11.8 Are there any other restrictions (besides the insert grade) for using one insert that covers all ISO insert grades?
11.9 Can the shape of an insert be selected independently of the tool holder? If so, explain.
11.10 Can an insert shape be selected based upon part shape alone? If not, what other factors should be considered?
11.11 What is the importance and effect of the clearance angle?
11.12 Discuss the statement, 'Ground inserts are not helpful on tolerances less than 0.07 mm'.
11.13 Discuss the factors to consider in selecting a clamping method for the insert on the tool holder.
11.14 A cut 5 mm deep and with a 0.4 mm/rev feed rate is used to produce a free surface of 9 μm Ra surface roughness.
(a) Recommend an insert shape.
(b) Recommend a clearance angle.
(c) Recommend a theoretical insert cutting edge length.
(d) Recommend an insert thickness.
(e) Recommend an insert type.
(f) Recommend a tool holder entering angle.
(g) Suggest an ISO code for an indexable insert (relevant digits).
(h) Suggest an ISO code for a toolholder.
11.15 The part shown in Fig. E11.1 is machined from a bar of free cutting alloy steel that passes through the spindle bore. Assume that one cutting pass is required.
(a) Select insert and toolholder for the job.
(b) Assign an ISO code for the selected insert and toolholder.

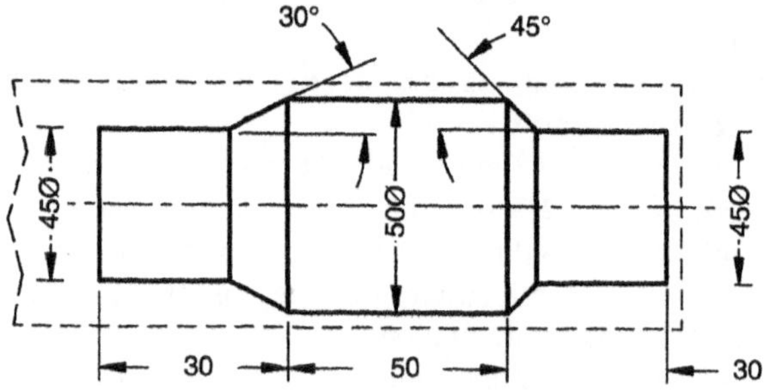

Fig. E11.1

11.16 The segment of a part as shown in Fig. E11.2 is to be machined.
(a) Select an insert shape and toolholder for the job.
(b) Assign an ISO code for the selected insert and toolholder.
(c) If L is modified to 50 mm, will it change your decision?
(d) If L is modified to 5 mm, will it change your decision?

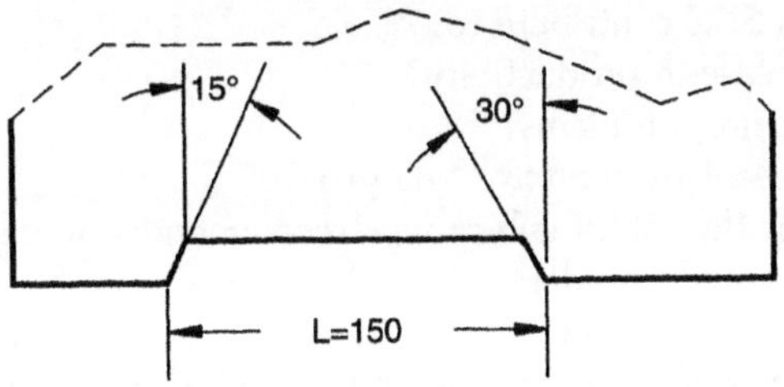

Fig. E11.2

11.17 The part shown in Fig. E11.3 is machined from cut pieces.
(a) Can this part be machined in one chucking? Show how.
(b) If this part is to be machined from one side and then from the other, then:
(i) Select the cutting location for each side.
(ii) Which side should be machined first and why?
(iii) What operations are needed on each side?
(c) Select insert and toolholder for the job.
(d) Assign the ISO codes for the selected insert and toolholder.

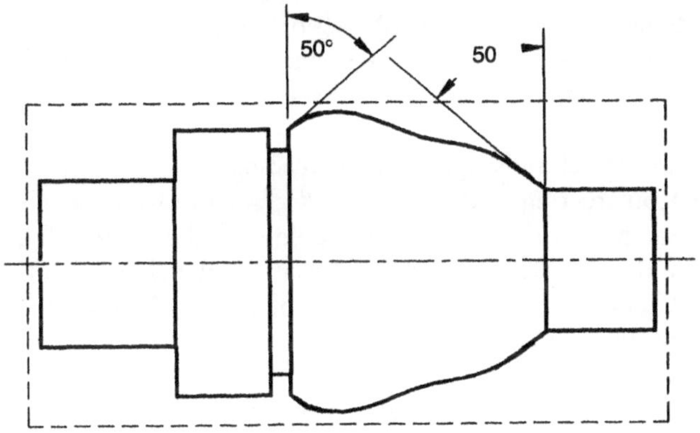

Fig. E11.3

CHAPTER 12

12.1 What is the difference between 'assignable causes' and 'natural causes'? To which cause would you identify machine run-out?
12.2 Can rejected parts be produced when practicing SPC? Explain.
12.3 Define the term 'process capability. Is it an absolute value?
12.4 Discuss the statement 'immediate feedback is the key to the success of any SPC system'.

12.5 How does SPC contribute to:
(a) an increase in productivity?
(b) prioritizing problems?

12.6 What is meant by the term 'sub-groups'?
(a) Why is the use of sub-groups recommended under SPC?
(b) How would you determine sub-group size and frequency?

12.7 Discuss the need to use 'mean', 'range' and 'standard deviation' in SPC.

12.8 What is the difference between range and average range?

12.9 Compute the standard deviation for a sample of five pieces and a value of $\bar{R}$ of 1.18.

12.10 Can SPC be implemented in any case? Discuss.

12.11 Discuss the purpose of control charts.

12.12 A dimension of 30 ± 0.1 is controlled by SPC, with a sub-group size of five pieces.
(a) The following readings have been recorded: 30.02, 30.06, 29.96, 30.04 and 29.94. What, if any, is the expected reject percentage?
(b) The following readings have been recorded: 30.07, 30.08, 30.05, 30.06 and 30.07. What, if any, is the expected reject percentage?
(c) Are the conclusions in (a) and (b) valid from a statistical standpoint?
(d) Compute the process capability for (a) and (b).
(e) Explain the differences between (a) and (b).

12.13 A dimension of 25Ø $+0.10/-0$ of a machine shaft is controlled by SPC. Throw two dies 100 times and record the sum of the double die results. Regard this reading as a dial indicator reading of the shaft, when it is zero to 25 000 mm.
(a) Construct a histogram of the readings.
(b) Construct an $\bar{X}$ chart and an R chart in case of a sub-group of 5 samples (units). Connect the points on the charts.
(c) Calculate $\bar{\bar{X}}$ and $\bar{R}$ and draw them on the charts.
(d) Compute upper and lower control limits of $\bar{X}$ and R and draw them on the charts.
(e) Compute C_p for the process.
(f) What is the expected reject percentage?

12.14 Examine the control chart as shown in Fig. E12.1 and draw your conclusion:
(a) Is the process stable?
(b) Should you investigate the process?
(c) What preliminary conclusions may be reached?

12.15 Examine the control chart shown in Fig. E12.2 and draw your conclusion:
(a) Is the process stable?
(b) Should you investigate the process?
(c) What preliminary conclusions may be reached?

12.16 Examine the control chart in Fig. E12.3 and draw your conclusion:
(a) Is the process stable?

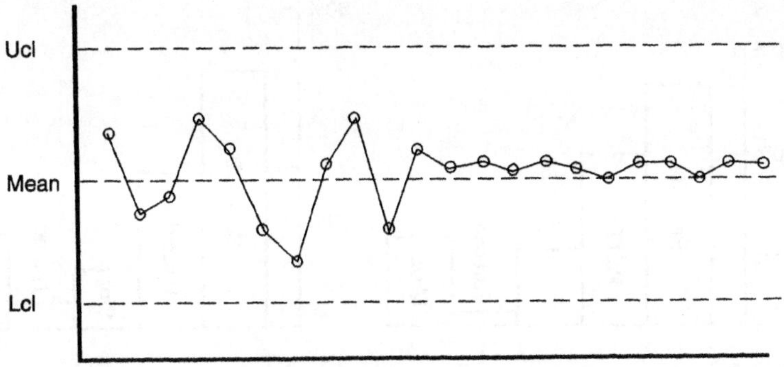

Fig. E12.1

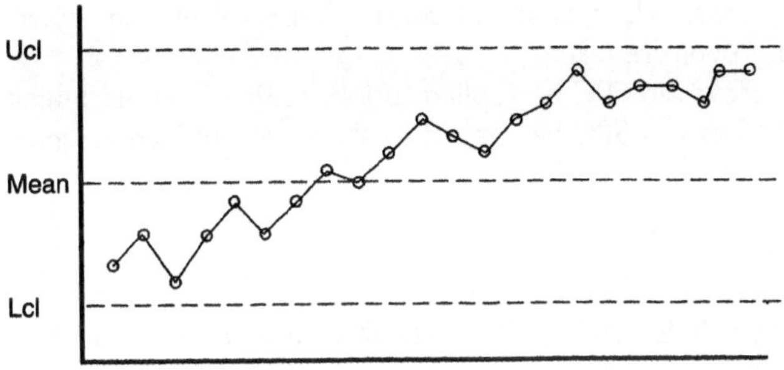

Fig. E12.2

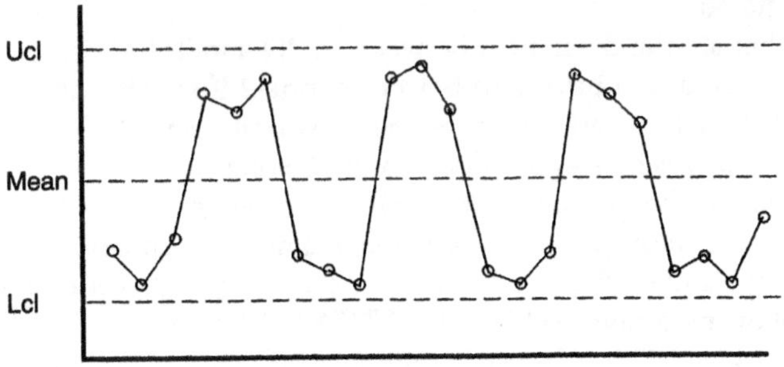

Fig. E12.3

(b) Should you investigate the process?
(c) What preliminary conclusions may be reached?

12.17 The management decide to take steps to reduce the number of defects of a product, an assembly of several parts.
(a) How can SPC (a Pareto chart) be employed?
(b) What conclusions can be made from the Pareto charts as shown in Fig. E12.4?

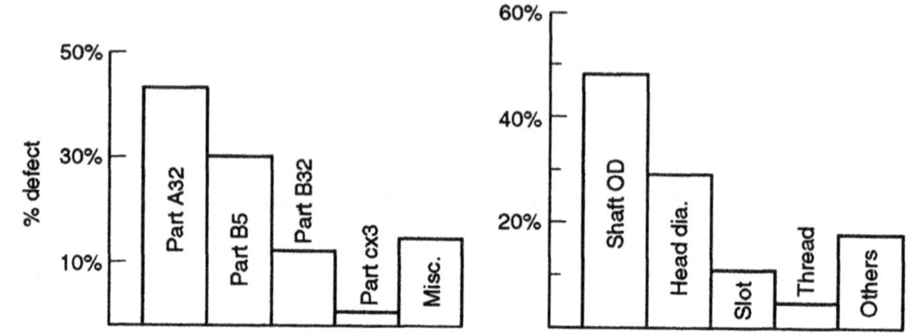

Fig. E12.4

12.18 (Optional) The validity of control limits calculation exists only if 20 subgroups are taken.
   (a) How can SPC be applied to jobs with a long machining time?
   (b) How can SPC be applied to short run production (small batches)?

CHAPTER 13

13.1 Why is hole making more difficult than other metal cutting processes?

13.2 What is the difference between part, center line and single hole optimization?

13.3 How are a_{min} and a_{max} used to determine if an intermediate operation is required?

13.4 What are the differences between MDD and TDD?
   (a) Can a tool belong to both categories? If so, give an example.
   (b) To which category does the twist drill belong?
   (c) To which category does boring belong?

13.5 When is the use of a center drill recommended?

13.6 What is meant by the expression 'diametric boundary'?

13.7 What is the difference between and the objectives of each one of the following arrays: TLSM, TLS, TLSW? Elaborate.

13.8 Can an MDD tool be used to produce holes that are located further than the MDH segment? If so, how?

13.9 In the case of a single hole, the exact dimension can be set when treating this hole. However, in the case of several holes on one center line, the exact dimensions are determined after treating all hole sizes. Why?

13.10 What is meant by 'blind hole', 'close hole' and 'open hole'? What purpose do they serve?

13.11 Why can a reamer not open a hole?

13.12 Why are the diameters and lengths that an insert drill can produce restricted?

(a) What are the benefits of using this tool?
(b) Can an insert drill be used as a boring tool as well?

13.13 What is the difference between an end mill and a disk mill?

13.14 Can a reamer improve location tolerance? If so, why and how?

13.15 Compute and compare the time to increase a hole from 37Ø to 38.5Ø for a length of 40 mm where the surface finish $R_a = 1.6\ \mu m$ and the material hardness is 225 HBN
(a) by boring;
(b) by end mill (assume cutting length > 40 mm).

13.16 Compute a process to produce an open hole of 8Ø ± 0.04, 10 mm long to a surface finish of 0.8 μm. List the steps and justify your decision.

13.17 Define a_{max} and compute its value for a hole of 30Ø ± 0.1 and 70 mm long with a surface finish of $R_a = 0.8\ \mu m$ in the following cases:
(a) Boring a finish pass.
(b) For a core drilling.
(c) For fine reaming.

13.18 Define a_{min} and compute its value for a hole of 20Ø ± 0.33 and 25 mm long with a surface finish of 3.2 μm produced by:
(a) Twist drill.
(b) Core drill.

13.19 The hole shown in Fig. E13.1 has to be produced. Propose a process to meet hole specifications, including exact dimensions of each stage, cutting conditions and machining time, for the following cases:
(a) $x = 0.5$; $y = 0.5$; $R_a = 8$; $D = 20 \pm 0.3$
(b) $x = 0.1$; $y = 0.5$; $R_a = 8$; $D = 20 \pm 0.3$
(c) $x = 0.5$; $y = 0.5$; $R_a = 0.8$; $D = 20 \pm 0.3$
(d) $x = 0.5$; $y = 0.5$; $R_a = 1.6$; $D = 20 \pm 0.08$
(e) $x = 0.5$; $y = 0.5$; $R_a = 8$; $D = 20 \pm 0.3$ and straightness of 0.03.

13.20 Three holes on one center line, as shown in Fig. E13.2, should be produced from a solid material of 220 HBN.
(a) List **all** combinations of possible processes, and compute the machining time for each case.

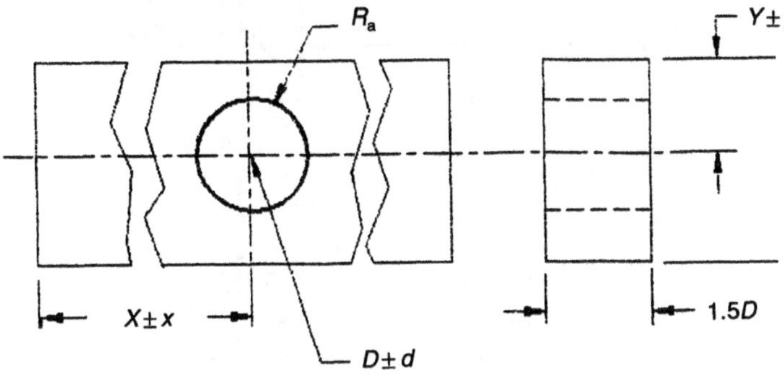

Fig. E13.1

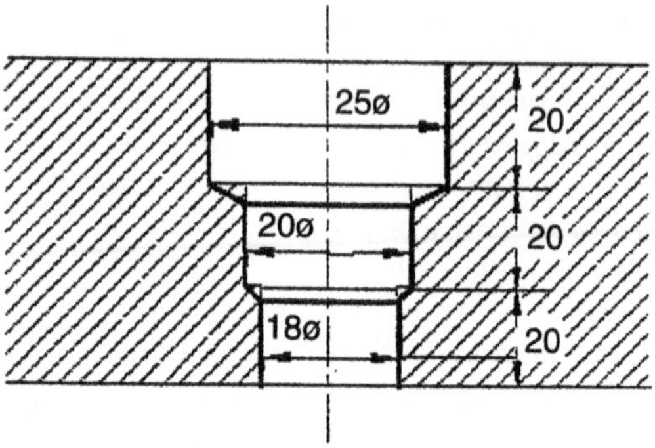

Fig. E13.2

(b) Adjust the process described in (a) for a 2 kW machine.

(c) Repeat (a) where segment #2 (20∅) calls for a surface finish of $R_a = 1.6\ \mu m$.

13.21 The part shown in Fig. E13.3 is machined from cut-to-length pieces. Only the holes have to be produced.

(a) Propose a process plan to machine these holes.

(b) (Optional) What should the machine accuracy be in order for there to be no rejected parts (to work with 9σ (k=3))? Prove your decision and show how you arrived at it.

13.22 (Optional) The hole, as shown in Fig. E13.1 with $x = 0.5$; $y = 0.5$; $R_a = 0.8$; $D = 20 \pm 0.1$ has to be produced. It is decided, due to the large quantity required, to produce this part from casting where a core hole will be cast.

(a) What should the cast hole dimension be, in order to minimize machining time and eliminate rejects?

(b) Propose a detailed process for producing this hole from the core hole.

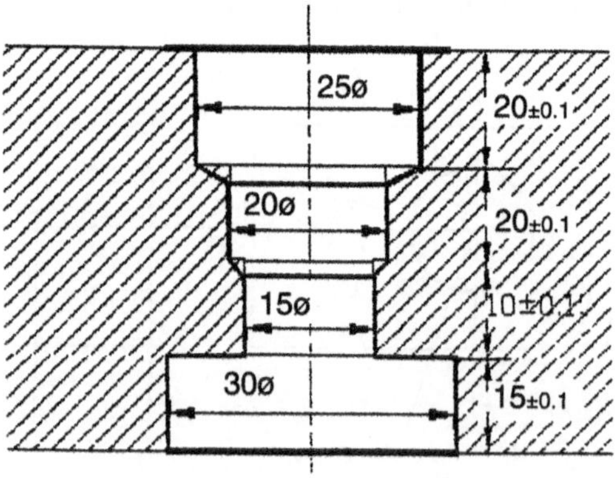

Fig. E13.3

CHAPTER 14

14.1 Why is the machining time equation different for milling and turning? Explain.
14.2 Why are cutting force equations different for milling and turning? Explain.
14.3 What are the difficulties in forming an equation to determine the optimum tool diameter?
14.4 When producing a surface, will a peripheral mill result in a shorter machining time than a face mill? Discuss.
14.5 Explain why face milling must overhang from the machined surface.
14.6 What does 'average chip thickness' mean? Why average?
14.7 Can a change in tool diameter affect the cutting forces? If so, why and how? Try to form a rule.
14.8 What is the main problem in machining a pocket?
14.9 What is the difference between a pocket and a semi-pocket?
14.10 Recommend an optimum process to machine surface A as shown in Fig. E14.1 in cases of: $R_{a1} = 1.6$ μm; Val = 0.1.
(a) Determine tool movement directions.
(b) Determine tool diameter (check catalog for standard tools).
(c) Determine number of cutting passes.
(d) Determine cutting conditions.
(e) Compute machining time and power.
(f) Check and adjust the process for use of a 10 kW machine.
14.11 Recommend an optimum process to machine surface B as shown in Fig. E14.1 in cases of: $R_{a2} = 12$ μm; Val = 0.2.
(a) Determine tool movement directions.
(b) Determine tool type and diameter (check catalog for standard tools).
(c) Determine number of cutting passes.
(d) Determine cutting conditions.
(e) Compute machining time and power.

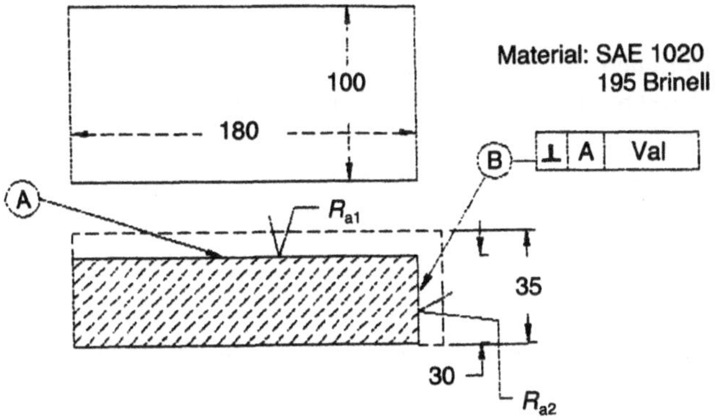

Fig. E14.1

14.12 Recommend an optimum process to machine surfaces A and B as shown in Fig. E14.1 where $R_{a1}=1.6\ \mu m$; $R_{a2}=0.8\ \mu m$; Val$=0.06$.
(a) Determine tool types and diameter (check catalog for standard tools).
(b) Determine number of cutting passes.
(c) Determine cutting conditions for each pass.
(d) Compute machining time and power.
(e) Determine machine accuracy for the job where K$=2$.

14.13 Recommend an optimum process to produce the slot shown in Fig. E14.2, on an aluminum body of 100 HBN. Determine tool type and diameter and the cutting conditions and compute the machining time for the following cases:
(a) $L=7$ mm; $R_a=12\ \mu m$.
(b) $L=10$ mm; $R_a=12\ \mu m$.
(c) $L=7$ mm; $R_a=1\ \mu m$.

14.14 Recommend an optimum process to produce the slot shown in Fig. E14.2, on an alloy steel body of 200 HBN. Determine tool type and diameter and the cutting conditions and compute the machining time for the following cases:
(a) $L=7$ mm; $R_a=12\ \mu m$.
(b) $L=10$ mm; $R_a=12\ \mu m$.
(c) $L=7$ mm; $R_a=1\ \mu m$.

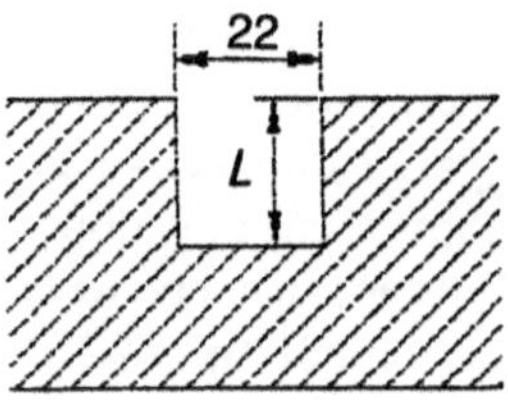

Fig. E14.2

14.15 Recommend an optimum process to produce the pocket shown in Fig. E14.3. The material is alloy steel 200 HBN and it is a rough pocket; $R_a=12.5\ \mu m$ on walls and bottom. Determine tool/s type and diameter, cutting conditions and length of cut and compute the machining time for the following cases:
(a) $r=3$ mm.
(b) $r=5$ mm.
(c) $r=15$ mm.

14.16 Recommend an optimum process to produce the pocket shown in Fig. E14.3. The material is alloy steel 200 HBN and $R_a=1.6\ \mu m$ on walls and bottom. Determine tool/s type and diameter, cutting conditions and length of cut and compute the machining time for the following cases:

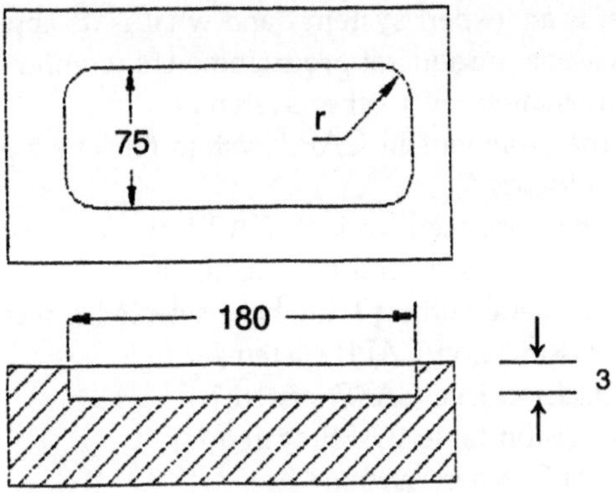

Fig. E14.3

(a) $r=3$ mm.
(b) $r=5$ mm.
(c) $r=15$ mm.

14.17 Compare and explain the differences between the results of exercises 14.15 and 14.16. Pay attention to the length of cut.

CHAPTER 15

15.1 Discuss the objectives and importance of having a CAPP system.

15.2 What are the difficulties in constructing CAPP systems?

15.3 Discuss the possible role of computers (not a system) in the work of a process planner.

15.4 What is the basic philosophy of variant CAPP systems?

15.5 Define the term 'family of parts'. Describe how they are created.

15.6 What side benefits can be achieved by creating a family of parts for use in process planning?

15.7 Are families of parts, created for process planning, used for other purposes (such as design, etc.)? Discuss.

15.8 Why should the length (number of digits) of a classification number be restricted? What subsequent restriction might there be on the coding system?

15.9 What is the difference between coding and classification?

15.10 What is a 'decision tree', and what is its application in CAPP? How does it operate and how is it programmed? What are its advantages over classification?

15.11 What is a 'decision table', and what is its application in CAPP? How does it operate and how is it programmed? What are its advantages over a decision tree?

15.12 What is an 'expert system'? and what is its application in CAPP? How does it operate and is it programmed for computer processing? What are its advantages over other systems?

15.13 Can the basic variant CAPP system retrieve a process for a compound part? Discuss.

15.14 Can the enhanced variant CAPP systems handle a compound part (which requires several processes)? If so, how?

15.15 How is metal cutting knowhow selected, collected and stored in the:
(a) Classification CAPP system?
(b) Decision tree CAPP system?
(c) Decision table CAPP system?
(d) CAPP expert system?

15.16 Is the data stored in the CAPP databases unique (the same in different plants)? Is it optimum? Discuss.

15.17 How is the input to a CAPP system (part specification) introduced to the system? Is it the desired method? Discuss.

15.18 What is GPP – generative process planning – and how is it different from the variant process planning system?

15.19 What are the difficulties in creating a GPP system?

15.20 Discuss the relationship between CAPP and CAD (computer aided design).

15.21 Why are the concepts on which CAD systems are based changing all the time?

CHAPTER 16

16.1 For the part represented below, prepare a complete process plan including the following:
(a) The table of the elementary operations with their machining conditions (see Exercise 2.5).
(b) The definition of associated surfaces and recommended jobs.
(c) The choice of appropriate machine tools.
(d) The table of the anteriority constraints.
(e) The anteriority matrix and the consecutive choice of jobs.
(f) Selection of positioning and clamping system (see Exercise 7.4) with justifications.
(g) The transfer of tolerances (see Exercise 2.4).
(h) The final editing of the process plan as shown in Fig. 16.6.

16.2 For the part shown in Exercise 2.4, prepare a complete process plan comprising all the steps as indicated in Exercise 16.1.

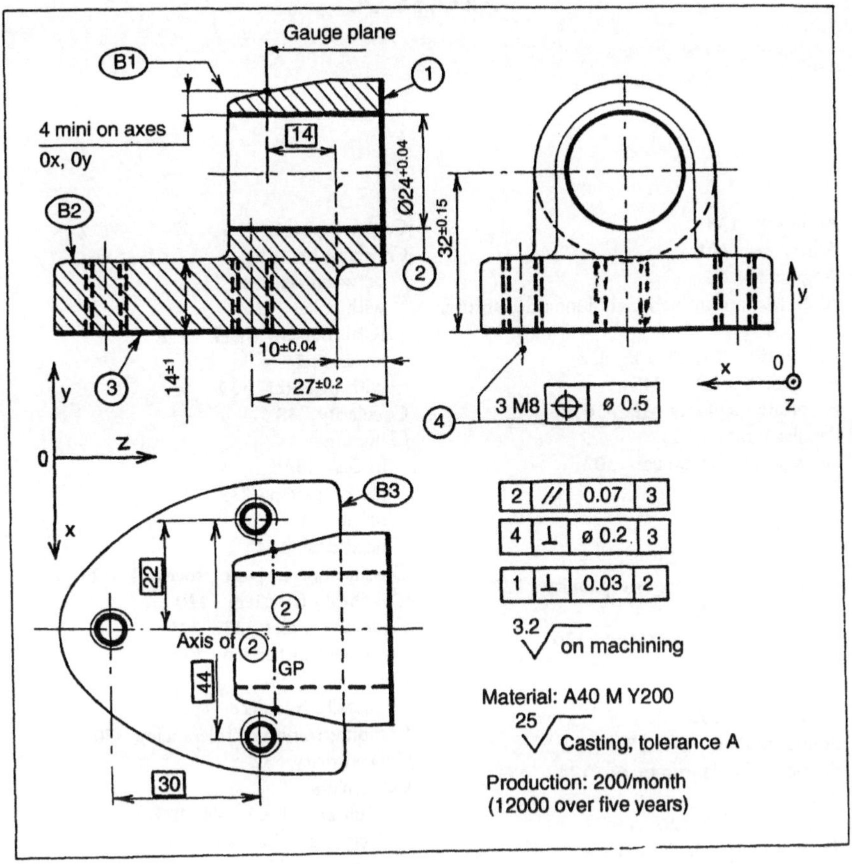

Fig. E16.1 Redrawn from Karr, J., *Methodes et Analyses de Fabrication Mecanique*, published by Dunod, Bordas, Paris, 1979.

Index

GPSR Compliance

The European Union's (EU) General Product Safety Regulation (GPSR) is a set of rules that requires consumer products to be safe and our obligations to ensure this.

If you have any concerns about our products, you can contact us on ProductSafety@springernature.com

In case Publisher is established outside the EU, the EU authorized representative is:

Springer Nature Customer Service Center GmbH
Europaplatz 3
69115 Heidelberg, Germany

Batch number: 10761599

Printed by Printforce, the Netherlands